T0140607

Advanced Sciences and Technologies for Security Applications

Indexed by SCOPUS

The series Advanced Sciences and Technologies for Security Applications comprises interdisciplinary research covering the theory, foundations and domain-specific topics pertaining to security. Publications within the series are peer-reviewed monographs and edited works in the areas of:

- biological and chemical threat recognition and detection (e.g., biosensors, aerosols, forensics)
- crisis and disaster management
- terrorism
- cyber security and secure information systems (e.g., encryption, optical and photonic systems)
- traditional and non-traditional security
- energy, food and resource security
- economic security and securitization (including associated infrastructures)
- transnational crime
- human security and health security
- social, political and psychological aspects of security
- recognition and identification (e.g., optical imaging, biometrics, authentication and verification)
- smart surveillance systems
- applications of theoretical frameworks and methodologies (e.g., grounded theory, complexity, network sciences, modelling and simulation)

Together, the high-quality contributions to this series provide a cross-disciplinary overview of forefront research endeavours aiming to make the world a safer place.

The editors encourage prospective authors to correspond with them in advance of submitting a manuscript. Submission of manuscripts should be made to the Editor-in-Chief or one of the Editors.

Hamid Jahankhani

Editor

Cybersecurity in the Age of Smart Societies

Proceedings of the 14th International Conference on Global Security, Safety and Sustainability, London, September 2022

 Springer

Editor
Hamid Jahankhani
London Campus
Northumbria University
London, UK

ISSN 1613-5113   ISSN 2363-9466 (electronic)
Advanced Sciences and Technologies for Security Applications
ISBN 978-3-031-20162-2   ISBN 978-3-031-20160-8 (eBook)
https://doi.org/10.1007/978-3-031-20160-8

This Springer imprint is published by the registered company Springer Nature Switzerland AG
The registered company address is: Gewerbestrasse 11, 6330 Cham, Switzerland

Committee

General Chair

Professor Hamid Jahankhani, UK
hamid.jahankhani@northumbria.ac.uk

Steering Committee

Professor Shaun Lawson, Northumbria University, UK
Dr. Arshad Jamal, QAHE and Northumbria University London, UK
Dr. Reza Montasari, University of Swansea, UK
Professor Stuart Macdonald, University of Swansea, UK
Professor Amin Hosseinian-Far, University of Northampton, UK
Professor Mohammad Dastbaz, Deputy Vice-Chancellor, University of Suffolk, UK
Dr. John McCarthy, Director, Oxford Systems, UK
Professor Ali Hessami, Director Vega Systems, UK
Dr. Sina Pournouri, Sheffield Hallam University, UK
Dr. Rose Cheuk Wai, QAHE and Northumbria University London, UK
Mr. Usman J. Butt, QAHE and Northumbria University London, UK
Dr. Mathews Z. Nkhoma, RMIT International University Vietnam, Ho Chi Minh City, Vietnam
Professor Babak Akhgar, Director of CENTRIC, Sheffield Hallam University, UK
Professor Gianluigi Me, Luiss University, Italy
Professor Vitor Sa, University of Minho, Braga, Portugal
Professor Bobby Tait, University of South Africa, South Africa

Programme Committee

Dr. Omar S. Arabiat, Balqa Applied University, Jordan
Mr. Konstantinos Kardaras, Technical Consultant, Greece
Dr. Chafika Benzaid, Computer Security Division, CERIST, University of Sciences and Technology Houari Boumediene, Algeria
Dr. Fouad Khelifi, Northumbria University
Dr. Ken Dick, Nebraska University Centre for Information Assurance, USA
Dr. Elias Pimenidis, University of West of England, UK
Mr. Ray Brown, QAHE and Northumbria University, UK
Professor Murdoch Watney, University of Johannesburg, South Africa
Dr. Mamoun Alazab, Macquarie University, Australia
Dr. Gregory Epiphaniou, Associate Professor of Security Engineering at University of Warwick, UK
Dr. Haider Alkhateeb, Wolverhampton Cyber Research Institute, University of Wolverhampton, UK
Dr. Carlisle George, Middlesex University, UK
Professor Gianluigi Me, University of Rome LUISS "Guido Carli", Italy
Dr. Ameer Al-Nemrat, University of East London, UK
Mrs. Sukhvinder Hara, Middlesex University, UK
Dr. Ian Mitchell, Middlesex University
Dr. Amin Hosseinian-Far, University of Northampton, UK
Dr. Reza Montasari, University of Swansea, UK
Dr. Ayman El Hajjar, Westminster University, UK
Dr. George Weir, University of Strathclyde, UK
Dr. Sufian Yousef, Anglia Ruskin University, UK
Dr. Alireza Daneshkhah, Coventry University, UK
Dr. Maryam Farsi, Cranfield University, UK
Mr. Stefan Kendzierskyj, CYFORTIS, UK

14th International Conference on Security, Safety and Sustainability, ICGS3-22 Cybersecurity in the Age of Smart Societies 07th and 08th September 22

About the Conference

Following the successful 13th ICGS3, we have much pleasure in announcing the 14th International Conference on Global Security, Safety and Sustainability.

2020 will be remembered as a year of COVID-19 pandemic with a profound impact on our lives and challenging times for governments around the world with technology as a backbone of tools to assist us to work remotely while in self-isolations and in total lockdown. Every individual and organization face the challenges of the adoption of technological change and understand that nothing will be the same for some time after and even when normality resumes. AI, blockchain, machine learning, IoT, IoMT and IIoT all are tested and applied at a very rapid turnaround as we go through the pandemic day after day. However, all this transformation can bring many associated risks regarding data, such as privacy, transparency, exploitation and ownership.

Since the pandemic in January 2020, governments and organizations, small or large, have seen an increase in cyberattacks. Hackers, through reconnaissance and some network intrusions, targeting nation-states medical research facilities, healthcare organizations conducting research into the virus, financial sectors and so on. Early in April 2020, the UK and US security agencies sound COVID-19 threat alert "…*Cybersecurity Infrastructure and Security Agency issued a joint warning that hacking groups associated with nation-state governments are exploiting the COVID-19 pandemic as part of their cyber operations…*"

The advancement of Artificial Intelligence (AI) coupled with the prolificacy of the Internet of Things (IoT) devices is creating smart societies that are interconnected. The expansion of Big Data and volumetric metadata generated and collated by these mechanisms not only gives greater depth of understanding but as change is coming at an unprecedented pace COVID-19 is driving the cultural changes and the public sector innovations to a new height of responsibilities.

Further development and analysis of defence tactics and countermeasures are needed fast to understand security vulnerabilities which will be exploited by cyber-criminals by identifying newer technologies that can aid in protecting and securing data with a range of supporting processes to provide a higher level of cyber resilience.

The ethical implications of connecting the physical and digital worlds, and presenting the reality of a truly interconnected society, presents the realization of the concept of smart societies in reality.

This Annual International Conference is an established platform in which security, safety and sustainability issues can be examined from several global perspectives through dialogue between academics, students, government representatives, chief executives, security professionals and research scientists from the UK and from around the globe.

The Two-Day Virtual Conference will focus on the challenges of complexity, the rapid pace of change and risk/opportunity issues associated with the twenty-first-century lifestyle, systems and infrastructures.

August 2022   Prof. Hamid Jahankhani
                                                          General Chair of the Conference

Contents

Security Risk Management and Cybersecurity: From the Victim or from the Adversary?

Jeimy J. Cano M

Abstract Cybersecurity and information security risks are generally addressed following the guidelines of current standards and best practices, which traditionally point to ISO 31000 risk management. This key exercise to ensure the "known" risks is carried out in a reactive perspective, since not only the risks are known and treated, but also the organization is prepared when one of them materializes. Thus, it is necessary to advance in cyber risk management from the adversary's perspective. That is, to understand both the intentions and their capabilities to get out of the comfort zone of standards, and mobilize the organization to another framework that seeks to detect, deter, delay, confuse and anticipate the adversary. Therefore, this article develops two perspectives of security and cybersecurity risk management (from the victim and the adversary), in order to tune efforts around the treatment of known risks and the development of capabilities to maintain a vigilant posture towards latent and emerging risks, allowing the organization navigate throughout the uncertainties and instability of the current international context.

Keywords Risk management · Adversary · Victim · Cybersecurity · Cyberrisk

1 Introduction

Cybersecurity and information security risks are generally addressed following the guidelines of current standards and best practices, which traditionally point to ISO 31000 risk management. This key exercise to ensure the known risks is carried out in a reactive perspective, since not only the risks are known and treated, but also the organization is prepared when one of them materializes.

In this sense, current practices such as that of the NIST cybersecurity framework (identify, protect, detect, respond, recover) [1] are placed in the perspective of the victim who must address the adverse event and where the attacker has many certainties about the functioning of its infrastructure and operations. This means, that the

J. J. Cano M (✉)
Law Faculty, Universidad de los Andes, Cra 1E No. 18A-10, Bogotá, Colombia
e-mail: jcano@uniandes.edu.co

© The Author(s), under exclusive license to Springer Nature Switzerland AG 2023
H. Jahankhani (ed.), *Cybersecurity in the Age of Smart Societies*,
Advanced Sciences and Technologies for Security Applications,
https://doi.org/10.1007/978-3-031-20160-8_1

1

application of standard practices inherently seeks to respond to adverse events in the infrastructure without considering the intentions or capabilities of the adversary, analysis that is advanced only if the attack was successful (if at all).

In this perspective, it would seem that the organization only has the option of receiving and handling the adverse events proposed by its attacker, leaving open the possibility of new and novel attacks from its potential adversaries. Security and cybersecurity frameworks seek to prevent attacks, usually known and documented, which complicates the traditional exercise of prevention, because in the real world, innovation, surprise and novelty are the essence of the adversaries' proposals.

For this reason, it is necessary to advance in cyber risk management from the perspective of the adversary. That is, to understand both the intentions and their capabilities [2] to get out of the comfort zone of standards, and mobilize the organization to another framework that seeks to detect, deter, delay, confuse and anticipate, to realize a more proactive and prospective management, where the organization will be more aware to its operational context, the protection of those critical assets that are interest to an adversary, people's behaviors and above all, a vigilant posture that maintains a strategic vision of the organization based on its risk appetite [3].

Therefore, this article develops two perspectives of security and cybersecurity risk management (from the victim and the adversary), in order to synchronize efforts around the treatment of known risks and the development of capabilities to maintain a vigilant posture against latent and emerging risks, allowing the organization to establish a trajectory in the midst of the uncertainties and instability of the current international context.

2 Dynamics of Cyber Risk. A Challenge from the Uncertain

Cyber risk or cybernetic risk is a risk that has different characteristics from traditional information and communications technology risks. While information technology risks are known and detailed, for which there is a set of established controls that are susceptible to formal inspection by the audit, cyber risks are uncertain, disruptive and emerging characteristics of a digital and interconnected context, where physical objects now have new ways of connecting and generating information flows that did not exist before [4].

Cyber risk is a risk that creates environments enriched with data, information flows and intelligent objects, where each new relationship between one object and another creates a new reality or experience for the client. In this sense, it is important to note that the mechanistic cause-effect perspective that is used to advance risk management to date is limited to identify and understand the cascade or domino effects that can be generated on account of the materialization of this type of risk.

Recognizing how closely connected objects are coupled, and how much interaction they have, requires recognizing the natural digital ecosystem in which cyber risk operates. To the extent that there is greater interaction and coupling, there will be greater propagation effects of risk materialization. Initially there will be recognizable and treatable effects from known practices, and then, because of their high connectivity and visible (and invisible) relationships, unexpected and unforeseen consequences will be revealed over time, which will lead the organization to address uncommon and possibly unexpected on account of these effects [5].

To understand cyber risk is to understand its systemic nature and recognize that there is an adversary or an attacker who has intentionality and a series of capabilities to produce instability and uncertainty. In this sense, unlike the information and communications technology risk, the fundamental perspective of its management is in how to identify the movements of the aggressor, how to detail its strategies in order to visualize it before it succeeds [2].

Thus, cyber risk necessarily involves navigating through uncertainty inherent to the digital and strategic initiatives of companies, in order to establish a framework that allows the organization to maintain a vigilant and proactive posture in the face of the adversary's disruptive proposals [6]. This implies establishing a set of tactical objectives specific to the management of this risk, which are finally articulated with a catalog of operational capabilities that must respond to the challenge of defense and anticipation in the face of unexpected events.

The tactical objectives and operational capabilities (to be called modes of operation) are detailed below:

Tactical objectives [7]:

- Implement elements of cyber risk adaptation and alignment with business needs and threat intelligence.
- Fully integrate security functions with business functions: information sharing and communication.
- Anticipate attacks and attacker behavior.
- Implement adaptation, learning and operational and organizational improvement based on the challenge of prior knowledge.
- Define and deploy cybersecurity as a transversal culture in the organization.
- Modes of operation [8]:
- Monitoring mode:

  Configure and update alerts and alarms that allow the company to develop preventive action maneuvers.
- Defense mode:

  Recognize the aggressor's intentions and capabilities to delay, deter or confuse the adversary.
- Crisis mode:

  Attend and overcome uncertain and unexpected events from the company's risk and control culture, activating coordination and communication procedures: playbooks.

- Radar mode:

  Intensively use data analytics to develop intelligence and reveal weak signals and unusual patterns.

3 Security Risk Management and Cybersecurity: From the Victim

This is the traditional perspective of risk management that seeks to locate a series of sensors in the organization (which we call controls) which are defined to generate alerts of possible adverse events and from there, initiate risk management based on the information generated on account of known risks [9].

This reading of risk management implies a reaction on the part of the company, to attend to possible events that are occurring and to advance the necessary actions for their treatment. Note that as long as alerts are not generated, the management system is not activated to advance in the knowledge of what is happening in the environment. Thus, the finer and better calibrated the control system is, the better alerts can be generated to advance in the identification and assurance of the possible risk.

In the specialized literature this posture is called passive defense [10]. An exercise based on a set of technological devices and mechanisms that monitor and update on threat and attack trends in order to alert the organization about these issues, and thus maintain the reliability of the infrastructure, based on known patterns, new techniques reported or signature updates in the appliance installed in the data centers or in the cloud.

When an organization sustains its risk management from the victim's perspective it is clear that it will be prepared to face the instability and the known adverse event. It will be able to mobilize activate the protocol identify, protect, detect, respond, recover [1], and the incident response team to identify, contain, update and eradicate the risk that has materialized, so that after a forensic examination of what has happened, it can capitalize on the lessons learned and make the adjustments of the case both in the infrastructure and in the procedures of the organization.

The management of cyber risk and information security in the perspective of "attacked or victim", reiterates the view of the quality framework known as PDCA (Plan, Do, Check and Act), which, although it allows to maintain a repeatable and predictable view on a scenario with known variables, is reactive to the emerging conditions of the environment, where it has no choice but to activate its "root cause analysis" protocol, in order to understand what has happened, how it has affected the organization and what corrective actions should be taken to incorporate into its current practice [11].

Advancing in risk management within the framework of quality can lead the organization to enter the "false sense of security" zone, where the adversary is able to develop reliable intelligence on the dynamics of the organization, as well as on the way in which it ensures concrete security risks in its different products and services. In this way, he/she manages to have extensive well-documented and consistent information from which he/she can quickly study how to create the distraction and deception to produce the necessary instability that will allow him/her to then carry out his pivot of action and deploy the concrete attack on a specific critical asset [12].

4 Security and Cybersecurity Management: From the Adversary

When shifting the risk management perspective now from the adversary, the organization must maintain what it knows how to do in the face of known risks and threats, and go out and explore new possibilities to better understand who its adversary is and what capabilities it has. While this will not limit or change the attacker's intentions, it is feasible to create a credible deterrent that allows them to move from one initially intended target to another [2].

While risk management from the victim responds to a mechanical cause-effect perspective, the perspective from the adversary is eminently systemic, that is, context-sensitive, interconnected, interdependent and with cascading effects. It is to recognize that the attacker develops an open and fine-grained view of the interaction and coupling of the different elements of the infrastructure, in order to produce maximum damage with minimum effort.

The adversary can produce an adverse event in cyberspace, such as the release of malicious code with unprecedented capabilities: multiplatform, polymorphic, encrypted, with self-destruct mechanisms (if detected) and hypercontagion, which reaches a technological infrastructure (either local or in the cloud) where it spreads in an accelerated manner, and ends up penetrating all nodes visible from that infrastructure, remaining active on some computers and passive on others, in order to give confidence to the incident response team in the treatment of the event, and subsequent invisibility to the adversary, knowing that he has available pivots to validate the behavior of the network and to be able to carry out a more silent, anonymous and invisible action to the radar of the installed controls [13].

In this context, risk management must go out and explore the environment, consult available data from current monitoring, use unsupervised artificial intelligence algorithms to find anomalies, recognize and reflect on security forecasts available to date, capture early warnings of emerging risks, and above all locate geopolitical tensions that are relevant to the development of the company's business [14].

With these inputs security and cybersecurity risk management should move forward in:

Detect: early warnings based on the identification of weak signals from the environment that must be crossed with consolidated trends and there find blank spaces from where adversaries can advance.

Deterrence: Incorporating moving target or deception technology in order to deteriorate the adversary's prior intelligence and thus increase his level of uncertainty, which forces the aggressor to rethink his strategy against an initial target.

Delay: Creating distraction zones for the adversary, where he can waste a lot of time trying to penetrate the infrastructure or reprioritize his plan of attack.

Confuse: Changing the infrastructure configuration dynamically, in order to create greater uncertainty in your risk model and thus, lead you to make missteps and expose your identity.

Anticipate: Having applied the previous steps, it is possible to see the path and movement of the adversary in the infrastructure, which will enable a concrete space to intercept him before he/she succeeds.

5 Security Risk Management and Cybersecurity—Victim and Adversary

If a complementary view of the two perspectives of risk management is established, it is necessary to create convergent conversation zone that allows visibility for the known risks inherent to the dynamics of the organization, as well as for emerging risks. This zone is referred to in the literature as scenario design and analysis. A collective construction exercise where the focus is not on the risks themselves, but on the context where the events may occur, for which it is necessary to recognize the challenges of the organization, the possible adversaries and the good practices that exist to date [11].

From the scenario practice it is possible to add different voices and perspectives of executives, tactical management and operation assurance, as a consolidated view that thinks about the possible events and effects they may have for the organization. From this perspective, a comprehensive view of risk is consolidated that goes beyond the technical zone and translates into possibilities that end up causing negative impacts on the organization. Cybersecurity and security translate into narratives that increase the situational awareness of the organization in the face of an increasingly BANI environment: Brittle, Anxious, Nonlinear and Incomprehensible [15].

When the two visions are combined: victim and adversary, a vigilant state is enhanced (which can be achieved to a variable degree) regarding the understanding and projection of risks and threats in the cyber environment and their evolution in the near future [16]. That is, an appropriation of risk at the corporate level, which led from the first level executives, manages to connect the organization at its different levels. None of this can be achieved without an openness to see things differently, without challenge previous knowledge and experiment uncertainty as an input to learn/unlearn.

Connecting the two visions allows not only to establish what is normal and which expected behavior of a technological platform, but also to recognize the behaviors and actions of people in the face of a threat, in order to create awareness and precautions against possible events that may be recognized as abnormal. The participation of multiple profiles and experiences in the development of the scenarios enables spaces for reflection and proactivity that demystify the security and cybersecurity risk, in order to place it in the daily operations.

To advance the design and analysis of scenarios is to enable a learning/unlearning platform where all perspectives add up, all possibilities are allowed, in order to build, from the company's strategic objectives and challenges, the required filter that defines the framework and relevant discussion to join efforts in the defense of the company against possible known and unknown adversaries, as well as against possible and probable contexts of instability where the organization must plot a course in the midst of a sea of uncertainties [11].

6 Conclusions

Advancing security and cybersecurity risk management is not a one-sided exercise. It is a dynamic that requires not only the traditional perspective of the victim preparing for an incident, but the challenge of stepping out of the comfort zone of standards, and placing yourself in the adversary's territory to increase uncertainty in your risk model.

Although risk management in the perspective of the identify, protect, detect, respond, recover framework has shown outstanding progress and achievements in different companies globally, it is necessary to mobilize efforts to complement the view from the capabilities and intentions of the adversary, as a way to maintain a vigilant and not only reactive posture, which drives the business dynamics and develops the necessary capabilities according to the risk appetite of the company.

Thus, it is feasible through the construction of scenarios to create a learning window, a "sandbox" where it is possible to experiment and explore possibilities located in the dynamics of the organization, in order to increase cyber situational awareness [16] in the daily life of the organization itself, without technological or specialized biases, where each person can recognize the impacts and effects of an adverse event as part of the way in which security and control risks should be understood and managed in the development of their activities.

When the two perspectives are added: the victim and the adversary, the organization advances in a position of organizational resilience, where it is able to define its operational thresholds, recognize its level of tolerance to risk, how much capacity it has to absorb the materialization of an adverse event and, above all, how it will maintain operations despite having been impacted by an unexpected situation [17]. The preparation and requirement to respond and recover that is acquired in the application of standards and good practices must be complemented with the necessary capabilities to recognize the adversary and face it on its own land.

Organizations that want to move forward through current tensions and face the uncertainty inherent to cyber risk must not only heed the recommendations and requirements of the standards, but also be open to taking risks in an intelligent and calculated manner, which basically implies "an ability to properly assess the probabilities and possibilities of the materialization of an undesired event and weigh it against perceptions, taking into account early warnings, consolidated trends and best forecasts, as well as the emotional aspects that may influence the final decision" [3].

References

1. NIST: Framework for Improving Critical Infrastructure Cybersecurity. National Institute of Standards and Technology. https://nvlpubs.nist.gov/nistpubs/CSWP/NIST.CSWP.04162018.pdf (2018)
2. Martin P (2019) The rules of security. Staying safe in a risky world. Oxford Press, Oxford, UK
3. Wucker M (2021) You are what you risk. The new art and science of navigating an uncertain world. Pegasus Book, New York, USA
4. Zukis B, Ferrillo P, Veltos C (2020) The great reboot. Succeeding in a world of catastrophic risk and opportunity. DDN Press, USA
5. Clearfield C, Tilcsik A (2018) Meltdown. Why our systems fail and what we can do about it. Penguin Press, New York, USA
6. Day G, Schoemaker P (2019) See soon, act faster. How vigilant leaders thrive in an era of digital turbulence. MIT Press, Cambridge, MA. USA
7. ISACA (2013) Transforming cyber security. https://bit.ly/36QGhie
8. Cano J (2021) Modes of operation of enterprise cybersecurity. Basic capabilities for navigating the digital context. Global Strategy. Global Strategy Report No. 44. https://bit.ly/3JEBsat
9. Kohnke A, Shoemaker D, Sigles K (2016) The complete guide to cybersecurity risk and controls. CRC Press, Bocaraton, Florida, USA
10. Cai et al (2016) Moving target defense: state of the art and characteristics. Front Inf Technol Elec Eng 17(11):1122–1153. https://doi.org/10.1631/FITEE.1601321
11. Cano J (2020) Rethinking the practice of security and cybersecurity in organizations. A systemic-cybernetic review. Global Strategy. Global Strategy Report No. 58. https://global-strategy.org/repensando-la-practica-de-la-seguridad-y-la-ciberseguridad-en-las-organizaciones-una-revision-sistemico-cibernetica/
12. Cano J (2021) The "false sense of security". The challenge of discomforting the certainties of standards and trying to "tame" the uncertain ones. In: SISTEMAS magazine. Colombian association of systems engineers—ACIS, pp 82–95. https://doi.org/10.29236/sistemas.n159a6
13. Forscey D, Batema J, Beecroft N, Woods B (2022) Systemic cyber risk: a primer. Paper. Carnegie Endowment for International Peace. https://carnegieendowment.org/2022/03/07/systemic-cyber-risk-primer-pub-86531
14. Reeves M, Ramaswamy S, O'Dea A (2022) Business forecasts are reliably wrong—Yet still valuable. Harvard Business Review. https://hbr.org/2022/03/business-forecasts-are-reliably-wrong-yet-still-valuable
15. Cascio J (2020) Facing the age of chaos. Institute for the future. https://medium.com/@cascio/facing-the-age-of-chaos-b00687b1f51d
16. Renaud K, Ophoff J (2021) A cyber situational awareness model to predict the implementation of cyber security controls and precautions by SMEs. Organ Cybersecur J: Pract Process People. https://doi.org/10.1108/OCJ-03-2021-0004
17. Denyer D (2017) Organizational resilience. A summary of academic evidence, business insights and new thinking. Report. BSI and Cranfield School of Management. https://www.cranfield.ac.uk/som/case-studies/organizational-resilience-a-summary-of-academic-evidence-business-insights-and-new-thinking

Comparison of Cybersecurity Methodologies for the Implementing of a Secure IoT Architecture

Nicolas Moreta, David Aragon, Silvana Oña, Angel Jaramillo, Jaime Ibarra, and Hamid Jahankhani

Abstract This research presents the collection of data from different cybersecurity methodologies used for securing IoT environments, of which a few were chosen for the validation of parameters. The methodology known as ENISA was used to filter, select and adjust to the needs of the architecture carried out in this project and was compared with the Hardening methodology. To review and validate the parameters of each one, a test bed environment was designed with several sensors and integrated systems (NodeMCU ESP8266 and Raspberry Pi). In addition, tests were carried out in three different scenarios and the validation of its security has been done using the Kali Linux distribution with tools like Nmap, Hydra, Wireshark, etc. The results were presented at the end in comparative tables attached to this document, which enabled the research to validate which methodology provides better resilience along with its applications to other architectures depending on the needs of the users.

Keywords Cybersecurity · Internet of Things · Hardening IoT Security

N. Moreta · D. Aragon · S. Oña · A. Jaramillo
Universidad de Las Americas, Quito, Ecuador
e-mail: nicolas.moreta@udla.edu.ec
URL: https://www.udla.edu.ec

D. Aragon
e-mail: esteban.aragon@udla.edu.ec

S. Oña
e-mail: silvana.ona@udla.edu.ec

A. Jaramillo
e-mail: angel.jaramillo@udla.edu.ec

J. Ibarra · H. Jahankhani (✉)
Northumbria University, Newcastle, UK
e-mail: hamid.jahankhani@northumbria.ac.uk
URL: https://www.northumbria.ac.uk

J. Ibarra
e-mail: jaime.ibarra@northumbria.ac.uk

© The Author(s), under exclusive license to Springer Nature Switzerland AG 2023
H. Jahankhani (ed.), *Cybersecurity in the Age of Smart Societies*,
Advanced Sciences and Technologies for Security Applications,
https://doi.org/10.1007/978-3-031-20160-8_2

1 Introduction

Kevin Ashton, professor at MIT, in 2009 used the expression Internet of Things (IoT) publicly for the first time and since then the increase and interest around the term has been increasing [1]. IoT is defined as the identification of digital entities and any physical object on the Internet. More and more people today understand that this trend will continue to grow as more and more simple devices (sensors, actuators, appliances, personalized medical devices) are connected to the Internet [2].

It is estimated that there will be approximately 50 billion IoT devices in 2030 [3]. This draws the attention of researchers, academics and industry. Thus, by connecting something to the Internet, applications can be easily created that allow adding value to the data collected by the objects. However, all IoT applications must be built securely [4], otherwise any device could be used to launch an attack.

Increased device connectivity and process integration exposes new vulnerabilities in IoT device security, data integrity, and system reliability [5]. In 2016 a major attack was detected on IoT devices, which consisted of a Distributed Denial of Service (DDoS) attack. This attack left Dyn without service, for this reason social networks like twitter, Spotify, pages like Amazon, GitHub and Reddit and several devices connected to the network were down. this attack left a considerable impact worldwide [6].

Cybersecurity mitigates attacks on heterogeneous devices that are part of the IoT infrastructure, which is why IoT-oriented cybersecurity architectures have been created, which are based on layered designs that require accessibility, confidentiality, integrity, availability and interoperability [7]. Likewise, methodologies have been generated that allow evaluating these infrastructures such as: Open Web Application Security Project (OWASP), National Institute of Standards and Technology (NIST) or Risk Management Framework (RMF) and thus be able to determine the risks and threats that are in evolution before making a design [8].

In [8] it is highlighted that an impact is divided into 3 levels of importance: Crucial, High and Medium. When the impact is crucial, the user must resolve it immediately, since sensitive information may be lost. On the other hand, if the impact is at a high level, the user must resolve it according to the damage it can cause. Finally, if the impact is medium, the user must be alert and take the necessary precautions. Figure 1 shows the most relevant impacts of IoT threats.

The provision of intelligent services has shown risks such as system penetration, data leakage and exposure to various cyber-attacks such as Man in the Middle (MITM). In addition, there have been several cyber-attacks and a growth in the number of threats such as: Hackers, Malwares, etc. [9]. Overcoming technical challenges requires a thorough evaluation of [10] IoT solutions.

In studies carried out by ESET due to SARS-CoV-2 (COVID-19), it was shown that 64% of Latin American companies believe that there is a risk of malicious code, while 60% consider the theft of information dangerous and 56% fear improper access to [11] systems. Additionally, another 2020 analysis found that weak passwords are a significant IoT security setback. Thus, it identified that the most used password is

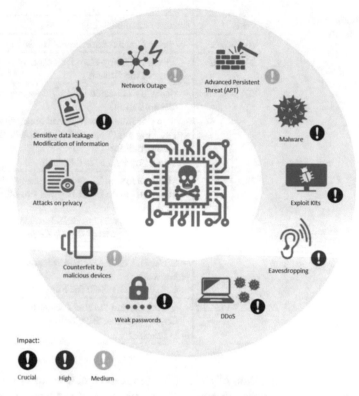

Fig. 1 Impact threats in IoT

"admin", followed by "root" and "1234". Consider that the analysis was performed on approximately 140,000 devices and found that 40% of routers used weak passwords and 34% of these were affected by at least one notable [12] vulnerability.

In [13] he explains the presence, trends and causes of security vulnerabilities in the source code of IoT operating systems (OS). In addition, a study is made of the most popular open-source OS in which traffic analysis tools are used (Cppcheck, Flawfinder and RATS). These results show the importance of applying protection methodologies to this type of infrastructure and avoid incorrect configurations or careless policies that can result in vulnerabilities and failures [14].

When designing the architecture of IoT protocols, long-range communication systems and protocols such as Low Power Wide Area Networks (LPWAN) should be considered. LPWAN is a type of wireless protocol that works at a low bit rate and supports long-range communication between IoT devices connected to a cloud system via [15] wireless gateways.

On the other hand, there are several areas that are expected to be directly affected initially including: healthcare, supply chain management, transportation system, agriculture and environmental monitoring, home life and many more as we move towards smart homes and cities [2].

Table 1 Timeline of IoT
security incidents

Years	Security incidents
2009–2013	– Puerto Rican Smart Meters Hacked
2013–2015	– Foscam IP Baby–Cam Hujacked – Target Data Breach
2015–2016	– BMW's Connected Drive Vulnerable (demostration) – Jeep Car Remotely Hijacked (demostration) – Trackingpoint's Smart Sniper Riflehack (demostration) – Vtech Toymaker Data Breach
2016–2017	– Mirai-DDoS on "Krebs on Security" Website – Mirai-DDoS on OVH Hosting Provider – Hajime – Mirai-DDoS on Dyn DNS Provider – DDoS on Building Block's Central Heating System – Mirai-DDoS on Deutshe Telekom – Network – Cloudpets DB Held for Ransom
2017-Present	– Cloudpets and "Meine Freundin Cayla"-Insecure Bluetooth – Brickerbot

Thus, the Internet connection of a smart home presents new security problems. For this reason, it is necessary to meet certain cybersecurity objectives such as: authentication, authorization, confidentiality, integration and availability. These goals must be met to prevent any kind of [16] attack. In addition, the behavior of IoT devices can be analyzed from the point of view of a user, therefore they can become victims of DDoS attacks directed from the outside. In [17] tests are carried out where individual attacks are made and the resistance of the IoT sensors against the DDoS attack is indicated.

Also, in smart houses, it is possible to develop a smart home network topology generator. More complex topologies inherit the threats faced by simpler ones. This is done using a graph-based model of malware propagation, where the effects of dynamic environment properties on specific security threats are effectively evaluated.

In 2021, an investigation of the Hardening methodology was carried out, where it was implemented in a layered architecture. The analysis was carried out through seven steps to perform security tests to reduce the risk levels in IoT systems, their vulnerabilities and the attack surface [18].

In 2017, the European Union Agency for Cybersecurity (ENISA) [8] considerably highlighted the increase in the number of attacks on IoT since 2009. In recent years, criticality has been presented in the penetration of architectures of IoT, due to the presence of smart devices in our daily activities. Regulatory agencies increasingly agree on the need for basic IoT security guidelines to ensure data protection [5]. This is due to the increase in incidents, as can be seen in the Table 1.

The ISO/IEC 27000 standards were created and managed by the International Organization for Standardization (ISO) together with the International Electronic Commission (IEC). For this reason, a wide dissemination and great recognition worldwide is guaranteed. These standards focus on improving practices in relation to different aspects related to information security management. ISO 27001 is an international standard which allows total confidentiality and integrity of data and information. In addition, it makes a risk assessment and the application of necessary controls to mitigate or eliminate them [19].

The ISO/IEC 30141 standard provides a common framework for designers and developers of IoT applications, which allows the development of reliable, secure, protected, privacy-respectful systems capable of dealing with interruptions due to natural disasters or cyberattacks [20].

The European Telecommunications Standards Institute (ETSI) specifies 65 security provisions for consumer IoT devices that are connected to a network. Its goal is to provide a relatively complete set of requirements to avoid restricting devices to a specific [21] technology or protocol.

NIST develops standards, guidelines, best practices and resources in different fields. In the field of cybersecurity, there are programs that seek to develop and apply practical and innovative security technologies and methodologies to address current and future challenges [22].

Considering that there is an increase in cyberattacks due to the COVID-19 pandemic, the aim is to generate a comparison between two cybersecurity methodologies for IoT infrastructures. In this study it will be possible to appreciate the weaknesses and strengths based on specialized standards and protocols. This will allow strengthening the research using criteria from other authors in order to provide a diagnosis of which aspects should be strengthened, with the purpose of establishing a study base for future implementations.

2 Analysis of Methodologies

After the investigation carried out, some methodologies were found in which their security can be verified based on important metrics such as: the type, the focus area, the implementation, the test configuration and the results. Below are some methodologies that have been applied in different IoT infrastructures.

A. **Cisco Security Framework**

Irshad [2] proposes a framework where the development of protocols, products and policy enforcement are used in operational environments for both applications and infrastructures. This framework has a communication layer that is responsible for authenticating, encrypting and verifying the integrity of messages using standardized mechanisms.

B. **Floodgate Security Framework**
   This security framework fully complies with the ISA/IEC 62443 standard, which specifies technical aspects associated with industrial cybersecurity processes. It protects IoT devices against cyber threats as well as future threats that will emerge during the lifetime of the device [2].

C. **Constrained Application Protocol (CoAP)**
   It consists of several modules to handle the security and trust issues of IoT environments. Due to severe power and memory limitations, devices rely on group communication, asynchronous traffic, and caching from millions of globally wirelessly connected smart objects [2].

D. **Object Security Framework for Internet of Thing (OSCAR)**
   It is based on the concept of object security because it protects against communication-related attacks by encrypting keys. It jointly raises end-to-end security and authorization issues [2].

E. **EBTIC Reference Architecture-based Methodology for IoT Systems (ERAMIS)**
   In 2019, Kearney and Asal [23] proposed a methodology that is based on the instantiation of a reference architecture that captures common design features. The research brings together best practices and good design security properties. Shows a clear vision for safety processes and services.

F. **Internet of Things Security Verification Standard (ISVS) de OWASP**
   Guzman and Bassem [24] specifies the security requirements for embedded applications and the IoT ecosystem in which they are found. While referencing existing industry accepted standards as much as possible. Security control requirements can be represented as a stack. This stack is made up of hardware, software and communication platforms, user space applications and IoT ecosystem.

   It has 3 Security Verification Levels, but not all ISVS requirements are applicable to all devices:

   1. ISVS Level 1: Provides a security baseline for connected devices where physical compromise of the device does not result in a high security impact [24].
   2. ISVS Level 2: Provide protection against attacks that go beyond software and target the hardware of the [24] device.
   3. ISVS Level 3: Provide requirements for devices where compromise must be avoided at all costs [24].

G. **IoT Security Compliance Framework**
   In 2020, the IoT Security Foundation [25] proposes to guide the user through a structured process and collection of evidence. This ensures that proper security mechanisms and practices are in place. This framework is intended to help all businesses make high-quality security decisions. The evidence collected in this study can be used to offer best practices to customers and other stakeholders.

   (1) Compliance Class: It uses a risk assessment that requires a significant compliance category, in order to lower certain levels of risk based on market

Table 2 Compliance class security features

Compliance class	Security features		
	Confidentiality	Integrity	Availability
Class 0	Low	Low	Low
Class 1	Low	Medium	Medium
Class 2	Medium	Medium	High
Class 3	High	Medium	High
Class 4	High	High	High

and application dependency. The checklist requirements range from class 0 to class 4 where compromise with generated data or loss of control results in escalating impact [25]. Table 2 shows the levels of integrity, availability and confidentiality (CIA triad) for each compliance class.

H. **Baseline Security Recommendations for IoT in the context of Critical Information Infrastructures**

This methodology provides information about IoT security requirements, mapping of critical assets, important threats, attack evaluation, detection of good practices and security measures. It aims to provide an adequate level of security to the different IoT architectures by evaluating the aspects of each [8] area.

I. **Security methodology for the Internet of things**

According to Surya Kant Josyula and Daya Gupta in [26] a new security methodology based on the Security Engineering Framework is proposed, which considers security measures that can be easily adapted during each stage of software development. This consists of the following phases:

Security requirements engineering phase: The security requirements for IoT systems are identified, since they are obtained, analyzed and prioritized.

Security design engineering phase: The design is analysed considering the limitations and the existing feasibility of the security mechanisms, to obtain the resolutions for the design.

Security implementation: It is based on the design and the functionality related to the security mechanisms is implemented. In this phase, the integration and documentation of the existing systems is carried out.

Security tests: The implemented software is tested and verified that it meets all security requirements. During the tests, vulnerabilities are repeatedly evaluated, possible attacks are tested and the results are analyzed, an evaluation and general security verification of the system is carried out. Finally, we proceed with audits and security reviews [26].

J. **Methodology for the assurance of sensors in the cloud**

In [27], a built-in security mechanism designed specifically for IoT is detailed, which addresses security with multiple parameters that must be met. It is important that you encompass the security of all components in the architecture. The objective of this methodology is that the implementation is easy, since if it is too complex it will be difficult for other organizations to apply it. Therefore, it is essential to know the IoT security requirements to comply with the security framework.

It is made up of 4 layers as shown below:

- Layer 1: Layer of things.
- Layer 2: Communication layer.
- Layer 3: Infrastructure layer.
- Layer 4: Data analysis layer.

K. **Cybersecurity Certification Framework for the Internet of Things**

In [28] uses the main ideas of ETSI, which consists of two components: security risk assessment and testing. The process determines the security objectives, the requirements are identified, the technicalities of the evaluated objective are analyzed and documented. Likewise, test planning establishes a strategy by defining the test phases and techniques [28].

The security assessment process includes risk assessment and testing:

- Risk identification
- Risk estimate
- Risk Assessment
- The security testing process comprises:
- Test design and implementation
- Setting up and maintaining the test environment
- Execution, analysis and summary of the tests

L. **Failure Modes and Effects Analysis Model (FMEA)**

It is a common tool within engineering, since it helps to prevent failures during certain processes. It is a methodology that is applied when implementing services or processes in order to be able to study their possible errors in the future through failure modes. This will serve to prioritize the most important bugs and be able to solve them, either due to their level of criticality or causing inconveniences to the user [29].

M. **Hardening methodology proposed at the University of the Americas**

In 2021, Echeverría et al. [18] conducted a study that focused on the analysis of IoT cybersecurity norms, standards, models and methodologies given by some international organizations (OWASP, ISO, NIST). The model handles a sequence of seven steps that allows reducing the attack surface by executing hardening processes, as shown in Fig. 2.

This study defined an IoT solution with a checklist that considers security aspects in all three layers (Application, Network, Perception). In addition, it established a risk matrix and vulnerability analysis to improve the level of security in the IoT ecosystem. They used as base standards: ISO/IEC 30,141, ISO/IEC 27,001, ISO/IEC 31,000 and ISO/IEC 25,010.

The methodologies presented above have been analyzed, in Table "A" of annex 1 the advantages and disadvantages of each one are exposed, in order to select the methodology that was compared with the Hardening methodology. Once the different

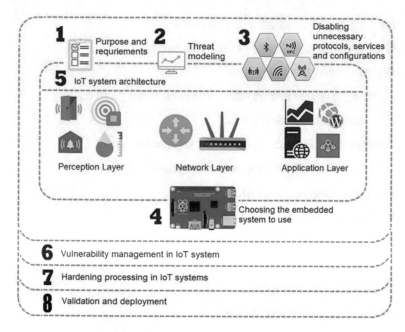

Fig. 2 Hardening reinforcement model

methodologies have been analysed, those parameters or security aspects that are considered relevant have been compiled below. To select the methodology, it was technically verified which of these met most of the parameters, as can be seen in the Table 3.

Table 3 Selection of methodology

Methodologies	P1	P2	P3	P4	P5	P6	P7	P8	P9	P10	P11	P12	Total
A	X	X		X			X		X			X	6
B	X	X				X	X				X		5
C				X								X	2
D	X			X			X		X				4
E			X								X		2
F	X		X	X			X	X	X	X		X	8
G	X	X	X		X		X	X	X		X	X	9
H	X	X	X	X		X	X	X	X		X	X	10
I	X			X		X	X	X	X		X		7
J				X			X		X			X	4
K	X	X	X				X				X		5
L							X						1
M	X	X		X	X	X	X	X	X	X		X	10

P1.- Use of encrypted communication protocols.

P2.- Identification and evaluation of risks and threats.

P3.- Hardware security.

P4.- Establish passwords that are difficult to crack, strong and secure.

P5.- Static and dynamic code analysis.

P6.- Personnel trained in secure software development.

P7.- Firmware version or most recent stable operating system.

P8.- Security of integrated systems.

P9.- Security requirements and mapping services (Authentication, confidentiality, integrity).

P10.- Network traffic analysis.

P11.- Privacy and security by design.

P12.- Secure data transfer and storage.

As can be seen in the Table 3, most of the chosen methodologies consider the use of encrypted communication protocols and the verification of the most recent firmware version or stable operating system. Which means that the minimum parameters must focus on secure communication through some type of encryption, in addition, that the IoT architecture uses updated systems and is as stable as possible. On the other hand, those methodologies that did not exceed six points were discarded, since these methodologies do not consider relevant security parameters such as: hardware security, network traffic analysis and secure code.

However, the methodologies that exceeded six points are those that met the highest safety parameters but had weaknesses in some criteria that the Hardening and ENISA methodologies take into account. Finally, the ENISA methodology was chosen, since it presented several precise parameters, which were adapted to the architecture and facilitated comparison with the Hardening methodology, since both complied with most of the proposed security parameters, obtaining a total of ten points, this allowed for a more sustainable comparison.

3   Assurance of Architecture

As can be seen in Figs. 3 and 4, the architecture of the IoT system consists of three layers: Sensors were used in the perception layer: temperature, humidity, gas, ultrasound and a relay module. In the network layer, the integrated systems Raspberry Pi 3 and NodeMCU ESP8266 were used and in the application layer, the Ubidots platform was used to visualize the data from the sensors.

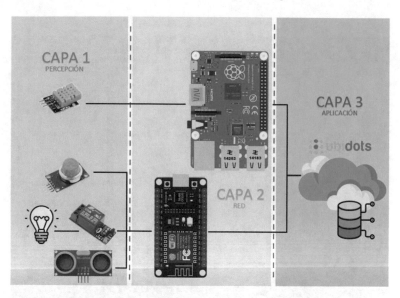

Fig. 3 Model for strengthening security in IoT systems

The union of IoT with the cloud offers services that minimize memory problems caused by handling large amounts of information [30]. IoT platforms can manage information, collecting it and sending it through a device. Likewise, they can store, analyze and display the information so that a user can easily view it [31].

In addition, the computing platforms proposed for IoT emerge on the market integrating different solutions and technologies that can be open source (open-source platforms) [30]. Next, some of the IoT platforms that were considered in this study will be named.

A. **Amazon Web Services IoT**

This platform can connect various devices to the Amazon Web Services (AWS) cloud and creating interactions between IoT devices. AWS IoT allows to connect sensors, actuators using HTTPS or MQTT. This platform provides ease to filter, transform and manage the information received from the devices. In addition, it offers security such as authentication and connection encryption, ensuring privacy and data integrity [31].

B. **Azure IoT Hub**

Azure IoT Hub enables you to manage secure, reliable, two-way communication between IoT devices and a backend. In addition, it can connect any device to the cloud [31]. It has built-in authentication and management for each device.

C. **Oracle Internet of Things Cloud Service**

This platform provides real-time connection to the cloud, allows you to analyze and integrate the data received with other applications. In addition, it has reliable

Fig. 4 Physical model of the IoT architecture

bidirectional communication and direct connection between the cloud and [32] IoT devices.

D. **Ubidots**

Ubidots can store and interpret sensor information in real time. In the same way, it displays information in a friendly way and allows creating applications for IoT in an agile, easy and creative way [32]. It is compatible with some devices including Arduino, Raspberry Pi, Android, Tessel and with those devices that can send HTTP requests [30]. It contains several libraries for API calls, it also has available languages such as: Python, Java, C, PHP, Node, Ruby [31].

The devices are displayed graphically on the platform panels. The API supports modifying, reading, deleting and sending the information that is stored through HTTP standards (GET, POST, PUT, DELETE) [32].

E. **Thinger.io**

It is an open-source platform, it provides a scalable cloud infrastructure ready to use, connect and manage IoT devices in a few minutes. Devices can be controlled from the Internet in an agile way, without worrying about the cloud infrastructure [30].

4 Case Study

With the collection of information and the selection of the ENISA methodology, a validation of the parameters of each methodology was carried out using compliance tables that are detailed below. In which the parameters were verified and a score was given with one (1) if the parameter was met and with zero (0) if the parameter was not met. Validation was performed by area or by layer according to each of the methodologies, this allowed the comparison of the following 3 scenarios:

A. **Scenario 1.- Without methodology**

For this scenario, certain parameters were considered to analyse both the Hardening methodology and the ENISA methodology, in order to verify the security status of the architecture. The percentage of compliance was calculated following the formula expressed below.

$$\% \text{ compliance } = \frac{\text{\# of parameters met}}{total\ of\ parameters} \times 100 \qquad (1)$$

In the first scenario, vulnerability tests were carried out in the base state of the architecture. The list of the parameters of the two methodologies was followed. These tests focused on reviewing the active services in the nodes through the Kali Linux distribution, the scan was carried out with the "Nmap" tool, which showed that the ports were in the "OPEN" state and in the number of default port, as can be seen in Fig. 5.

Then brute force attacks were carried out where it was possible to obtain the credentials that allow access to the device. This represents a weakness in the architecture, since it was found that the password and the user were established by default, as seen in Fig. 6.

The validation showed that the configurations of the active services were in the default ports. Users and passwords also presented a low level of security according to the recommendations, since they were vulnerable to attacks such as brute force. The tools used to check some of these results were: Nmap, netstat, IPTraf, Wireshark, etc., from the Kali Linux operating system (OS). This showed that the architecture had a low level of security as shown in table "B" of annex 1.

B. **Scenario 2.- Hardening Methodology**

In this scenario, we proceed with the implementation of the Hardening Methodology carried out at UDLA. This has 17 parameters which will be validated and scored in table "C" of annex 1. The parameters of security measures and good

Fig. 5 Scanning ports with nmap

Fig. 6 Hydra attack on IoT architecture

practices were analyzed, performing a validation in the communication protocols. In addition, passwords were strengthened according to security criteria that explain that a strong password must have at least 12 alphanumeric characters with a combination of uppercase, lowercase and special characters. In the same way, the change of default ports was carried out to ensure the implemented services as seen in Fig. 7. Network traffic was monitored to check for encrypted communication protocols.

Once the passwords and ports that were in default configurations had been changed, the brute force attack was executed with the "hydra" tool. This attack had no significant effect on password leaks, as can be seen in Fig. 8.

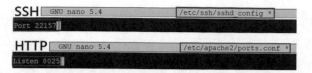

Fig. 7 Change default ports

```
                          root@kali:/home/kali                    _ □ ×
 File  Actions  Edit  View  Help
 6
      ●       /home/kali
   hydra  -L /home/kali/Desktop/user.txt  -P /home/kali/Desktop/password.txt  -t 6 ssh://192
 .168.100.105
 Hydra v9.1 (c) 2020 by van Hauser/THC & David Maciejak - Please do not use in military or
 secret service organizations, or for illegal purposes (this is non-binding, these *** igno
 re laws and ethics anyway).

 Hydra (https://github.com/vanhauser-thc/thc-hydra) starting at 2021-12-03 17:44:13
 [DATA] max 6 tasks per 1 server, overall 6 tasks, 50 login tries (l:5/p:10), ~9 tries per
 task
 [DATA] attacking ssh://192.168.100.105:22/
 1 of 1 target completed, 0 valid password found
 Hydra (https://github.com/vanhauser-thc/thc-hydra) finished at 2021-12-03 17:44:38

      ●       /home/kali
 ▌                                                                              2 ◎
```

Fig. 8 Attack with hydra to the IoT architecture after applying the second methodology

The verification of the parameters showed that the Hardening methodology obtained an improvement in the security level, since it was possible to contain some vulnerabilities such as brute force attacks, DDoS, MITM, etc. This showed that the architecture presented an acceptable level of security as shown in table "C" of annex 1.

C. **Scenario 3.- ENISA Methodology**

This methodology has 83 available parameters. Through a comparative analysis with the Hardening methodology, a list with 30 parameters was obtained in total, which is observed in table "D" of Annex 1, these are divided into areas with a certain number of criteria to be evaluated.

In the ENISA methodology, security aspects of privacy by design were evaluated, where parameters focused on privacy impact assessments, protection and password strengthening are considered, as shown in the Fig. 9.

Thus, authorization aspects were also taken into account to create protection policies and avoid manipulation, both of hardware and software. In the same way, it was verified that the communication through the different protocols is safe and reliable. On the other hand, network traffic was controlled to reduce vulnerabilities such as credential exposure. This ensures that devices on the network do not compromise one another and leak the data they save and send.

In Fig. 10, the port scan was carried out with the "Nmap" tool in order to know the active services and to be able to change the default port number and close the services that are not necessary. In addition, in this methodology the brute force attack was re-launched with the "hydra" tool, to verify its security, in

Fig. 9 Create and change default username and password

Fig. 11 thanks to the change of ports and not using passwords and default users, it presents a higher level of security.

Table "D" of annex 1 shows that the validation of parameters of the ENISA methodology increased the security of the architecture since it includes the strengthening of passwords, the change of default users and ports, showing an acceptable level of security.

5   Discussion of Results

To know the risks that an IoT architecture can have, 5 criteria were obtained, which are: NULL, LOW, MEDIUM, HIGH AND CRITICAL. To obtain the percentage of each criterion and the risk ranges in which the IoT architecture is found in each methodology, as seen in the Table 4. In formula (2), the risk number (nr) is used, which goes from HIGH to NULL, numbered from 1 to 4, respectively, and the total risks (cr).

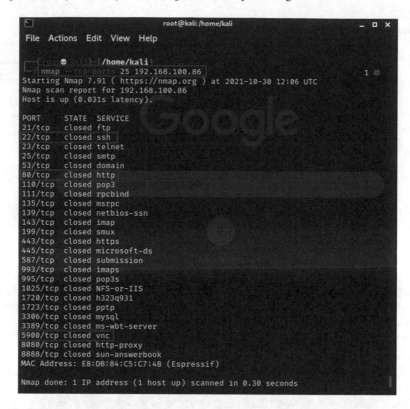

Fig. 10 Scanning ports with Nmap after applying the ENISA methodology

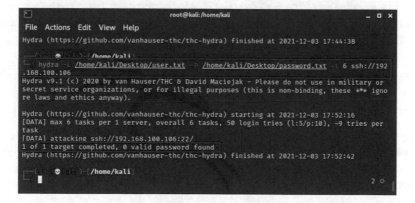

Fig. 11 Hydra attack on IoT architecture after applying the third methodology

Table 4 Risk percentages

Risk	Percentage (%)	Range
Null	100	x = 100
Low	75	$100 < x \leq 75$
Medium	50	$75 < x \leq 50$
High	25	$50 < x \leq 25$
Critical	0	$0 \leq x < 25$

$$\% \, risk = \frac{nr}{cr - 1} \times 100 \tag{2}$$

To present the results of the first scenario, table "B" of Annex 1 was made in which parameters were validated to verify the level of security presented by the IoT architecture.

It was evidenced that there were some vulnerabilities since the compliance percentage is 23.53%, which means that it presented a CRITICAL level of risk validating the parameters of the Hardening methodology. In the same way, the parameters of the ENISA methodology were verified, which yielded a compliance percentage of 30%, which represents a HIGH level of risk according to the Table 4.

Table "C" of Annex 1 shows the parameters met, reaching a total of 70.59% security in the architecture, using the Hardening methodology. Likewise, table "D" of annex 1 shows the parameters met, reaching a total of 73.00% security in the architecture, using the ENISA methodology.

Once verified that without methodology there is a high level of risk due to the vulnerabilities found. It was verified that by implementing the methodologies, ENISA had a higher security percentage than the Hardening methodology. Therefore, it was possible to state that said methodology complies with more security aspects for IoT architectures.

6 Conclusions

Over the years, cybersecurity has become a very relevant issue for society, since the protection of user data is given priority and it is necessary to apply different methodologies to correctly secure them. With this comparison of parameters that are presented between the selected methodologies, it will help to measure a considerable level of security and will facilitate the selection of the best methodology for the protection of users and their data.

When validating the information obtained from each methodology and using the correct pen testing tools when verifying compliance with the parameters in the 3 scenarios, it was possible to observe that both methodologies exceeded 70% compliance after being applied, respectively, which means that there are no considerable differences in the security they provided to the IoT architecture in which they were tested, but both methodologies managed to obtain an acceptable level of security according to the tables attached to this document.

By means of a mathematical analysis carried out in compliance with the parameters that are needed to grant an adequate level of security to the IoT architecture. It can be noted that an ideal maturity level was validated in the Hardening methodology. However, compared to the ENISA methodology, Hardening does not cover certain security aspects that are considered important to have a secure ecosystem, according to various international organizations.

In conclusion, it was confirmed that the IoT base architecture was insecure, since it had a low percentage of compliance that compromised the aspects of confidentiality, integrity and availability. Therefore, it is currently necessary to apply cybersecurity methodologies so that the information that circulates on the Internet is not compromised, either due to ignorance or due to not correctly applying the appropriate security measures.

The results indicate that both the Hardening methodology and the ENISA method ology contain minimum parameters to ensure an IoT architecture. However, it depends on the IoT ecosystem to be implemented, since each one requires knowing the business approach, the sensitivity of the data that is going to be handled in the network, and the needs of the user.

Hyperlink Annex

https://udlaec-my.sharepoint.com/:f:/g/personal/nicolas_moreta_udla_edu_ec/EpB c75mf3c5Cu-3f7odqpZcByMVXn8vJhdVux0-PZca6Ng?e=ZrHt5m.

References

1. El origen e historia del Internet de las Cosas (IoT). http://www.bcendon.com/el-origen-del-iot/
2. Irshad M (2017) A systematic review of information security frameworks in the internet of things (IoT). In: Proceedings—18th IEEE international conference on high performance computing and communications, 14th IEEE international conference on smart city and 2nd IEEE international conference on data science and systems, HPCC/SmartCity/DSS 2016, pp 1270–1275
3. Vailshery L (2021) Número de dispositivos conectados en todo el mundo 2030—Statista, Jan 2021. https://www.statista.com/statistics/802690/worldwide-connected-devices-by-access-technology/
4. Acharya V, Hegde VV (2020) Security frameworks for internet of things systems—A comprehensive survey. In: Proceedings of the 3rd international conference on smart systems and inventive technology, ICSSIT 2020, no. Icssit, pp 339–345
5. Brass LT, Carr M, Elsden M, Blackstock J (2018) Standardising a moving target: the development and evolution of IoT security standards. In: IET conference publications, vol 2018, no. CP740, pp 1–9
6. Arcos E (2016) Mirai e internet de las cosas responsables del ataque DDoS de octubre 21, Oct. https://hipertextual.com/2016/10/mirai-ddos-internet-cosas
7. Investigación sobre ciberseguridad de internet de las cosas (iot): una revisión de los temas de investigación actuals. IEEE Internet of Things J 6(2): 2103–2115 (2019)

8. E. U. A. for Cybersecurity (ENISA), Baseline Security Recommendations for IoT in the context of Critical Information Infrastructures (2017), November. https://www.enisa.europa.eu/public ations/baseline-security-recommendations-for-iot
9. Khatoun R, Zeadally S (2017) Cybersecurity and privacy solutions in smart cities. IEEE Commun Mag 55(3):51–59
10. A survey on IoT architectures, protocols, security and smart city based applications. In: 8th International conference on computing, communications and networking technologies, ICCCNT 2017 (2017)
11. Introducción conclusiones hallazgos ciberseguridad en tiempos de pandemia preocupaciones incidentes controles. https://www.welivesecurity.com/wp-content/uploads/2021/06/ESET-sec urity-report-LATAM2021.pdf
12. ESET: Threat Report Q2 2020. Comput Fraud Secur 2020(8):4 (2020)
13. Al-Boghdady KW, El-Ramly M (2021) The presence, trends, and causes of security vulnera-bilities in operating systems of iot's low-end devices. Sensors 21(7). https://www.mdpi.com/1424-8220/21/7/2329
14. Pacheco J, Tunc C, Hariri S (2019) Security framework for IoT cloud services. In: Proceedings of IEEE/ACS international conference on computer systems and applications, AICCSA, vol 2018-November, pp 1–6
15. Ichi.pro. Estándares y protocolos para soluciones de IoT. https://ichi.pro/es/estandares-y-pro tocolos-para-soluciones-de-iot-37517846919484
16. Coates M, Hero A, Nowak R, Yu B (2002) Internet tomography. IEEE Signal Process Mag. May, to be published
17. Huraj L, Šimon M, Horák T (2020) Resistance of iot sensors against DDoS attack in smart home environment. Sensors 20(18). https://www.mdpi.com/1424-8220/20/18/5298
18. Echeverría, Cevallos C, Ortiz-Garces I, Andrade RO (2021) Cybersecurity model based on hardening for secure internet of things implementation. Appl Sci 11(7). https://www.mdpi.com/2076-3417/11/7/3260
19. Peciña K, Estremera R, Bilbao A, Bilbao E (2011) Physical and logical security management organization model based on ISO 31000 and ISO 27001. In: 2011 Carnahan conference on security technology, pp 1–5
20. Núñez R (2018) Primera Norma internacional ISO para Internet de las Cosas IoT world online. https://www.iotworldonline.es/primera-norma-internacional-iso-para-internet-de-las-cosas
21. Cyber. En 303 645 - v2.1.0 - cyber; cyber security for consumer internet of things: baseline requirements (2020). https://www.etsi.org/deliver/etsien/303600303699/303645/0201.0030/en303645v020100v.pdf
22. Webb J, Hume D (2018) Campus IoT collaboration and governance using the nist cybersecurity framework. In: Living in the Internet of Things: cybersecurity of the IoT—2018, pp 1–7
23. Kearney P, Asal R (2019) ERAMIS: a reference architecture-based methodology for IoT systems. In: Proceedings—2019 IEEE world congress on services, Services 2019, vol 2642–939X, pp 366–367
24. Guzman, Bassem C (2021) Using the isvs - owasp isvs (pre-release 1.0rc). https://owasp-isvs.gitbook.io/owasp-isvs-pr/usingisvs
25. Iot SF (2020) IoT security compliance framework release 2.1. https://iotsecurityfoundation.org
26. Josyula SK, Gupta D (2017) A new security methodology for internet of things. In: 2017 international conference on computing, communication and automation (ICCCA), pp 613–618
27. Rahman FA, Daud M, Mohamad MZ (2016) Securing sensor to cloud ecosystem using internet of things (iot) security framework. In: Proceedings of the international conference on internet of things and cloud computing, ser. ICC '16. Association for Computing Machinery, New York, NY, USA. https://doi.org/10.1145/2896387.2906198
28. Matheu SN, Hernandez-Ramos JL, Skarmeta AF (2019) Toward a cybersecurity certification framework for the internet of things. IEEE Secur Privacy 17(3):66–76
29. Asllani AL, Lari N (2018) Strengthening information technology security through the failure modes and effects analysis approach. Int J Quality Innov 4(1):1–14

30. Ortiz Monet M (2019) Implementación y Evaluación de Plataformas en la Nube para Servicios de IoT. Oct 2019. https://riunet.upv.es/handle/10251/127825
31. Martinez Jacobso R (2017) Comparativa y estudio de plataformas IoT. https://upcommons.upc.edu/handle/2117/113622
32. Pires PF, Delicato FC, Batista T, Barros T, Cavalcante E, Resumo MP. Capítulo 3 Plataformas para a Internet das Coisas

A Critical Review into the Digital Transformation of Land Title Management: The Case of Mining in Zimbabwe

David V. Kilpin, Eustathios Sainidis, Hamid Jahankhani, and Guy Brown

Abstract The mining sector is worth trillions globally and the cornerstone of many developing nations wealth. National and corporate interests sit at the heart of negotiating fair value when pursuing profit while at the same time attempting to balance equality and sustainability of the sector. While the mining sector is of significant value to the economy, there are challenges in managing land administration. This research aims to provide a strategic template for the Zimbabwean government to apply the technological benefits of digital transformation to bring transparency and accountability for stakeholders, including the people and partners of Zimbabwe. The government would respond to implement technological advances such as Blockchain, Self-Sovereign Identity (SSI) and Artificial Intelligence (AI). The fundamental importance of land and its management lies at the core of this study as strategic research for Zimbabwe. Therefore, this research investigates how digital transformation can support mining land title management to create transparency and accountability. From a myriad of available options, this research proposes a technological solution based on Blockchain's Self-Sovereign Identity (SSI) to address the lack of Transparency and Accountability in mining land title management.

Keywords Blockchain · Self-Sovereign Identity · SSI · E-Government administration · Land registry · Mining administration · Cadastre · Digital transformation · Zimbabwe · Decentralised identifiers · DID · Transparency · Accountability · Smart contracts

1 Introduction

The fundamental importance of land and its management lies at the core of this study as strategic research for Zimbabwe. Therefore, this research investigates how digital transformation can support mining land title management to create transparency and accountability.

D. V. Kilpin · E. Sainidis · H. Jahankhani (✉) · G. Brown
Northumbria University, Newcastle Upon Tyne, UK
e-mail: Hamid.jahankhani@northumbria.ac.uk

H. Jahankhani (ed.), *Cybersecurity in the Age of Smart Societies*,
Advanced Sciences and Technologies for Security Applications,
https://doi.org/10.1007/978-3-031-20160-8_3

Major weaknesses associated with the manual system of title administration provided for by the Mines and Minerals Act of Zimbabwe can be categorised into: (i) too much discretionary power; (ii) lack of commonly-shared information base among the various vertical levels of decision-makers; (iii) failure to conduct adequate due diligence on assets and title awardees; (iv) human errors of licensing authorities; (v) high burden of administration of a manual system and high cost of compliance; and (vi) frequent disputes between title holders due to overlapping or coincident titles.

Excessive discretionary powers emanate from lack of clearly specified criteria or 'reasonable grounds' as well as a distinct lack of judicial review of certain decisions by administrators. Examples of these weaknesses are manifested in Section 20 of the Mines and Minerals Act (subsections 3 and 4) pertaining to refusal to grant a prospecting license, Section 27(1)(a) on permission for a prospector to dispose of minerals for assay purposes, Section 85 on re-award of a Special Grant, and Section 89 with regard to refusal to grant an EPO on the grounds of inadequate financial standing or national interest.

The important objective in exposing these weaknesses is to clearly indicate the need for a digital transformation of the manual mining title management system as outlined in the Mines and Minerals Act, and hence significant reform of the Act itself. Discretionary powers that are not subject to judicial review or appeal are breeding grounds for corruption, bribery, extortion, and various other forms of unethical practices by both administrators and industry participants. A computerised system that hosts the various clearly specified decision criteria or reasonable grounds and links them to particular geographical areas and ranges of possible optional decisions could help deal with these problems and risks as shown in Fig. 1.

In any case, where a decision is final, the need to base such decisions on the most accurate information and data is obvious.

RISK ASSESSMENT OF MINING LAND TITLE MANAGEMENT (MLTM) IN THE CONTEXT OF TRANSPARENCY AND ACCOUNTABILITY. (AS DEFINED BY ZIMBABWE'S MINES AND MINERALS AMENDMENT BILL - 2016) In answer to the question: "Are self-sovereign identity (SSI) and Blockchain feasible tools to enhance the transparency and accountability of mining land title management?		
RISK NUMBER	**RISK DESCRIPTION**	**RISK LEVEL (LOW, MEDIUM, HIGH)**
1	In practice, there is no due diligence capability during Title applications to verify claims regarding their capacity, financial resources or integrity - such as past lawful conduct and compliance.	HIGH
2	Title award processes unlikely to follow set procedures as legislated with auditable authorisations at each stage.	HIGH
3	Mining Land Titles and licence information will not be publicly known or available in real-time.	HIGH
4	Mining Titles can be altered, transferred, terminated or modified without miner's awareness. Little opportunity to respond before changes take lawful effect requiring difficult and expense reversals.	HIGH
5	Speculation by exploiting weak manual systems and stock piling Mining Titles without production intentions	HIGH
6	Opaque Mining Title Ownership contrary to legislation, regulation or policy.	HIGH
7	First come, First Served', Mining Title applications can be circumvented or tampered.	HIGH

Fig. 1 Risk assessment of mining land title management

2   Deployment of the Cadastre System in Zimbabwe

The Zimbabwean government has recently initiated the process of establishing a mining cadastre in line with the prevailing trend in Africa, has created an online mining cadastre portal [33]. The Ministry of Mines and Mining Development is currently going through a process of data capture, data cleaning and verification, and at a later date is expected to declare that all data have been verified. The current process apparently depends on the Ministry's own records and the input from title holders, applicants and other stakeholders who can contribute through verifying the records in the developing cadastre and alerting the Ministry of any errors or omissions.

In order to engage with the Ministry through the portal, the various stakeholders are required to register. Existing rights holders or those with pending applications need to download an MCO01 Registration form which they complete and submit in person together with supporting documents. Prospective rights applicants do self-registration completely online.

If administrative discretionary powers are not curtailed by law, then the implementation of the CMC and its ability to achieve the lifeblood of the mining sector (protection of mining titles) will be seriously impeded.

3   Blockchain Technology and Its Implications for Land Title Management

A blockchain is a growing list of records, or blocks, that are linked using cryptography [35, 40]. A cryptographic hash is an algorithm that maps data of an arbitrary size to a bit array of a fixed size. Each block contains a hash of the previous block, or a form of transactional data. The purpose of having information about the previous block means that each additional block is reinforced by the preceding one. Blockchains are typically managed in a peer-to-peer network for use on a distributed ledger, where nodes collectively adhere to a protocol to communicate and validate new blocks. The fundamental advantage offered by blockchains which may be useful for their relevance to land title management is that that are secure by design as data in any block cannot be retroactively altered without altering the subsequent blocks [24]. The chain that joins the two blocks provides an iterative process and ensures the integrity of other blocks, and the integrity of the data is ensured after the block is digitally signed [26].

What this means is that blockchain provides a fool-proof means of storing and retrieving data that is not possible in the manual storage of land titles in many developing countries.

One question that then arises is about the digital identity of those who hold lands. In the digital space, identities are constantly negotiated rather than being fixed. According to SSI, there are entities whose identities are generated which correspond to the former. The identities, at the same time, form the basis of attributes or identifiers. Thus, we can notice how SSI ensures anonymity at the same time as ensuring that trust is established. This is because one party often provides credentials to another party and the recipient that they can trust and the trust of the verifier are then transferred to the credential holder [42].

Decentralised identifiers (DIDs) are often used to generate SSIs. These DIDs enable a verifiable, decentralised identity.

Properties have traditionally been dealt with through cadastres in their capacity to provide a comprehensive recording of the real estate. Cadastral surveys document the boundaries of a property. A cadastral parcel is referred to as land that has a unique set of property rights [10]. What makes cadastral maps particularly open to politicking or manipulation is the paperwork they are generally accompanied by when compared to the seamless process of authentication offered by blockchains. There are specific provisions related to the jurisdictions that shape how the documents are managed and authenticated. These jurisdictions shape the stakeholders who are required to sign and authorise the document. While in principle the same mechanism can take place through blockchains via decentralised identities, traditional cadastral regulations, often make it a tiresome process requiring multiple signatures, also creating problems of storage.

As has been shown in case of post-colonial societies, these documents are accompanied by paperwork which can be manipulated to make claims to land, such that the government's claim may vary from the individuals [23]. There is also an element of surveying which is shaped by legislation that determines the form of these competing claims. While there have been calls to digitise and introduce blockchains to reduce corruption, [27] has made the case that individuals should possess the right to choose whether they prefer traditional registries of blockchains, arguing from a shift from cadastre bodies as having sole monopoly on the market. Scholars have referred to the use of blockchains in land regulation as real estate tokenisation [27]. Moreover, it has been argued that it is important to have a cross-blockchain protocol designed to support free choice and transferability, as blockchains offers a disadvantage, namely related to intolerance to retroactive transactions, as mentioned above.

In many contexts, there is a suspicion toward centralised government land management systems given longstanding colonial histories that still impact national land management systems in many countries in Africa. In such a context many scholars have referred to blockchains as anti-corruption tools, with Aarvik [1] suggesting that the introduction of blockchain directly impact many areas of governance and trust. Aarvik [1] has suggested that blockchain is likely to work more effectively in areas where trust in codes is likely to be greater than trust in persons yet has made the case that digital literacy is necessary before the implementation of blockchains.

4   The Applicability of Blockchain and SSI Technology to Mining Title Management: Adaptation or Adoption?

Existing research on mining in Africa has shown that private companies have a far greater power over local government in shaping mining regulations, and this is a situation that has resulted in the marginalisation of local owners. Due to the incapacity to monitor and implement regulations, it is possible that the result may by the bypassing of government regulations by private entities, who in the past have also co-opted armed groups to protect their installations. Land use conflicts between large scale miners and community members has been a frequent occurrence around the world. As a result, it has become increasingly difficult for miners who demand a significant area to co-exist with community members, especially as community members rely on these lands for their livelihoods [18] (Fig. 2).

It has also been noted that government involvement for the purpose of dispute resolution is often miniscule. Thus, in the past it has been recommended that if governments play a central role in arbitrating conflicts, it can possibly allow for improved dispute resolution. As Castro and Neilson [7] have shown, natural resource conflicts are often "severe and debilitating, resulting in violence, resource degradation, the undermining of livelihoods, and the uprooting of communities (229)".

Often in the utilisation of natural resources like mines, parties tend to try and maximise their own interests and, in the process, end up defeating the interest of the other [37]. Yet, among developing countries in Africa, such as Zambia, mining contributes to 90% of the country's export earnings and provides employment to 15% of the country's workforce, meaning that it is often in the government's interests to incentivise mining companies over the interests of the communities [32].

Fig. 2 Spatial characteristics and risk factor identification for land use spatial conflicts. *Source* [48]

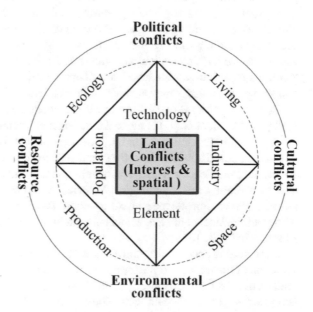

In many countries, mining efforts have also meant that entire communities have effectively come to live under a country's mining project, as in the case of the Northern Brazilian state of Para where 400,000 individuals have come to live seeking to tap into the residents' resources, 100,000 of whom directly live inside the gold mine [29]. Another way in which mining effects a local community is when demographic shifts caused by mining result in increasing the cost of living, which in turn results in indigenous communities rapidly seeking to derive a livelihood in other areas [12]. Not only is there a possibility of conflict between landowners and owners of mining companies, but also between miners and, say, herders, as mining introduces new ideas of capital ownership that local communities are not familiar with.

Countries like Ghana are characterised by a history of conflict between small miners and large scale mine management. Often individual mine lands are located illegally within land occupied by large-scale mining companies which is often the product of familial and cultural relations with land.

There are many challenges of the land mining sector, especially as private companies conduct mining activities with the help of government issued contracts over publicly owned lands. In some countries, both federal and state laws apply to mining, depending on whether the mine is situated on land where both surface and subsurface rights are held by private owners. When there is a split estate, the owner of the mineral rights often has the right to develop the mineral resources.

This situation forces us to consider what possibilities and challenges exist in relation to the implementation of blockchains given the concept of split ownership. To answer this question, it would first be useful to consider the challenges of digitisation alongside the opportunities it creates created for actors. Partial digitisation has been more time consuming as it appears technically complicated and therefore requires process-specific training to technical and non-technical employees [11]. Information systems targeted at enhancing learning and knowledge management have mixed effects, often depending on organisational and other specifics [20, 43]. For example, access to knowledge management systems strengthens the association between innovation culture and innovation outcomes but diminishes the association between an autonomous culture and innovation outcomes [13]. Relatedly, using IS to enhance knowledge capabilities may enable continuous innovation [25].

Given the problems of digitisation without adequate training in much of the developing world, Alam et al. [2] make a case for incremental digitisation or the implementation of blockchains in land title management in Bangladesh. They propose a blockchain based solution that offers data synchronisation and transparency, ease of access, immutable records management as a faster and cheaper solution. They state that "considering the technological knowledge and capacity of the people and the government, we introduced a phase-by-phase Blockchain adoption model that starts with a public Blockchain ledger and later gradually incorporates two levels of Hybrid Blockchain" ([2], 1319).

The resulting decentralisation of control in combination with the immutable representation of the transfer of possession provides a unique opportunity to build collaborative, multi-sided, 'trustless systems' [14]. Ølnes et al. [39] enumerate the benefits and promises of blockchain technology in government as a means of information sharing. In general, blockchain technology allows for client-defined chains of entries, client-side validation of entries, a distributed consensus algorithm for recording entries, and mechanism for ensuring security. The data is secured by publishing it in an encrypted form in the immutable, distributed ledger [36]. The storage of critical information in the decentralised data layer protects the system from malfunction, obsolescence, and deliberate or involuntary record changes. All critical data records are secured by a "proof of existence" in the blockchain [29] (Fig. 3).

Every essential step in the registration process that qualifies as a high "standard of care" can be published to the blockchain. This provides data security and long-term data stability by making the entire data set immutable via notarisation. In addition, publishing data during each step of an instrument's recording process provides irrefutable "proof of process", which in turn protects the submitted records and establishes accountability. It is important to note that blockchain technology does not help address the accuracy of the land titles [39].

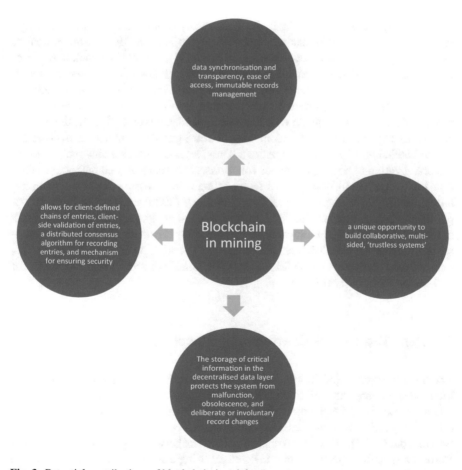

Fig. 3 Potential contributions of blockchain in mining

5   Discussion of Applicability of Blockchain and SSI Technology to Mining Title Management

Over the past decade, the mining industry has also integrated digital certification methods across mineral supply chains, which has added to a raft of high-level regulations and initiatives to bring a measure of accountability to an industry that has been tarnished by human rights violations, child labour, and minerals used to fund wars. As the most significant of these, blockchain technology entered the lexicon of the mining industry with the promise of eliminating the need to rely on third-parties for verification, verification, and auditing of supply chain information. Researchers have critically reassessed the subject-object dichotomy in extractive process by considering the cultural, social and biophysical value ascribed to natural resources.

As a result of technological advancements and security measures, new forms of monitoring and surveillance are necessitated. This is because blockchains enable technocratic governance that alienates or marginalizes people without the technical know-how about its uses ([8], 202). As Käll observes, it is precisely the "security" supposedly afforded by blockchain technology that leads to "locked-down control over the digitalized worlds we inhabit" (2018, 134). It therefore makes sense to have a more nuanced approach toward blockchain claims of openness, democratization, and accessibility, especially as these technologies cause a wide range of profound changes at both structural and mundane levels. In light of attempts to create blockchains, one must contend with the new dependencies and relations that will be created as a result, the private companies or semi-private partnerships oversee the use of blockchains in land management. Thus, arguably blockchains may result in the privatization of some governance functions. Some have argued that actors organized outside these institutions that face obstacles in accessing "closed accounting blockchains" will be displaced, to follow Çalışkan's (2020) actor-based taxonomy of public and private blockchains. By taking a political geography perspective, we can measure whether and how blockchain-based digital extraction produces new landscapes, which may result in social and economic exclusion in mining economies, unequal data production and validation, and inequitable distribution of incentives for end-to-end traceability (Fig. 4).

6   Steps Toward the Creation of a Smart City

The most intriguing applications among the numerous intriguing ones based on the blockchain concept are smart contracts, which have programmable code that runs when specific criteria in the blockchain network are satisfied. "Smart Contract" is an outstanding idea to systematically execute contracts between the linked parties, and it was first put forth by Nick Szabo in [45]. A contract between two or more parties that follows a specified course of action is created using a smart contract. In an effort to liquefy these hard assets into a liquid economy, blockchain is being actively pursued

Pros	Cons
Transparency	Unequal data production
reduction in corruption and illicit flows	Unequal distribution of incentives for end-to-end traceability
prevention from deliberate entry changes	Social and economic exclusion based on digital literacy
Traceability	

Fig. 4 Blockchains and mining

[28]. Additionally, it will eliminate corruption, produce tamper-proof land register records, strengthen the mortgage market, and produce a liquid economy. This section discusses possible and implemented blockchain-based management strategies for various city departments and their administration (Fig. 5).

The majority of the discussed proposals in this field respond to the challenge of urban space sustainability and advance governance and e-government [6, 15, 16]. Other levels of government, like regional or national government, may be able to use this governance.

Citizen transactions often have a severe lack of trust. Most often, they also demand expensive monitoring, time-consuming reputation checks, or third-party mediation [4]. The majority of these difficulties have been addressed as a result of the advent of blockchain in government. The blockchain-based smart city solutions offer a standard platform to integrate other city services and increase their benefits. Other technologies have also been crucial to this development. The Internet of Things (IoT), cloud computing, and mobile computing are highlighted in this series. Other crucial technologies include 5G, IA, and big data.

Mobile computing in this context mostly alludes to the possibility of smart devices and other wearables in individuals' hands, which have become commonplace in contemporary society. City administrators can offer new citizen-centric tools to make it easier for users to interact with the city and access its services through new Apps and mobile services. Additionally, the majority of the services provided by blockchain technologies are typically accessed via mobile devices.

IoT is a new paradigm that can give things communication, processing, and sensing capabilities. These "objects" or "devices" are employed to collect data from the urban environment and give city officials useful information to make decisions [34]. Security and privacy concerns become a crucial issue in a smart city since it collects and processes information about people's lives and environments that is sensitive to privacy. In order to achieve this, the IoT and blockchain combination improves security for accessing data collected by things and for controlling devices.

One of the most cutting-edge approaches to societal adoption of information technologies is the cloud computing paradigm. This paradigm gives city services universality and processing capability. The cloud has evolved into a location where

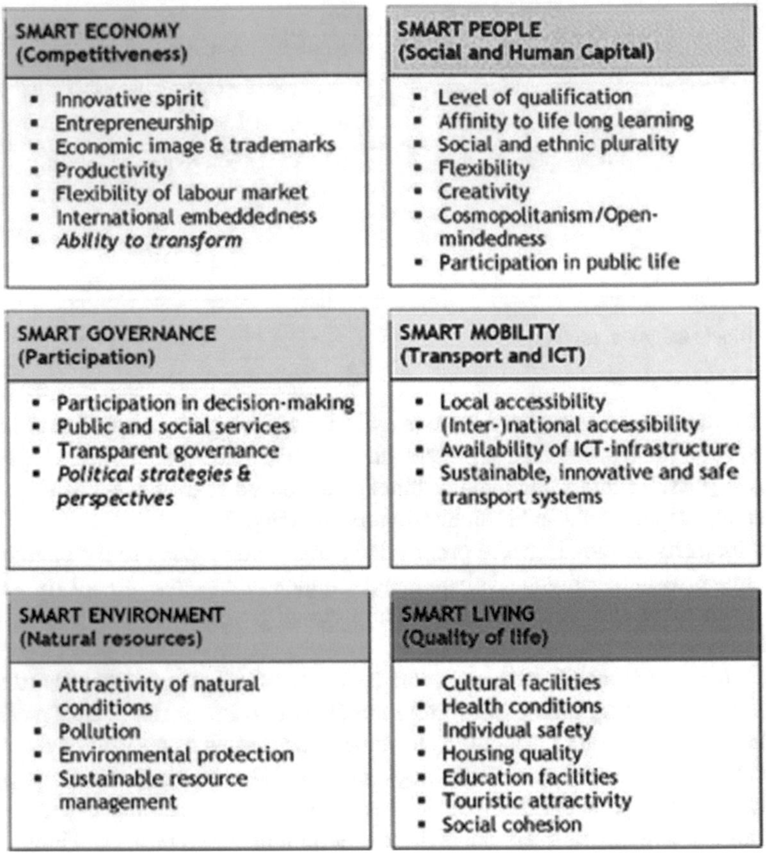

SMART ECONOMY (Competitiveness)	SMART PEOPLE (Social and Human Capital)
• Innovative spirit • Entrepreneurship • Economic image & trademarks • Productivity • Flexibility of labour market • International embeddedness • *Ability to transform*	• Level of qualification • Affinity to life long learning • Social and ethnic plurality • Flexibility • Creativity • Cosmopolitanism/Open-mindedness • Participation in public life
SMART GOVERNANCE (Participation)	SMART MOBILITY (Transport and ICT)
• Participation in decision-making • Public and social services • Transparent governance • *Political strategies & perspectives*	• Local accessibility • (Inter-)national accessibility • Availability of ICT-infrastructure • Sustainable, innovative and safe transport systems
SMART ENVIRONMENT (Natural resources)	SMART LIVING (Quality of life)
• Attractivity of natural conditions • Pollution • Environmental protection • Sustainable resource management	• Cultural facilities • Health conditions • Individual safety • Housing quality • Education facilities • Touristic attractivity • Social cohesion

Fig. 5 Smart environment and mobility characteristics, Orecchini et al. [38]

data is sent for additional processing and where city services are hosted. In this situation, blockchain offers confidence to verify the accuracy of user data and provide easy data access [9].

The idea of "e-government" refers to the role of public administration that is limited to the formalisation of processes and digitization of resources [4]. The use of ICT technology to uphold good government ideals and accomplish policy objectives is referred to as the new paradigm of the Digital Government (DG).

E-government is a potent tool for implementing international sustainability goals and tackling the aforementioned difficulties. The majority of the solutions are used in smart city settings, where the benefits they offer are integrated with other features of the digital society to enhance citizens' quality of life. The primary and most prominent solutions in this area, involving a number of city management departments. These concepts can frequently be linked with other digital services offered by the city to work in tandem with them and reach a wider audience.

Real estate transactions are lengthy and complex, and they are frequently complicated by fraud or other irregularities (such as double-selling, forged documents, and collusion). Blockchain technology can be used to help prevent fraud, improve efficiency and transparency, enable strict ownership protection, and shorten transaction times and costs. A permanent record of real estate transactions can likewise be created using blockchain technology [46]. Current bidding systems can be replaced with blockchain-based ones, which would enhance the process and increase data accuracy and transparency. Some governments have already begun using Blockchain technology for title tracking [31].

Another example is land administration, which covers all tasks related to establishing, preserving, and disseminating data on ownership, value, and land use. In a manner similar to the real estate industry, blockchain technology can be utilised in land registry and management as a transparent and decentralised public platform for property rights and historical transactions, which is a persistent issue worldwide, particularly in emerging nations [47]. Poor governance, a lack of transparency, corruption, and limited access by appropriate jurisdictions, particularly in regional areas and following the occurrence of natural catastrophes that damage data centres and transactions, are problems in land administration.

According to [5], a few benefits of blockchains include (a) Title deed registration, (b) time-stamped transactions; (c) multi-party transparent governance tools; (d) tamper-proof recording system; (e) disaster recovery system; and (f) restitution and compensation in post-conflict zones. Asymmetric information and transparency, when information about properties is not easily available or is concealed, are two problems the market faces.

Managing title deeds and sharing data are significant issues for which blockchain can offer a solid answer. Building confidence among co-owners is a challenging process, in particular. With the help of Blockchain, the real estate market can be freed from these limitations and challenges and function more like the stock market. This would cause a tectonic shift with potential seismic repercussions for other upstream or downstream markets, including banking, insurance, mortgage, and construction [17].

Blockchain has a strong foundation for authenticating and registering every action on digital platforms in incredibly fine detail, including creation, modification, and deletion [22]. This eliminates any potential for data exploitation. In addition, Blockchain enables the establishment of multi-layer data access controls to thwart unlawful or damaging activities on the dataset. By doing this, centralised agencies or organisations that once owned, regulated, and controlled data are rendered obsolete.

The information could be securely kept in a manner that guarantees and protects its authenticity and acts as a firewall to prevent unauthorised access [21]. Finally, blockchain technology and the ecosystem that surrounds it are developing quickly and perhaps producing more questions than they are providing. One of the major concerns that need more investigation is the requirement for laws and rules to develop a transparent governance framework in order to guarantee the life of the Blockchain.

7 How Blockchain-SSI Technology Can Facilitate the Adoption of Transparency and Accountability Initiatives (TAIs)

Our current understanding of the technology limits its application in academics and management. Blockchain may bring supply chain transparency to a new level, but it is difficult to implement. The impact of Artificial Intelligence and Blockchain technology has been dramatic, and most people's daily routines have been impacted by these technologies. Such technologies provide various reliable services, which are being trusted by all the users. Thus, it provides an effective means of using artificial intelligence and blockchain technologies for governance transparency. Since the traditional mechanisms utilized in governance need to be reformed in respect of various parameters, such as the availability of information to users, as well as information asymmetries between users [3]. Similarly, the speed of transfer of data and the security of sensitive information are being improved. As a result, through blockchain technology, a check and balance mechanism can be applied to each and every aspect of governance, thus eliminating the possibility of corruption [30]. As blockchain technology has recently emerged, it can enable things that seemed impossible in the past, such as recording assets, allocating value, and most importantly registering and monitoring the footprints of transactions. without any central It provides transparency, integrity, and traceability of information and data which can be validated and verified by trustless parties without the need for a central authority. By providing a distributed database capable of certifying records and transactions—or "blocks"—without using a central database and in a way that cannot be altered, tampered with, or altered, blockchain can raise transparency and governance levels to an unprecedented level. By using this, we can be assured that this system has an unprecedented level of integrity, confidentiality, and reliability, and that there is no single point of failure. In addition, as artificial intelligence (AI) is increasingly deployed across applications and multiple sectors for decision-making and facial recognition, the concern is across the transparency and responsibility of comprehensively testing or verifying advanced or AI-powered algorithms from time to time. It has the potential to bring a lot of value to our society today and in the future, but it needs to be controlled thoroughly by establishing norms, strategies, and procedures that govern how AI applications are built, implemented, and used. For AI-based algorithms to succeed, optimization will be crucial to capturing the value intended for their deployment. Government use cases are particularly relevant in these areas because they have implications for the adoption of BC infrastructure. It can be an issue of the national government, such as non-action or resistance from regulators, or an issue of the environment, such as digital transformation laws or disadvantaged persons' and officials' personal data. Additionally, to create a sense of trust in a learning process, it is crucial to capture the perspectives of various participants from different perspectives. As one use of federated learning that has become known recently, the authors discussed a joint AI application. Through the proposed work, multiple aspects of trust are addressed in the domain of AI training by leveraging Blockchain technology, but for providing transparency in governance, some governing parameters are still required in relation to the use of AI and Blockchain technology [41].

As non-technical issues need to be addressed, creating transparency and account-ability might be more challenging than expected. Several key issues have been discussed, such as digital IDs, privacy, interoperability, connectivity, and technology awareness. efficiency and storage size. Among the important considerations in developing a transparent and accountable e-Government system are the following: data validity, check and control mechanism, digital signature, algorithm transparency, law and regulation support, and dispute resolution. Considering the practical realities in China and the potential applications of blockchain technology in e-government, [19] finds that blockchain technology can lead to positive effects such as: (1) improved quality and quantity of government services, (2) more transparent and searchable government information, (3) more collaborative efforts within different organizations, and (4) assisting the development of an individual credit system in China. However, information security, cost and reliability are still major problems in application. For promoting and applying blockchain technology in e-government, it is essential to develop an application platform and management standards. Using blockchain technology to improve government services is a great idea, but standardizing the management system, processes, and responsibility for the application is necessary to promote it in the future. Some articles have examined the current implementation of blockchain-based applications in Estonia [44] and Chinese e-Government [19]. The majority of articles presented conceptual frameworks, as can be seen from this overview. Most articles did not present empirical data. The adoption of blockchain technology in the public sector represents a theoretical view, rather than a practical one or based on empirical evidence. There is a lack of academic research in this field, as evidenced by an analysis.

8   Conclusions

The presentation has considered the specific problems posed by land titles in the mining industry, especially in developing countries. It has considered how mining practices bring to fore conflicts between local owners, governments and mining companies over ownership over surface and subsurface mines. In the context of developing countries where land titles are often handled through excessive paper work and bureaucracies, the competing claims of ownership can be resolved through the introduction of blockchains. Blockchains typically rely on a cryptographic hash which is an algorithm that maps data of an arbitrary size to a bit array of a fixed size. Each block contains a hash of the previous block, or a form of transactional data. The purpose of having information about the previous block means that each additional block is reinforced by the preceding one. The chain that joins the two blocks provides an iterative process and ensures the integrity of other blocks, and the integrity of the data is ensured after the block is digitally signed. Yet just as the introduction of blockchains have the potential of resolving long-standing conflicts, it also creates new digital territorialities, as access to knowledge may differ and the difficulty as a block cannot be retroactively altered without altering the subsequent blocks, creates new forms of challenges and contestations.

References

1. Aarvik P (2020) Blockchain technology to prevent corruption in Covid-19 response: how can it help overcome risks? Blog post. https://www.cmi.no/publications/7259-blockchain-technology-to-prevent-corruption-in-covid-19-response-how-can-it-help-overcome-risks. Accessed 25 Oct 2022
2. Alam KM, Ashfiqur Rahman JM, Tasnim A, Akther A (2020) A Blockchain-based land title management system for Bangladesh. J King Saud Univ Comput Inf Sci
3. Alkalha M, Al-Zoubi & Alshurideh (2012) Investigating the effects of human resource policies on organizational performance: an empirical study on commercial banks operating in Jordan. Eur J Econ Finan Admin Sci
4. Allessie D, Sobolewski M, Vaccari L (2019) Blockchain for digital government: an assessment of pioneering implementations in public services. https://doi.org/10.13140/RG.2.2.34874.85449
5. Anand A, McKibbin M, Pichel F (2016) Colored coins: Bitcoin, blockchain, and land administration. In: Proceedings of annual world bank conference on land and poverty, Washington DC
6. Bibri S, Krogstie J (2017) Smart sustainable cities of the future: an extensive interdisciplinary literature review. Sustain Cities Soc 31. https://doi.org/10.1016/j.scs.2017.02.016
7. Castro A, Nielsen E (2001) Indigenous people and co-management: implications for conflict management. Environ Sci Policy 4:229–239. https://doi.org/10.1016/S1462-9011(01)00022-3
8. Cavanagh CJ, Benjaminsen TA (2017) Political ecology, variegated green economies, and the foreclosure of alternative sustainabilities. J Polit Ecol 24(1):200–216. https://doi.org/10.2458/v24i1.20800
9. Cha J, Singh SK, Kim T, Park J (2021) Blockchain-empowered cloud architecture based on secret sharing for smart city. J Inf Secur Appl 57:102686. https://doi.org/10.1016/j.jisa.2020.102686
10. Dale P, McLaughlin J (2011) Land administration. https://doi.org/10.1007/978-94-007-1667-4_15
11. Deshpande A (2019) An inquiry into the impact of digitization and customized ERP applications on twin engineers overall efficiency—An empirical study approach
12. Dorian J (1996) Minerals and mining in the transitional economies
13. Durcikova A, Fadel KJ, Butler BS, Galletta DF (2011) Knowledge exploration and exploitation: the impacts of psychological climate and knowledge management system access. Inf Syst Res 22(4):855–866
14. Glaser F (2017) Pervasive decentralisation of digital infrastructures: a framework for blockchain enabled system and use case analysis. HICSS
15. Grover BA (2021) Blockchain and governance: theory, applications and challenges, blockchain for business: how it works and creates value. Scrivener Publishing LLC. https://doi.org/10.1002/9781119711063.ch6
16. Guarda T, Augusto MF, Haz L, Díaz-Nafría JM (2021) Blockchain and government transformation. In: Rocha Á, Ferrás C, López-López PC, Guarda T (eds) Information technology and systems. ICITS 2021. Advances in intelligent systems and computing, vol 1330. Springer, Cham. https://doi.org/10.1007/978-3-030-68285-9_9
17. Hastings EM, Wong SK, Walters M (2006) Governance in a co-ownership environment: the management of multiple-ownership property in Hong Kong. Prop Manag 24:293–308

18. Hilson G (2002) The environmental impact of small-scale gold mining in Ghana: identifying problems and possible solutions. Geogr J 168(1):57–72
19. Hou H (2017) The application of blockchain technology in e-government in China, pp 1–4. https://doi.org/10.1109/ICCCN.2017.8038519
20. Huber GP (2001) Transfer of knowledge in knowledge management systems: unexplored issues and suggested studies. Eur J Inf Syst 10(2):72–79
21. Huckle S, Bhattacharya R, White M, Beloff N (2016) Internet of things, blockchain and shared economy applications. Proc Comput Sci 98:461–466
22. Huh S, Cho S, Kim S (2017) Managing IoT devices using blockchain platform. In: Proceedings of advanced communication technology (ICACT), 2017 19th international conference on, pp 464–467
23. Hull MS (2012) Government of paper: the materiality of bureaucracy in urban Pakistan. University of California Press, Berkeley (California), 320 pages
24. Iansiti M, Lakhani K (2017) The truth about Blockchain. Harvard Bus Rev 1–11. https://enterprisersproject.com/sites/default/files/the_truth_about_blockchain.pdf
25. Joshi KD, Lei C, Datta A, Shu H (2010) Changing the competitive landscape: continuous innovation through IT-enabled knowledge capabilities. Inf Syst Res 21(3):472–495
26. Knirsch F, Unterweger A, Engel D (2019) Implementing a blockchain from scratch: why, how, and what we learned. EURASIP J Inf Secur. https://doi.org/10.1186/s13635-019-0085-3
27. Konashevych O (2020) Cross-Blockchain Protocol for Public Registries (April 24). The paper is accepted for publication in the International Journal of Web Information Systems, Emerald Publishing. This is a pre-print version under Creative Commons Attribution Non-commercial International Licence 4.0 (CC BY-NC 4.0). https://doi.org/10.1108/IJWIS-07-2020-0045. Available at SSRN: https://ssrn.com/abstract=3537258 or http://dx.doi.org/10.2139/ssrn.3537258
28. Kundu D (2019) Blockchain and trust in a smart city. Environ Urban ASIA 10(1):31–43. https://doi.org/10.1177/0975425319832392
29. Lemieux A (1997) Canada's global mining presence. In: Canadian minerals yearbook. Natural Resources Canada, Ottawa
30. Mamoshina P, Ojomoko L, Yanovich Y, Ostrovski A, Botezatu A, Prikhodko P, Izumchenko E, Aliper A, Romantsov K, Zhebrak A, Ogu IO, Zhavoronkov A (2017) Converging blockchain and next-generation artificial intelligence technologies to decentralize and accelerate biomedical research and healthcare. Oncotarget 9(5):5665–5690. https://doi.org/10.18632/oncotarget.22345
31. Mashatan A, Roberts Z (2017) An enhanced real estate transaction process based on blockchain technology, 2018. In: A policymaker's guide to connected cars.
32. Mbendi (1999) Information for AfricaFAfrican mining statistics. Africa's leading business website. http://www.mbendi.co.za/indy/ming/mingaf.htm#Overview
33. Ministry of Mines and Mining Development (2022). http://www.mines.gov.zw/
34. Mora H, Mendoza-Tello J, Varela-Guzmán E, Szymanski J (2021) Blockchain technologies to address smart city and society challenges. Comput Hum Behav 122:106854. https://doi.org/10.1016/j.chb.2021.106854
35. Morris D Z (2016) Leaderless, blockchain-based venture capital fund raises $100 million, and counting, Fortune (magazine). http://fortune.com/2016/05/15/leaderlessblockchain-vc-fund/)
36. Nakamoto S (2008) Bitcoin: a peer-to-peer electronic cash system. https://bitcoin.org/bitcoin.pdf
37. Ochieng Odhiambo M (2000) Oxfam Karamoja conflict study: a report. Oxfam, Kampala
38. Orecchini F, Santiangeli A, Zuccari F, Pieroni A, Suppa T (2018, October) Blockchain technology in smart city: a new opportunity for smart environment and smart mobility. In: International conference on intelligent computing & optimization, pp 346–354. Springer, Cham
39. Ølnes S, Ubacht J, Janssen M (2017) Blockchain in government: Benefits and implications of distributed ledger technology for information sharing. Gov Inf Q 34:3
40. Patil P, Narayankar P, Narayan DG, Meena SM (2016) A comprehensive evaluation of cryptographic algorithms: DES, 3DES, AES, RSA and Blowfish. Proc Comput Sci 78:617–624. https://doi.org/10.1016/j.procs.2016.02.108

41. Schelter S, Lange D, Schmidt P, Celikel, F Biessmann, A Grafberger (2018) Automating large-scale data quality verification. Proc VLDB Endow 11:1781–1794. Posted: 2018
42. Sedlmeir J, Smethurst R, Rieger A et al (2021) Digital identities and verifiable credentials. Bus Inf Syst Eng 63:603–613. https://doi.org/10.1007/s12599-021-00722-y
43. Srivardhana T, Pawlowski S (2007) ERP systems as an enabler of sustained business process innovation: a knowledge-based view. J Strateg Inf Syst 16:51–69. https://doi.org/10.1016/j.jsis.2007.01.003
44. Sullivan C, Burger E (2017) E-residency and blockchain. Comput Law Secur Rev 33. https://doi.org/10.1016/j.clsr.2017.03.016
45. Szabo N (1997) The idea of smart contracts, Nick Szabo's Papers and Concise Tutorials, vol 6
46. Veuger J (2018) Trust in a viable real estate economy with disruption and blockchain. Facilities 36:103–120
47. Vos J (2017) Blockchain and land administration: a happy marriage? Eur Prop Law J 6:293–295
48. de Zhou, Lin Z, Lim S (2019) Spatial characteristics and risk factor identification for land use spatial conflicts in a rapid urbanization region in China. Environ Monitor Assess 191. https://doi.org/10.1007/s10661-019-7809-1

Harnessing Big Data for Business Innovation and Effective Business Decision Making

Umair B. Chaudhry⬤ and M. Abdullah Chaudhry

Abstract Innovation and other decision-making processes in businesses are very critical to their growth and survival in an environment of ever-increasing competition and technological advancement. Large corporations have the resources to invest in an expensive Big Data and Analytics infrastructure or platform, however, using business intelligence technologies is a challenging effort for start-ups owing to hefty license costs, the need to create an integrated data warehouse, and a lack of Business Intelligence capabilities. This paper offers an in-depth look at how businesses can take advantage of big data when making business decisions or embarking on any form of innovation be it process, product, service or strategy. A comprehensive evaluation of literature discerns that an intricate rationalisation of key implications, use and benefits are widely evaluated. In addition, for the instigation of comprehensive unstructured data analysis, the paper shows how, instead of having to invest in a sophisticated and expensive infrastructure, Microsoft Power BI allows start-ups to reap the advantages of large-scale big data tools and allows new businesses to obtain access to data analytics to see a direct effect on their bottom line by detecting trends and patterns, as well as new possibilities, success rates, and user preferences.

Keywords Big data · Analytics · Power BI · Business decision making · Data mining

U. B. Chaudhry (✉)
Northumbria University, Middlesex Street, London 1 7HT, UK
e-mail: u.b.chaudhry@qmul.ac.uk

Queen Mary University of London, London 1 4FZ, UK

M. A. Chaudhry
Lahore College of Arts, Science and Technology, GCUF, Lahore, Pakistan

1 Introduction

Improved processes, new and better products and services, more efficiency, and, most importantly, increased profitability is all dependent on a company's ability to effectively utilize new insights and ideas. The competitiveness of the local, regional, national, and worldwide markets is increasing. Competition has increased as new technologies have become more generally accessible and as the internet's commerce and information-sharing capacities have improved. Consequently, as time passes, businesses are increasingly confronted with complex situations that require rapid decision-making in order to promote organizational development and success [1]. It is true that Big Data Analytics has played a significant part in company operational dynamics, helping companies not only with comprehensive market and production research, but also with the prediction of the best move using rigorous data and analytics [2]. Information systems connected to the internet, cloud computing, mobile devices, and the Internet of Things (IoT) have produced enormous quantities of data. There are three types of data in it: structured and semi-structured. It also includes unstructured and real-time data. Innovative methods to leveraging the potential of big data have been developed by several businesses and academia, respectively. For decision-making, large datasets may be a useful source of additional information.

Big data has been described in a variety of ways by various academics. Big data is a cultural, technological, and intellectual phenomenon, according to [3], whereas the ocean of information is big data, according to [2]. "Big data" refers to a huge quantity of both structured and unstructured data [1]. Large datasets are referred to as "big data" [4] because traditional data processing techniques cannot handle them. The Big Data revolution is much more successful than previous research, as shown by the comparisons. Instead of making decisions based on gut feelings, managers gain from utilizing big data. Data gathered by organizations has surpassed its usefulness, and big data now helps firms make better predictions and smarter decisions. Management procedures worldwide have benefited from big data usage by corporate executives. Some areas, including commerce, social media, and supply chain big data, have had a variety of studies done on them. The consequences of big data analytics vary depending on the business decision environment. They call it conventional business machine produced and social data, according to [5] 'Big data' refers to the massive quantities of structured and unstructured information that companies must deal with on a daily basis. Data volume, on the other hand, isn't the most important factor to take into account.

What is really important is how all businesses can make use of this enormous data. Big data insights may help you make better decisions and drive businesses forward. As [6] points out, for many individuals, "big data" simply means millions of data points that may be analyzed using various technologies. Real-world big data is the appropriate use of data collected via various means, such as technology, in any given area. During the first decade of the twenty-first century, internet and start-up companies embraced big data. There is a new kind of data speech, text, logs, pictures, and videos have emerged [7]. When big data is used correctly, it may lead to a variety of applications that aid in decision-making. Unstructured data will be rationalized for certain chosen companies, and unstructured big data will be evaluated for business possibilities and assist identify their aid in decision making for the selected enterprises, in order to assess these elements.

This paper will provide a substantial outlook to the use of unstructured big data and how it can be harnessed for business innovation and strategic decision-making. The paper is arranged in 5 sections. Section 1 presents the introduction and the background which is followed by Sect. 2 having a detailed literature review. Section 3 presents the data, experimentation and results. Section 5 presents the discussion and recommendations and concludes the research.

2 Literature Review

2.1 Unstructured Data and Its Importance

There is a clear need to keep ahead of the curve and produce breakthrough insight when technology and business concepts change. This necessitates looking outside of SAP Applications data. By being proactive rather than reactive in the face of market shifts, businesses may become more flexible and adaptable.

According to [8], unstructured data is rising at a rate of 55–65% each year. Users now have access to AI-powered unstructured data analytics tools that were expressly designed to get insights from unstructured data. Because unstructured data makes up the majority of today's data, companies must develop methods to manage and analyse it to make critical business choices. As per [9] Algorithms based on artificial intelligence now assist in the automated extraction of meaning from the massive amounts of unstructured data generated on a daily basis. By integrating unstructured data from a wide range of sources, businesses can have the intelligence they need to make quick decisions that improve relationships with customers. Decision-makers may discover which goods or services are most enticing for their target market by scanning large datasets fast and finding patterns in consumer behaviour. This is critical for product development as well as determining which marketing efforts are the most effective. Organizations must break down data silos and replace them with a scalable data hub in order to take use of unstructured data's potential. Organizations may ultimately discover the immense commercial potential of unstructured data if they have the tools in place to store, analyse, and report data from many sources and to share it with corporate decision-makers.

Organizations should concentrate on developing organizational agility in order to adapt appropriately and quickly to a broad range of environmental business changes according to [10]. Process agility is particularly important since it refers to how fast and easily organizations can retool their processes in response to changes in the market environment. Firms are progressively investing in different technologies and integrating them into their business operations in order to take advantage of insightful information that can yield quality decision making. Information System (IS) may help companies make timely choices, provide them a competitive edge, spur innovation, and help them handle unpredictability in their environment. It is envisaged that IT-enabled information would play an essential role in the development of organizational capacities because of the hypercompetitive nature of current corporate settings [11]. Companies are also always looking for insights from the growing amount, variety, and speed of data in order to make sense of the information and enhance decision-making [12]. Firms are also looking for strategies to solve previously unknown issues, such as emerging trends.

Business Data Analysis

A growing interest in business data analysis (BDA) can be attributed to the opportunities that come with data and analysis in various organizations. Operational and strategic decision-making are aided by the use of BDA in operations management, which improves performance as per [13]. According to [10], the IT business value literature has long emphasized the ability of "information systems (IS) to inform decision-making and improve firm performance". It's been shown in studies of business performance that IS may help companies make timely choices, provide them a competitive edge, spur innovation, and help them handle unpredictability in their environment.

2.2 Big Data Analytics and Its Various Facets in Business Practice

As per [14], the use of big data analytics has grown in importance as a decision-support tool for managers. By using big data discovery methods, managers may get new insights on previously undiscovered results. Data storage and processing capabilities were severely constrained before the introduction of computers. As a general rule, analytics may be broken down into one of three types depending on their intended purpose of usage. Through reports and dashboards, descriptive analytics describes a phenomenon using historical data and aids in comprehending what has transpired. Predictive analytics aids in our comprehension of what may occur [15]. Using prior data, correlations, and trends, it helps make predictions about the future. Executive decision-making may be aided by prescriptive analytics, which is yet another useful tool. So that various results may be better understood under varied circumstances, it's helpful. It includes a variety of tools, including optimization, simulations, and what-if-analysis scenarios that vary the initial set of parameters that are supplied.

   According to [16], Managers can make an informed judgment and prepare for the worst-case scenario if they know what to anticipate. Companies are increasingly using big data analytics to gain a competitive edge and enhance their performance. This trend has lately gained substantial attention. Technology and procedures for storing, processing, and displaying data are necessary to take full advantage of ever rising data volumes and velocity and diversity [17]. But there has been considerably less emphasis paid to how organizations may embrace these technologies for further progress. Currently, big data analytics is the most popular way for analysing large volumes of raw data because of its greater capacity to collect vast amounts of data and apply the finest analytical procedures to it. There are three main components to big data analytics: the data, the analytics, and the presentation of findings in a manner that enables companies and their customers to create commercial value. Big data analytics may offer a lot of advantages, but many firms are choosing not to use them. Companies that have failed to use business intelligence are more likely to have this problem.

Some company leaders may wonder whether big data analytics is really different from traditional business intelligence and data mining, or if it's just a new skill that requires a lot of money to implement. It's critical for policymakers and business practitioners and academics to know the answers to these issues because they invest in new data analytics initiatives. For starters as per [18], the writers look at the differences between conventional business intelligence and big data analytics. Despite the fact that the big data age barely began in 2005, the amount of big data is expanding at a rapid rate of roughly 50% every year [19]. Surprisingly, unstructured data makes up a significant portion of this increase, including video, photos, social media postings, user comments, and any other sort of data that is difficult to categorize into repeating areas. Large-scale datasets that are difficult for enterprises to acquire and manage in a timely way need the use of the most powerful data management methods important to information processing, which are known as "Big Data" [20]. Big data analytics has been hailed by some as a quantum leap forward from traditional business intelligence methodologies, yet it may still be a mystery to social scientists. Business intelligence, on the other hand, has been shown to be a critical component of many big data analytics programs.

There are many new possibilities for organizations to use big data and analytical methodologies to get a better understanding of their market position and make better choices to remain competitive and expand their share of the market. Using big data analytics, companies have been able to develop suitable plans via the lens of data, increasing their efficacy and efficiency. Data analytics has become an essential part of agile businesses' decision-making processes.

2.3 Source of Big Data Analytics and Its Role on Decision Making

Big data comes from a variety of sources, "including conventional information systems", "social network sites", "cloud applications", "software", "social influencers", "Data warehouse appliances", "public network technologies", "legacy documents", "corporate applications", "meteorological data", and "sensor data". Russom showed that just a small fraction of people is familiar with terminology like predictive analytics, advanced analytics, and big data analytics. In order to analyse large amounts of data, businesses need a variety of different technologies. These include relational databases and data warehouses as well as data mining techniques such as clustering and association. Big data is very useful for decision-making since it provides companies with useful data and enhances business analytics in particular. Despite the advantages, big data analytics for decision-making faces few obstacles. Obstacles often include a lack of qualified personnel to handle decision-supporting sophisticated analytics as well as issues with database software.

Use of Decision Support Systems

Conventional solutions for supporting business choices were based on data produced by transaction processing systems like ERP and supported internal business decisions which are driven by the use of specific analytical tools such as Tableau. As time went on, more and more complex supply and demand mechanisms were added (SRM and CRM) as per [21]. With the use of these decision support tools, internal operations and tactical choices can be made in an innovative way. This data, aided internal decision-making by increasing the precision and speed by means of these innovations. According to [22], the system as a whole includes a transactional core database and a data warehouse for storing and classifying extracted data. These datasets may be used to generate business insight with the help of additional data mining technologies. The analysis and identification of patterns, correlations, and association rules made use of data mining from amassed data.

Benefits of Using Big Data on Decision Making

There are a range of conventional datasets as well as huge datasets derived from a variety of sources, whether organized, semi-structured, or unstructured. As per [23], using these datasets for strategic, tactical, and operational decision-making is possible in a number of ways. When business transaction data is processed for clustering algorithms, crucial insights about goods purchased together or forecasting demand for certain things are revealed for decision-makers to use. Demand forecasting aids in better preparedness for catastrophic natural events.

Increasingly, IT providers and consultants like IBM are pitching "Big Data" as a solution in an attempt to achieve agility. Companies have been using data analysis to boost efficiency for more than a decade due to underdeveloped methods for managing data that prevented them from going any farther. Data connected to the manufacturing process has a direct relationship to the deployment and usage of data management and processing systems as well as the demands of an organization [17].

Data formats and processing methods have radically changed with the emergence of Big Data Analytics (BDA), making this a game-changer for businesses. This implies that more can be done using analytical approaches to get value from the sheer amount of data, creation speed, and variety of sources from which it is gathered. Using BDA to transform industrial procedures and increase performance adds to the commercial value of IT publications. Because of the change to BDA-supported performance indicators, decision-makers may use more data to assess

Fig. 1 Conceptual model of using big data for decision making [24]

alternative strategies for achieving their objectives [19]. Researchers found, in line with past operations literature research, that increasing the use of business process analysis (BDA) helps organizations better anticipate and improve process performance. It is thus a win–win situation for businesses since they can save costs while improving operational efficiency and customer service while also reducing inventory levels and organizing the operations better. Figure 1 shows the flow from data gathering till decision making. Once the organization has a solid analytics approach in place, we suggest having a senior executive oversee this department [20]. As a result, the insights gathered through data analysis will be put to good use in making decisions. Analytical tools like dashboards, scorecards, and KPIs may all be used in conjunction with other performance assessment systems. Big data plays a key role in analytics, which in turn provides insights for decision making, according to the literature assessment.

Using domain expertise and data analysis, several chances for analytics initiatives may be discovered [19].

2.4 Evaluation of Data Analytical Tools and Its Implication on Unstructured Data Analysis

Businesses of all sizes, from start-ups to multinational giants, are increasingly turning to data to create a story. Only through analysing data and drawing relevant conclusions can any value be derived from it. It's no surprise that many large corporations across the world have already made considerable investments in big data and data analytics and are now starting to enjoy the advantages [7]. Large corporations have the resources to invest in an expensive Big Data and Analytics infrastructure or platform, but start-ups and small firms often do not. Using business intelligence technologies is a challenging effort for start-ups because of hefty license costs, the need to create an integrated data warehouse, and a lack of Business Intelligence capabilities. They are aware of the benefits of data-driven initiatives in business intelligence. Sadly, they do not have the resources to recruit data scientists or invest in expensive infrastructure improvements as a result. What can small businesses do to take advantage of the promise of big data in light of this?

As mentioned by Benoit [13], Microsoft Power BI is the answer. Instead of having to invest in sophisticated and expensive infrastructure, Power BI allows start-ups to reap the advantages of large-scale big data tools. Simple setup allows new businesses to obtain access to data analytics and see a direct effect on their bottom line.

Power BI helps companies create complete dashboards using the current data sources. Platform-independent availability is a possibility for these dashboards. There are some additional benefits to using Power BI for start-ups. For example, firms may use Power BI to produce, examine, and monitor numerous reports, trends, graphs, and so on, all from a single place, taking many elements of the company into account. As a result, stakeholders have a better understanding of the overall performance of the organization's numerous processes. To assist management, detect any problematic areas and take proactive actions for mitigating risks, many real-time warnings may be created as a component of the dashboard [8]. An additional problem for start-ups is funding training for complicated business intelligence (BI) products. As a result, Power BI is a product that even non-technical users can use (and ultimately master). Any company user may examine and comprehend the data using the reports and dashboards' easy user interface. Power BI Designer and Power Pivot, for example, allow advanced users to do complex analyses.

Start-ups may utilize Power BI to analyse data gathered from a variety of sources, including databases, files, and online services. Using a single application, even non-technical individuals may combine data from many sources and analyse it. Additionally, Power BI offers a broad selection of report visualizations. With the use of these visualizations as [25] have implied, the data may be presented in a variety of forms, such as Pi-charts, bar graphs, and more, which aids management in making better decisions and determining where to concentrate. Since Power BI is now an open-source project on GitHub, it helps start-ups simply generate unique visualizations of high quality with minimum expenditures.

Microsoft Power BI can easily be connected with other Big Data sources to gather and analyse massive volumes of data, however, as has been shown. Microsoft's Power BI contains Power Query and Power Map, both of which may be used in conjunction with Office 365's Big Data analytics as per [26]. Through its capacity to extract data from Microsoft Excel and further use and breakdown as required, Power Query aids in boosting the BI experience. Start-ups, on the other hand, may use Power Maps to visualize and analyse geographic data in three dimensions.

3   Experimentation

From the comprehensive evaluation of the above literature, it can be discerned that an intricate rationalization of key implications, use and benefits are widely evaluated however the literature does not contextualize any actual data-driven findings presenting the analytical use of Big Data and its gradual use on business future decision making. This can be further addressed by a systematic use of data analytical tool in specific business contexts to stimulate how harnessing unstructured data can aid to business decision making.

To carry out this work, the scientific approach of positivism is deemed effective as it offers a comprehensive picture of empirical evidence and hypotheses. According to this philosophy, harnessing unstructured data (Big Data) may assist with innovation and business decision-making because of positivism's designative and positive interpretation [27]. When positivism is used to create interactions between social entities by changing a single independent component, it may also entail the distortion of reality, which is what we're looking at in this research: the practical use of Big Data for commercial decision-making. Hence, we have implemented an intricate methodological prospect through which it aids in accumulating organizational based Big Data and its associated metrics and rationalize and evaluate the analytics of such data collected. Considering that the research is striving to ascertain gradual use of the data analytics on business decision and its practical benefit, it can be understood that the following research contextualizes a substantive quantitative data evaluative framework where large sums of unstructured data associated with different selected businesses will be rationalized based on the analytics gained from the market and how such data aids in channeling the designative data towards determining the right production or marketing of products of the concerned businesses which in turn contextualizes the key importance, role and practicalities of harnessing unstructured (Big Data) for business innovation and decision making.

Quantitative method to data collection was utilized, and the acquired data is then subdivided into distinct data and market trends. For the instigation of comprehensive unstructured data analysis, the research has opted towards using Microsoft Power BI as the key data analytical tool on data from three companies namely Soma chocolate company, Tesla, and Ubisoft. Exploring and analysing enormous chunks of data to find relevant patterns and trends is what data mining is all about. Insight from the explored data can then be utilized by businesses in pursing any innovation plan or making any other critical decision. Figures 2 and 3 show the steps and stages involved.

The results of data mining are used to develop analytical models and with the help of analytic tools like Microsoft Power BI, insight can be produced graphically that can aid in decision-making and other commercial activities.

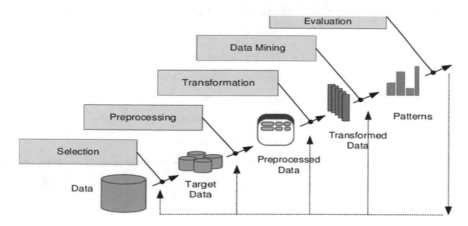

Fig. 2 Steps in data mining

Fig. 3 Stages in data mining

4 Data Analysis and Mining

Microsoft power BI has been used as the data analytic tool on three data sets gained from Soma chocolate company, Tesla, and Ubisoft [28–30] to show how big data can be harnessed towards effective decision making.

Figures 4 and 5 are sample extractions from Soma suggesting the country where we can launch our product and from where we can acquire the beans. From the graphical representation it can be seen that the rating of the company Soma is highest among all and the most interesting observation that the analytics get is the percentage of cocoa is also high in the product of soma can be seen a kind of linear structure between the rating and percentage of cocoa. So, it can be concluded that consumers want cocoa in their product. Moreover, from a country-based analysis it is seen that mainly in USA, France, Canada and UK product market rating is very high and especially people of USA and France have the liking of cocoa in their product. Thus, from this analysis it can be discerned that type of product that chocolate consumers are attracted of and the place where the consumer demand is very high and the research also gives an idea of most successful chocolate company, Soma secret behind success.

Figures 6 and 7 are mined from Tesla's data set. It can be observed that Tesla is holding the highest battery pack fast charging and overall efficiency. Further, from these analytics it can be implied that all the brands can't offer body styles. Thus,

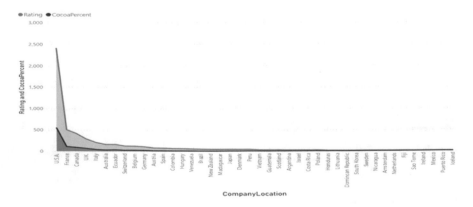

Fig. 4 Analyzing Cocoa percent and rating by company location

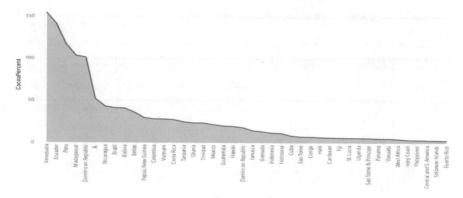

Fig. 5 Analyzing Cocoa percent against broad bean origin

from this analysis a point can be made that in terms of electric vehicle market tesla is the best from all points but though tesla brings this era in automobile industries, other brands like Audi Porsha, Nissan etc. are also gradually updating and coming highly in this competitive market.

Figures 8 and 9 are extracted Ubisoft. Data set of different games including their sales value in different region of the world is taken into account. The data analytics rakes account of different genre of games on different market and evaluates their sales and business performance on specific genre in specific markets. This analysis helped in categorizing the best genre of game for the specific market which may aid Ubisoft in future decision making regarding new game developments.

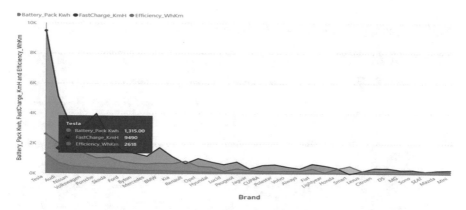

Fig. 6 Analyzing battery, charge time and efficiency by brand

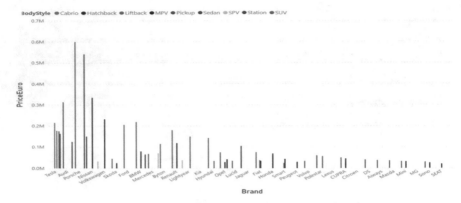

Fig. 7 Analyzing sales by brand and body style

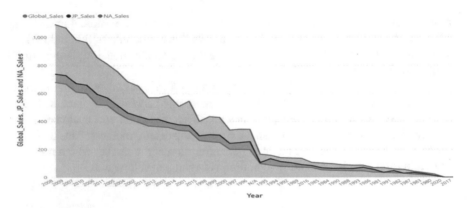

Fig. 8 Comparing global sales, North America and Asian sales

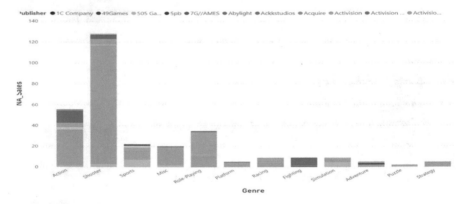

Fig. 9 Analyzing North American sales by genres and publishers of games

5 Discussion and Conclusions

From a comprehensive outlook into the data analytics carried out in the research, it can promptly be evaluated that the use of unstructured data can be of substantive use in the modern business world. Not only accumulation of these unstructured data aided in acquiring extensive market-oriented big data for carrying out analytics in specific business prospects but also provided substantive insights into how such analytics plays a crucial role in supporting organizational decision making by means of extensive innovation and ultimately contribute to performance enhancement.

The data analysis carried out by this research thus conforms to the definitive fact that Big Data analytics and the use of unstructured data can be of utmost significance for the modern business prospects. Nevertheless, by putting emphasis on these analytics, it can promptly be evaluated that Big Data Analytics by means of unstructured data evaluations and analytics is has enormous potential for businesses to succeed in the market. As the conducted data analytics showcases, the big data analytics provides a wider outlook on the market, which insights into how competitors are becoming successful, prospects of product sales and demands of specific businesses on specific markets. Hence, considering the prospects of Soma, Tesla and Ubisoft, it is clear that implementation of strategic initiatives driven by the insights gained from the Big Data analytics would provide these businesses with a more definitive outlook into market insights and through it help formulating more promising business decisions. By this, businesses can take more comprehensive approaches to future initiatives that would aid companies to earn more benefits in the market by means of consumer engagement and more product sales. These benefits would eventually contribute to the financial performance of the businesses. After getting a substantial knowledge through the carried data analytics, the selected businesses can be provided with definitive data-driven recommendations for future growth.

References

1. Kitchin R (2014) Big data, new epistemologies and paradigm shifts 1(1). https://doi.org/10.1177/2053951714528481
2. Fan J, Han F, Liu H (2014) Challenges of big data analysis. Natl Sci Rev 1(2):293–314
3. Boyd D, Crawford K (2012) critical questions for big data 15(5): 662–679. https://doi.org/10.1080/1369118X.2012.678878
4. Waller MA, Fawcett SE (2013) Data science, predictive analytics, and big data: a revolution that will transform supply chain design and management. J Bus Logist 34(2):77–84
5. Dubey R et al (2015) The impact of big data on world-class sustainable manufacturing. Int J Adv Manufact Technol 84:1. Springer, pp 631–645. https://doi.org/10.1007/S00170-015-7674-1
6. Jeble S, Kumari S, Patil Y (2018) Role of big data in decision making. Oper Supply Chain Manag 11(1):36–44
7. Chang V, Lin W. How big data transforms manufacturing industry: a review paper
8. Nisar QA, Nasir N, Jamshed S, Naz S, Ali M, Ali S. Big data management and environmental performance: role of big data decision-making capabilities and decision-making quality

9. Shamim S, Zeng J, Shariq SM, Khan Z (2019) Role of big data management in enhancing big data decision-making capability and quality among Chinese firms: a dynamic capabilities view. Inf Manag 56(6)
10. Lee M, Kwon W, Back KJ (2021) Artificial intelligence for hospitality big data analytics: developing a prediction model of restaurant review helpfulness for customer decision-making. Int J Contemp Hosp Manag 33(6):2117–2136
11. Awan U, Shamim S, Khan Z, Zia NU, Shariq SM, Khan MN (2021) Big data analytics capability and decision-making: the role of data-driven insight on circular economy performance. Technol Forecast Soc Change 168
12. Gupta S, Justy T, Kamboj S, Kumar A, Kristoffersen E (2021) Big data and firm marketing performance: findings from knowledge-based view. Technol Forecast Soc Change 171
13. Benoit DF, Lessmann S, Verbeke W (2020) On realising the utopian potential of big data analytics for maximising return on marketing investments 36(3–4):233–247. https://doi.org/10.1080/0267257X.2020.1739446
14. Roper S. Ethicality of data mining and predictive analytics introduction to practices
15. Nygård R, Mezei J (2020) Automating lead scoring with machine learning: An experimental study. In: Proceedings of Annual Hawaii International Conference System Science, vol 2020, January, pp 1439–1448
16. Steff R, Soare S, Burton J. Emerging technologies and international security machines, the state and war
17. Sadati N, Chinnam RB, Nezhad MZ (2017) Observational data-driven modeling and optimization of manufacturing processes. Expert Syst Appl 93:456–464
18. Joshi M, Biswas P (2018) An empirical investigation of impact of organizational factors on big data adoption. Smart Innov Syst Technol 79:809–824
19. Ma S, Zhang Y, Ren S, Yang H, Zhu Z (2020) A case-practice-theory-based method of implementing energy management in a manufacturing factory 34(7–8): 829–843. https://doi.org/10.1080/0951192X.2020.1757154
20. Babu MM, Rahman M, Alam A, Dey BL (2021) Exploring big data-driven innovation in the manufacturing sector: evidence from UK firms. Ann Oper Res 1–28
21. Maroufkhani P, Wagner R, Wan Ismail WK, Baroto MB, Nourani M (2019) Big data analytics and firm performance: a systematic review. Information 10(7):226
22. Shujahat M, Sousa MJ, Hussain S, Nawaz F, Wang M, Umer M (2019) Translating the impact of knowledge management processes into knowledge-based innovation: the neglected and mediating role of knowledge-worker productivity. J Bus Res 94:442–450
23. Iqbal A, Latif F, Marimon F, Sahibzada UF, Hussain S (2019) From knowledge management to organizational performance: modelling the mediating role of innovation and intellectual capital in higher education. J Enterp Inf Manag 32(1):36–59
24. When smart gets smarter: how big data analytics creates business value in smart manufacturing. https://publications.hse.ru/en/chapters/251951534. Accessed 08 Jun 2022
25. Ghasemaghaei M, Turel O (2021) Possible negative effects of big data on decision quality in firms: the role of knowledge hiding behaviours. Inf Syst J 31(2):268–293
26. Fredriksson C (2018) Big data creating new knowledge as support in decision-making: practical examples of big data use and consequences of using big data as decision support 27(1):1–18. https://doi.org/10.1080/12460125.2018.1459068
27. Research methods: bloomsbury business briefing Peter Stokes red globe press. https://www.bloomsbury.com/uk/research-methods-9780230362031/. Accessed 08 Jun 2022
28. Chocolate Bar Ratings | Kaggle. https://www.kaggle.com/datasets/rtatman/chocolate-bar-ratings. Accessed 08 Jun 2022
29. Cars Dataset with Battery Pack Capacity | Kaggle. https://www.kaggle.com/datasets/divyanshugupta95/cars-dataset-with-battery-pack-capacity. Accessed 08 Jun 2022
30. Video Game Sales | Kaggle. https://www.kaggle.com/datasets/gregorut/videogamesales. Accessed 08 Jun 2022

Building a Resilient Cybersecurity Workforce: A Multidisciplinary Solution to the Problem of High Turnover of Cybersecurity Analysts

Babatunde Adetoye and Rose Cheuk-wai Fong

Abstract The high turnover of Security Operations Centre analysts, (SOC analysts), is a current issue being discussed globally, especially after the COVID-19 pandemic and the recent global widespread of cybersecurity breaches. The gaps in security operations, due to inconsistency in the availability of cyber security workforce exposes the organisation to the possibility of costly security breaches. Although there are diverse views on the factors responsible for the high turnover of SOC analysts, whatever the reasons, high turnover of SOC analysts, in many ways, constitute a serious risk to organisations. Employers have used financial incentives to address this problem, however, the problem still persists. This research therefore explores a multidisciplinary approach to finding comprehensive and sustainable solutions to this problem, through seeking the views of the stakeholders; the SOC analysts and cybersecurity leaders, including the views of experts from the fields of Human Resources (HR) and psychology. This research was conducted as a case study of the two types of Security Operations Centres (SOCs): in-house and Managed Security Service Providers (MSSP), between the months of March and May 2022. Semi-structured interview was used as the qualitative method of gathering the necessary data. The key findings from this research suggest that most SOC analysts consider other factors than money in deciding whether to leave or stay in an organisation. These factors are mainly informed by the characteristic needs and motivations of the young generation of SOC analysts; hence, cybersecurity leaders should be tactical in finding ways to engage and retain their critical workforce.

Keywords High turnover · SOC analysts · Multidisciplinary solutions

B. Adetoye · R. C. Fong (✉)
Northumbria University, London, UK
e-mail: rose.fong@northumbria.ac.uk

B. Adetoye
e-mail: tundetoye@ymail.com

© The Author(s), under exclusive license to Springer Nature Switzerland AG 2023   61
H. Jahankhani (ed.), *Cybersecurity in the Age of Smart Societies*,
Advanced Sciences and Technologies for Security Applications,
https://doi.org/10.1007/978-3-031-20160-8_5

1 Introduction

According to World Economic Forum [89], the cybersecurity industry has the most pronounced workforce skills gap, with evidence of about 350% growth in the number of unfilled cybersecurity jobs between 2013 and 2021 [56]. In the face of the ever-increasing, sophisticated, and widespread threat landscape [26], organisations are now facing high turnover of SOC analysts [5]. The combination of the existing skills gap and high turnover of SOC analysts constitute a massive vulnerability that could endanger organisations' security posture [12].

SOC analysts are the frontline cybersecurity employees responsible for the security of the organisation's IT infrastructure and information asset [50]. They are responsible for round-the-clock monitoring and rapid response to incidents for remediation and disaster recovery. Evidence shows that the existing cybersecurity skills shortage is putting a lot of pressure on the few SOC analysts who must handle all the workload [74]. Some critics are suggesting that this workload pressure is responsible for the high turnover that is now prevalent in SOCs globally, however, others suggest that high turnover is due to the desires to find new employers that can pay higher salary. In any case, high turnover not only creates problems for hiring managers and HR in terms of the time and the money that go into finding and onboarding new talent [49], the time it takes to fill any vacant cybersecurity position is also a window of vulnerability that attackers could exploit [10].

1.1 *People, Process, and Technology*

Although, the challenges in cybersecurity are always around three key areas: people technology and process. This research, however, seeks to explore the people aspect of cybersecurity because people drive processes and technology [39]. In addition, people can either be the weakest link when they are overlooked [1], or they can be the force that build effective technologies and processes when given adequate support [88].

1.2 *Research Aim and Objectives*

To explore the role of the digital leaders in facilitating the development of comprehensive and sustainable solutions that can reduce the turnover rate, foster a highly engaged and resilient cybersecurity workforce that can ensure a consistently robust security operation.

1. To discuss the key factors that are contributing to the current high turnover of SOC analysts.

2. To consider the risk implication and potential impact of the high turnover of SOC analysts on organisation's security posture.
3. To determine the role of digital leaders in facilitating the implementation of new and unconventional solutions to the problem of high turnover.
4. To explore possible multidisciplinary solutions that can help engage and retain SOC analysts for a more stable security operation.
5. To formulate and recommend comprehensive solutions for a resilient cybersecurity workforce for a consistently robust organisational security posture.

2 Literature Review

2.1 The Need for Security Operations in Contemporary Organisations

A Security Operations Centre (SOC) is a centralised place where an organisation's networks, devices and endpoints are monitored for malware infection or Indicators of Compromise (IoC) to ensure immediate detection and eradication before they cause any adverse effect [51]. Although, some organisations may not be able to build and manage their in-house SOC for financial reasons or lack of necessary skillset, they may therefore decide to employ the services of a Managed Security Service Provider (MSSP) to handle their security operations [59], whichever option an organisation chooses, the SOC needs highly engaged SOC analysts to run an effective security operations [59].

One key factor of security operations is the need to consistently operate round the clock to protect, detect and quickly respond to any threat [62]. SOC analysts are therefore expected to keep a round-the-clock watchful eye on the whole organisation's infrastructure.

2.2 The Existing Skills Gap Problem in Cybersecurity and SOCs

The field of cybersecurity is in shortage of skilled professionals who can effectively plan, manage, and optimise security technologies and processes. Although, critics argue that the talent gap is a myth, suggesting that organisations who cannot find talent are either not paying market rate or have unrealistic list of requirements for the roles they wanted to fill, as some of them even ask for advanced-level certifications for entry-level roles [70].

On an average, statistics indicate that it takes about 21% more time to fill cybersecurity roles than any other roles in IT [80]. Although, according to the 2021 (ISC)[2] Cybersecurity Workforce Study, the 2021 Cybersecurity Workforce Gap decreased to 2.72 million compared to 3.12 million in the year 2020. Nevertheless, the same report suggests that the global cybersecurity workforce still needs a 65% growth to adequately protect organisations' information assets and infrastructure as the global threat landscape keeps expanding. While the skills gap in cybersecurity has been so widely discussed as a persistent problem for organisations over the years [34], organisations are now facing additional problem of high turnover of SOC analysts due to more reasons than one [61].

2.3   The Current High Turnover Rate of SOC Analyst

According to XpertHR (2019), the average turnover rate in the field of technology was 18.3%, ranking third highest in all the various industries [58].

Although, experts claim that any organisation with less than 15% should not see their turnover rate as unhealthy, as this is common to most organisations [58]. Therefore, in the light of this claim, 18.3% turnover rate in technology field can be classified as unhealthy, especially when this figure is even higher for SOC analysts [4]. Although, some management consultants suggest that employee turnover should be seen as an unavoidable reality in the contemporary business world [30]. In their views, employee turnover appears good for the industry, as new challenges, and new opportunities that new environments bring seem to help employees keep their skills fresh and organisations remain competitive [85]. Such proponents argue that low employee turnover is not necessarily a good thing for organisations [58]. However, when the rate or frequency at which employees leave organisations are high, especially in critical operations such as the SOC, this inconsistency in staffing might constitute a significant risk to organisations security posture in the face of ever-increasing threat landscape [5].

2.4   The Cocktail of Reasons for Employee Turnover

Employee turnover is caused by many reasons that can either be classified as the 'pull' turnover, a situation where better opportunities attracted the employee out of the organisation, or the 'push' turnover where the employees walk away from unpleasant experience at work [15]. Figures indicate that contemporary employees move from one organisation to another on an average of four-year cycle to either improve their skills or attract better salary [54]. For example, 27% of people who changed roles claim they did so for increase in salary, followed by career progression, at 24% [21]. Experts claim that, as long as the market keeps rewarding job switching, the issue of high turnover will only increase [75].

In addition, people are also more confident to change jobs with the slightest opportunity or inconvenience because they know that the demand for cybersecurity professionals is high [41]. Although the same research found that 68% of respondents indicated that they would rather stay with an organisation where their opinions are valued. On an average, evidence shows that about 66% of cybersecurity professionals change roles within two years [21] and less than 20% have stayed within the same organisation for four years and above. This confirms how dynamic the cybersecurity industry is. Reports show that, on an average, SOC analysts stay within an organisation for between one to three years [43]. Some might argue that this is so short a time due to the consistency that the SOC team needs to be effective.

2.4.1 Pressure from Recruiters (a 'Pull' Factor)

Some critics blame the high turnover on recruiters, as evidence shows that about 46% of cybersecurity professionals claimed they get weekly offers from recruiters to leave their organisation for another [83]. Evidence suggests that recruiters put so much pressure on cybersecurity professionals to move from one organisation to another [41] as many cybersecurity professionals are constantly contacted several times with attractive offers.

2.4.2 The Pressure of Heavy Workload (a 'Push' Factor)

Reports claim that SOC analysts leave one organisation for another due to the constant heavy workload that cybersecurity threats generate. For example, more than 50% of security alerts generated by the security monitoring devices are false alerts, causing alert fatigue for SOC analysts [13]. The recent global COVID-19 pandemic has also added to the workload as workers used their home network and personal computers to connect to organisation's network, thus increasing the volume of alerts that SOC analysts have to deal with every day [63].

The World Health Organisation [90] has also linked the issue of high turnover to burnouts. In addition to alert fatigue, burnout may be responsible for the risk of missing Indicators of Compromise in the network [69].

Working in a SOC demands that SOC analysts repeatedly operate in emergency and alert mode on a daily basis to safeguard the organisations network against the ever-present threats. This fire-fighting mode of detecting, triaging, and responding to threats can cause chronic mental health issues if not managed properly and on time [72]. Although work-related stress and burnout are not only peculiar to SOC analysts in this always-connected, always-on, and high-paced contemporary work culture [29].

A recent Ponemon Institute survey suggests about 65% of SOC analysts have considered quitting their jobs due to work-related stress [82]. The repetitive and meticulous nature of going through huge amount of data to find Indicators of Compromise (IOC) is akin to finding a needle in the haystack. This very monotonous activity can be mind-numbing causing a concept called '*boreout*', the situation where SOC analysts do not feel challenged enough by the sedentary tasks, they have to carry out every day [82]. Both these issues of '*burnout*' and '*boreout*' caused by the alert-overload and the monotonous nature of the task, can become a '*push*' factor for the SOC analysts.

2.4.3 The Frustration for Lack of Clearly Defined Career Path and Half-Life of Skills

Evidence suggests that about 65% of cybersecurity professionals leave their jobs because they do not see clearly defined career paths in the organisation they work for [74]. Others complained about lack of formal training opportunities for personal growth and career development [24]. While they have to regularly study to ensure they are able to catch up with changes in the threat landscape [7], they also have to regularly reskill as the technology and products they use keep changing or being replaced by new technologies [3]. HR and learning think-tank practitioners refer to this phenomenon as the half-life of skills. They agree that a lot of skills in the contemporary world have a "half-life" of five years, however, many technical skills, like the skills needed to work in cybersecurity, have a half-life of less than two years because these vendor-related or platform-related tools are frequently being updated [3]. In addition to this, employers continue to expect their staff to fully embrace the "always-learning" mindset [17], although, the workload and the round-the-clock working hours barely leave room for personal and professional development. A post pandemic study by Exabeam [28] indicated that 62% of the 432 cybersecurity professionals who responded to the survey claimed that their work is stressful [18], suggesting that such experience may be responsible for disengagement or turnover.

2.5   The Risk Implications of High Turnover of SOC Analysts on Organisations

A recent survey conducted by the Centre for Strategic and International Studies (CSIS) reported that 71% of IT decisionmakers claimed the talent gap in cybersecurity is a significant risk to their organisation [47]. For example, this claim is substantiated by the recent report of ransomware attack on Georgia Pipeline during the time that they had two unfilled cybersecurity leadership positions [44]. These reports suggest that the absence of security professionals to protect the organisation in the face of ever-present threats to organisations is a serious vulnerability that threat actors always exploit. In a recent ISC^2 [42] research, 60% of cybersecurity professionals suggests staffing shortage is a key factor that makes it difficult for organisations to meet their cybersecurity goals and this puts the organisation at risk. Already, research suggests that 59% of organizations say that they are either extremely or moderately at risk due to their staff shortage [38]. In addition, when team members leave, this could affect the morale of the others who remain [60].

Although the SOCs that report high employee retention suggests that workplace benefits and high salary help keep their staff, however, the persistent high turnover rate suggests that it takes more than salary increase to retain employees [68].

2.6 Finding Solutions to the High Turnover (a Multidisciplinary Approach)

Since high turnover of SOC analysts is a people issue, it therefore requires a people approach. HR and psychology, as disciplines that study people's behaviours and motivation could therefore help throw some light on how to get SOC analysts to engage more and stay longer for the sake of consistency in cybersecurity operations. Although, most leaders in the cybersecurity fields may not be adequately knowledgeable in handling issues that are psychological and behavioural in nature, however, seeking the expertise of HR practitioners and psychologists for solutions should pose no problems as leaders in the digital age are known to embrace non-convectional approaches to solving problems [76].

2.6.1 What HR Could Contribute

The HR department is tasked with the management of employees in terms of engagement and retention [15]. It is also usually the first and the last point of contact for employees and they are in a very strategic position to influence the opinions of employees about the organisation and their engagement with the organisation [84]. HR leaders are also in the best position to advice the organisation on certain policies that may affect employees and help create the culture that fosters security mindedness [36]. They could also work closer with cybersecurity leaders in setting up an effective cybersecurity workforce development programme, not just only during recruitment but throughout employee tenure in the organisation [9]. Although, realistically, SOC analysts will eventually have to leave the organisation for one legitimate reason or the other, HR and the SOC leadership could come up with succession plan, talent pipeline and exit strategy, to ensure the exit of a SOC analyst is well managed without affecting the security operation of the organisation.

2.6.2 What Psychologist Could Contribute

Evidence shows that working in a SOC is unavoidably demanding because cyberthreats are ever-present and ever-increasing [67]. Since the cyberthreats that generate the heavy workload for SOC analysts are not going away, and constantly moving from one organisation to another is not the solution, as the problem is everywhere. It is therefore important for SOC analysts to build resiliency to cope with the challenges of the contemporary workplace [29]. Personal resiliency of employees is now considered an essential factor if the individual and the organisation must survive and thrive in the contemporary world [14]. Although some argue that employers might abuse the term 'resilience' as a smokescreen to blackmail employees to endure unhealthy organisational culture [37].

This concept of resilience has its root in behavioural psychology, and it is defined as the ability to bounce back and even thrive when overwhelmed by unavoidable challenge [19]. It is a soft skill that can be learnt, especially in this Volatile, Uncertain, Complex and Ambiguous world [2]. Evidence shows that few employers know how to help their employees be more resilient [87]. Hence, experts from psychology could help in providing practices to support the development of resilient qualities for workforce engagement and retention.

3 Research Methodology

As this research seeks to investigate a people-issue as a priority over technology and processes or natural science, it adopts an interpretivist's perspective to have adequate insight into the complex world of people at work, and to seek their views regarding the issue facing them [73]. People issue can be too complex to be subjected to the positivist's philosophy, which assumes that the result produced from research can only be objectively validated.

The diversity of views on the reasons for high turnover, the complexity of people's experience, and the contexts in which SOC analysts work demands an inductive approach to be properly understood (Bryman and Bell 2011). Although, the inductive approach works with a small sample, however, this research ensures that diverse and relevant views are represented in the sample to derive useful themes [73]. Since business situations can be very complex, dynamic, and unpredictable, an interpretivist perspective is appropriate for this research [73].

Since this research seeks to explore the views and feelings of the stakeholders, as stated in the research objectives, this type of data will be descriptive data, meaning, the research will be collecting qualitative data [25]. The collected narrative data, the views, experiences, and expectations of the stakeholders, especially, the SOC analysts on the issue at hand, will help to inductively generate theory from research [73]. The choice of the qualitative method has, therefore been dictated by the research objectives and not just the personal preference of the researcher (Bryman and Bell 2011).

Case study is the most appropriate strategy to investigate a contemporary issue like the high turnover of SOC analysts within the context of security operation of an organisation [73]. In order to have a credible research data for a case study, this research has chosen to factor in the two types of SOC environments. The first is an in-house SOC of a popular insurance company and the second is an MSSP providing security operations for many organisations. Both have operations and workforce dispersed globally. A case study strategy allows for greater depth of exploration of the research objectives [45], and it benefits from triangulating different views of people in producing useful correlation of data collected [73], especially for research that seek to take a multidisciplinary approach to finding solutions to the problem at hand, by comparing and contrasting the views of the participants (Bryman and Bell 2011).

3.1 Sample Selection and Data Collection

This research seeks to consider the views of SOC analysts working in both an in-house SOC and an MSSP environments, therefore, the representative sample were selected equally from these two environments (Fig. 1). In addition, the participants were also chosen based on their years of experience as SOC analysts, which range from four to seven, as this range is high enough for them to provide credible views regarding working in SOCs. Choosing years of experience below three years would not have given the participants enough experience to provide rich and sufficient data [45]. In addition, the selection also included women to be represented in the sample, although, one-third of the sample were women, this is closer to the representative ratio of women to men in cybersecurity, which is estimated at 20% [57].

Interviews are very effective for explorative research such as a case study [31]. Interviews are the most appropriate to collect data about the feelings and views of people [22]. In addition, interviewing lends itself more to triangulation of views [6] especially where different opinions regarding the cause of high turnover already exist, as the literature review revealed. Triangulation of views from empirical data against established theories makes the case study analysis credible [46].

A semi-structured interview was used to collect empirical data. This approach made use of a list of themed questions from the research objectives (Bryman and Bell 2011). Although, an in-depth interview would have given a richer data, however,

	Participants	Type of SOC environment	Yrs. of experience	Gender
1	Security Analysts 1	In-house	6	Male
2	Security Analysts 2	In-house	4	Female
3	Security Analysts 3	MSSP	7	Male
4	Security Analysts 4	MSSP	5	Male
5	Security Analysts 5	In-house	7	Male
6	Security Analysts (SOC lead)	MSSP	8	Female
7	Cybersecurity Leader (Architect)	MSSP	11	Male
8	HR Director		12	Female
9	Psychologist (Wellbeing expert)		20	Male
	Total participants			**9**

Fig. 1 The list of the participants (interviewees)

allowing the participant to talk freely throughout an interview may waste time and lead the interview into discussions that are not relevant to the research topic. The semi-structured interview is more flexible and open to potentially unexpected insights, thus, allowing the researcher to adapt and rearrange the questions on the fly, depending on the flow of the interview [73].

An email was sent to the potential participants to inform them about the aim of the research and arrangements were made to conduct online Zoom interviews as soon as they were available. Zoom interviews made it convenient for the participants rather than a face-to-face interview. The interviews were recorded for accurate capture of responses, however, the respondents were notified and promised the confidentiality of their responses.

3.2   Data Analysis

The possible methods of analysing qualitative data include content analysis, thematic analysis, and narrative analysis [73], however, for this research, thematic analysis was the most appropriate because it gives room to realise emerging patterns or themes from diverse sources of data. The audio recording of the interviews was first transcribed using the Microsoft Word speech-to-text feature in readiness for data processing, sorting, coding, and analysis. The initial level of analysis and coding of the qualitative data from the interviews made use of the computer-assisted qualitative data analysis tool, NVIVO, to reduce the tedious manual coding process (Bryman and Bell 2011).

This research employed the deductive (concept-driven) coding approach, where the researcher starts data analysis from predefined set of codes from research questions, and literature review [52]. This is justifiable since the case study employed semi-structured interview based on research objectives. This approach was chosen over the inductive coding, which does not use any predefined code frame, in order to keep the research within the context of the set research objectives and to avoid the possibility of unintentional detour [11]. This research was however opened to the possibility of unexpected but important emergent ideas (inductive codes) that could enrich this multidisciplinary research or may inform future research, even if they were outside of the scope of this research.

4   Data Presentation, Findings and Analysis

4.1   The Coding Framework

Figure 2 shows the list of some codes derived from literature review and the research questions, which were used as the initial deductive framework to guide the process

	Nodes/codes (Deductive coding)
1	Awareness of high SOC analysts' turnover
2	Impact of high SOC analysts' turnover on organisation's security operations
3	The leadership openness to recommendation of solutions
4	The multidisciplinary solutions from experts
5	Repetitiveness of tasks
6	Alert fatigue
7	Work related stress
8	Employee wellbeing
9	Criticality of the workload
10	Competitive salary
11	Talent shortage
12	Increased job opportunities in the marketplace
13	Pressure from recruitment agencies
14	Leadership & management response to employee needs
15	The role of the experts (HR and Psychologists)
16	The concept of resiliency

Fig. 2 The coding table/framework (A Priori)

of looking for patterns and correlations within the data towards thematic analysis [11]. However, the researcher was opened to the possibility of newly inductively generated codes during the preliminary data analysis.

4.2 Data Coding, Categorisation, and Thematic Analysis

NVIVO, a qualitative data analysis tool, was used to code the nine different interview transcripts. The process involved going through the data, line-by-line to objectively capture the essence of the data [35]. This was an iterative process that involved continuously applying the codes as the researcher moved back and forth across all the interviews transcripts as a form of triangulation of views and code categorisation based on their similarity to find useful themes.

The Case Study Results Presentation

As the first step in the presentation of result and research findings, the Word Cloud feature in NVIVO was used on all the empirical data to determine and visualise the most frequently used words. '*People*' emerged as the most frequently used word (Fig. 3). The implication and significance of this will be discussed in the discussion chapter.

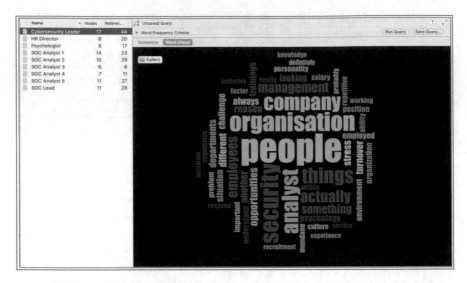

Fig. 3 The word cloud generated in NVIVO

Views on Awareness of High Turnover of SOC Analysts

Participants were asked regarding their awareness of the current high turnover of SOC analysts, and everyone indicated that they are aware of the situation. For example, the Cybersecurity Leader said,

> Looking at the issue of high turnover in SOC teams from a global point of view, the problems are not local; it's a worldwide issue with retention.

This was also confirmed by the SOC Lead, who said, *"Yes, in my view of our organisation, there is a fairly high turnover"*, although SOC Analyst 2 added by saying, *"I've seen a number of articles relating to this topic and I guess it's becoming more familiar to those in the security workforce to transfer from one work to another in such a short span of time"*.

Nevertheless, there were diverse reasons given regarding why SOC Analysts leave. For example, SOC Analyst 4 said,

> I could say that the high turnover in SOC is because there are loads of opportunities with a lot of new and cool companies that can give you new challenges and can set new goals for every individual that is keen to raise their level of experience and probably receive also a higher compensation and gratification in return.

SOC Analyst 5 added a different opinion by saying,

> Analysts may leave a job because they have gained experience or certification while at a SOC, and the company's HR is unable to give the analyst a proper raise that matches the analyst's upgraded value.

The SOC Lead, again added,

Employers don't have the reward for loyalty, which is something that drives me crazy, but that's another matter. Anyway, although, not the only reason, people leave for financial gain, and I also think potential lack of development opportunities.

The HR Director said,

Mostly, the pandemic has created the opportunity for people to rethinking their lives and their careers and coming up with changing plans to their lives.

Although, she added a different but important view regarding the term 'turnover', she said,

I would describe it as a fluid employment market much more than turnover.

SOC Analyst 1 is of the opinion that,

It would be nice to get more money, but I don't think that's the main factor, I think it's the repetitive role of SOC analysts. You are going to get bored, especially if you are doing the same thing every day at work. People want a bit of variation in their role.

SOC Analyst 2 said,

There's a high demand within the security workforce, so, based on that, companies are offering a much higher or competitive salary than the other company to attract candidates.

She further added,

It can be stressful working with critical incidents and if the management will not listen to all those concerns, then, that's a big factor for an employee to leave an organisation immediately.

Views on Risk Implication of High Turnover of SOC Analysts

When asked about the impact of high turnover on the organisation, the Cybersecurity Leader said,

Losing SOC staff, who are responsible for detect-capability means that you are blind to what's happening in your environment and therefore, you are exposed to a lot of risk, and your ability to respond quickly is also impacted, so, it has detrimental security impact on the organisation's ability to defend itself.

The SOC Lead added,

The major risk is the obvious one, it is about a potential leak. A leak in information of structure, credentials, set up, how people in the team go about things, potential vulnerabilities that are there and all those kinds of things. The more turnover of staff you have, the more it widens or enhances the risks and the likelihood.

SOC Analysts 5 said,

The morale of the SOC analysts that are left behind may deteriorate. A lower morale is a poison for any department, and with high turnover rate, the poison could spread to other members in the SOC and even members in other departments that work closely with the SOC.

In addition, SOC Analyst 4 said,

Another risk for an organisation is that a new hire will need time to learn and understand the new environment; during that time frame, the company will surely be exposed to threats or attacks while the new employee is still catching up with tools and procedure, creating a hole in the security posture.

These responses indicated that the participants are aware of the many security risks that the current high turnover may introduce to the security posture of the organisation.

Views on the Role of the Leadership in Facilitating Workforce Engagement and Retention

When asked about the role of leadership in facilitating workforce engagement and retention, the Cybersecurity Leader said,

As leaders, we can't succeed without considering the fact that we are dealing with human beings at the end of the day, our staff are human beings. They have lives outside of work. It is the totality of themselves that they bring to work, and we need to make sure that we think very carefully about that.

He further gave examples of what the leadership of their organisation has been doing to engage and retain SOC Analysts by saying,

We have an HR business partner that partners with our organisation and we're always looking at these in terms of the ranking level, in terms of the compensation and remuneration. We are also looking into wellness, mindfulness, and psychological safety, hence, we as leaders are going through many trainings to ensure that we provide a platform where people feel psychologically safe and can communicate their thoughts without fear of repercussion.

He also said,
However, in the view of SOC Analyst 2,

If the management will not listen to employee concerns, then, that's a big factor for an employee to leave an organization immediately. It affects you psychologically, if you don't feel like what you're doing is that important, or there is a way that you get feedback that your views are being heard by the executives. I think that's something that definitely boosts the morale of employees. Although the problem with an MSSP is that there's no channel for open communication between the actual employees on the SOC side and the client's management side.

The HR director revealed,

One of the things that we have done in our organisation, for example, is that we're going to start a major programme for our 600 plus employees, so, we've divided them amongst the leadership team, so, they can have regular coffee sessions, where they can find out how they're doing with their wellbeing and things like that. And we've also introduced executive conversations, a free time for employees to come and have conversations with us.

Views of HR and Psychology Experts on Building a Resilient Workforce

The HR director said,

It's about, how do we ensure that people find work meaningful? How do they connect to the values of the organisation, much more than things like pay? How do we ensure that we're taking care of the wellbeing of the people in our organisation?

In addition, she said,

You know, there used to be times when we used to have exit interviews, but in addition to exit interviews, we have introduced stay interviews. So, if you think you're not feeling it anymore, just come out, have a chat with us.

In the psychologist's view, he said,

Even in the recruiting or induction process, it is important to help people understand their personality traits and incorporating some sort of personality assessment into that process. It should promote self-awareness for people to play to their strengths and there's quite a lot of literature on that.

He further explained,

A personality test is practically powerful in understanding the individual and in understanding the way they work best. Some roles might look like an opportunity with a nice price tag to it, but it will actually be a threat to their wellbeing because they are not so wired for it and it's only going to affect their productivity. For some people, the honest question is, maybe retrain and go do something else or even change to a totally different industry.

4.3 Other Emergent Codes from Empirical Data

Although, the interview data used deductive coding to highlight key data based on literature and research questions, however, there were new and useful themes that emerged which were not directly part of the research questions.

The New Generation of Employees and Implications of their Value and Behaviour

While discussing the reasons behind high turnover, the SOC Lead alluded to the influence of the new generation of employees on the issue at hand. She said,

I think this is a generational thing as well. The 'generation snowflake' are coming into the role now and they're delicate, as in, you can't say anything to them without them breaking or taking offense; if you don't have someone holding their hand, that's it, they throw their toys out the pram and go.

Similarly, the HR Director said,

You know, with the young people, job security is not an issue for them, with a little challenge, they just move back home, (laugh) so, they don't think too much about it before they leave a job. They have a totally different perspective of the way that they respond, where we, older generation, would exercise perseverance, they can't stand anything like that, so, we're quite watchful in that space.

SOC Analyst 2 also said,

The younger generation move quickly because they have little tolerance to challenges. when they do experience stress or a little challenge, maybe the work is not favourable to what they want for an outcome; maybe they see the income to be too little for an entry-level experience, then, that would be a good enough reason for them to leave immediately.

Mentoring Programme for Employees

Another interesting point that inductively emerged is mentoring. The HR Director revealed that,

> Another thing that we've done is to introduce the 'reciprocal mentoring' programme, whereby we have employee mentoring by senior managers and the executives, and people have responded to that quite positively.

Similarly, the Cybersecurity Leader also said,

> Part of the initiatives for the more senior roles is to actually mentor the ones that are coming behind them. The junior ones will want to grow, because they're involved in interesting projects, and they are much more likely to be sticky with the organisation.

The Psychologist also said,

> Mentoring is something that must be built into the system, particularly for people that are just entering into the system. There are certain courses they can take or subscribe to mentoring from somebody who is more senior.

Career Development Path for SOC Analysts

The Cybersecurity Leader also mentioned the need for establishing clearly defined career path for SOC analysts as a way to engage and retain them; he said,

> Many organisations actually don't have a growth or promotion path to more senior roles for these analysts. There has to be an active plan to ensure growth and progression for the SOC folks, to have that growth trajectory that you can see through to more senior roles.

The SOC Lead also said,

> Well, you can't make anyone stay, however, I think a clear, structured opportunity path would enable people to have a clearer focus of where they might want to head, so they don't necessarily have to move up in SOC, but they could move across, and I think people join SOC because it's an easier way to get in, and then they worry about where they want to go next.

The Psychological Environment for Psychological Wellbeing

Another point raised is about the psychological safety and wellbeing. For example, the HR Director said,

> We're giving our employees the psychological safety-space much more than remuneration safety-space, although we try to match the median of any pay scale through job evaluation, but we found it more effective to have the psychological contract really strengthened with the employee, and we found it more relevant to connect with the emotions of people. It's really about creating safe spaces for them to explore what they really want to explore, in terms of their careers and things like that. That's an area that I've been able to push quite a bit in our organisation, and it's been well received.

Similarly, SOC Analyst 2 mentioned that *"......it has to be a psychological thing more than money"*.

The psychologist added another view by saying,

There are other things that employers can do in terms of the physical working environment to make it conducive to mental wellbeing. They should explore Ecotherapy, using the environment to promote wellbeing. Music therapy can also be used to create an atmosphere to lift moods and boost performance.

Peer Support and Organisational Culture

Regarding challenges or work-related stress that could 'push' SOC analysts out of the organisation, the psychologist said,

Peer support is something that should be encouraged and be incorporated into the system at all levels. Sometimes, people underestimate their ability to cope with stress because they see themselves as loners; alone in trying to make sense out of a stressful situation. So, helping people to know there are support systems outside of themselves; they should not feel intimidated or shy to put up their hands and ask for help.

SOC Analyst 1 said,

I think it's definitely a culture thing when people get stressed. If you're working within the company that's got a good culture and good people that support you, then I think you'll be fine.

SOC Analyst 2 also said,

I get to speak with other colleagues or other employees that are not within our team. People stay longer in an organisation if they are comfortable with who they are, and with whom they're working with, and it feels like they have become a family. I can say that our culture in the Philippines is more like a family, that's why we tend to stay longer in companies.

SOC Analyst 4 said something similar,

Pressure in Cybersecurity may be high sometimes, so, the best way to cope with this is to share and speak with colleagues. In this way, new ideas and new ways of dealing with issues may come up or it could help you release the stress.

Likewise, the Cybersecurity Leader said,

A toxic culture, where you know the more senior folks feel superior and the junior ones feel less appreciated, doesn't cultivate an atmosphere where people feel like they would want to be there for a long time. So, you actually see a lot of people just leaving the organisation.

5   Discussion

5.1   *There is an Awareness of the Current High Turnover of SOC Analysts and the Potential Risk Implications to organisation's Security Posture*

From the empirical findings, the views of the participants agree with literature that there is high turnover of SOC analysts globally and its risk implications to organisations. Literature suggests it is an existential threat to organisations that find it difficult to retain cybersecurity employees [74]. These risks include data leakage by employees leaving the organisation, reduced morale of the employees that are left behind, and the inability of the organisation to detect threats, defend itself with the possibility of security breaches [64, 81]. Although empirical findings suggest that turnover is an indication of a 'fluid' employment market, which may lead to redistribution of talent that could benefit organisations if well managed, however, it can be said that the risk associated with this phenomenon cannot be eliminated but managed [66].

5.2 Reasons for the Current High Turnover of SOC Analysts Are Many, Thus, Needing Tactical Solutions, Because There is no 'One-Size-Fits-All' Solution

Literature review highlighted a number of reasons why SOC analysts move from one organisation to another so rapidly, which include money [54], stress caused by alert fatigue [29], burnout, repetitive work, lack of clearly defined career path [24], unsupportive management and lack of development opportunity in the organisation [71]. Likewise, the interviews also confirmed that the reasons are varied, including the recent changes that the COVID-19 pandemic introduced. For example, the remote working has even led to increasing level of 'overemployment', where IT professionals can now work for more than one organisation in order to earn more money [16]. Although, this phenomenon seems to debunk the argument that people leave due to stress. If the work with an organisation was stressful, it would not make sense to take on another role in another organisation.

5.3 The Role of Digital Leaders is Critical to Finding and Implementing Sustainable Solutions to the High Turnover of SOC Analysts

It was interesting to note that the Word Cloud in NVIVO highlighted 'people' as the most prominent word in the empirical data (Fig. 3). Literature suggests that organisations that adopt a 'human-centric work model', which emphasises employee wellbeing in their policies and operations are more likely to engage and retain their employees [40]. The Cybersecurity leader emphasised this in the interview, that their success will be guaranteed when they see and treat their employees as human beings with lives outside of work. Some key points that emerged from empirical data, agrees that digital leaders have the responsibility of creating an enabling culture for sustainable high performance and retention of employees [55]. Literature also suggests that leaders must create the culture that is able to meet the deep human need to have autonomy and direct their own lives [65]. This is also in alignment with self-determination theory which suggests that people become self-determined to do well when their needs for autonomy, competence and connections are met for their psychological wellbeing [48]. The Human Capital theory suggests that investing in the training of individuals is a resource that is more important than capital, technology, and natural resources. SOC analysts are the human capital of an organisation and digital leaders must ensure proper investment in their continuous improvement for efficient security operation [81]. Literature also suggests that the best cybersecurity leaders are digital leaders who are humble enough to accept they do not have all the answers to the issues facing their organisation and are thus open to advice from experts from other fields to enrich their own knowledge and practices [8].

5.4 Understanding the Needs and Motivation of the New Generation of Employees is Critical in Employee Engagement and Retention

It also emerged from the interviews that the issue of high turnover is a generational matter. Evidence shows that the contemporary generation called the 'Gen Z', are characterised as a hypercognitive generation, who collect and cross-reference diverse information online and on-demand to achieve their aims [32]. They can see what many organisations are offering to employees and can make their career decisions based on these information. The influential power that they have on socioeconomic issues is confirmed by the response of the SOC Lead and HR Director in Sect. 4.4.1. Therefore, organisations should consider this when they are looking at how to engage and retain this cohort of employees. For example, they are easily bored and want variety without which it is almost impossible to engage them for long [23]. Literature indicates that trust is a key factor that can engage this generation, irrespective of financial rewards or perks (Hall, 2019). As far as this cohort is concerned, loyalty to an organisation is based on trust and it must first be demonstrated from the organisation's side as the bigger entity in the employment contract [86].

6 Conclusion and Recommendations

This research started with the aim of exploring the role of the digital leaders in facilitating the development of sustainable solutions that can reduce the turnover rate of SOC analysts. This aim was broken into five objectives. The first was to discuss the key factors that are contributing to the current high turnover of SOC analysts. The second was to consider the risk implication and potential impact of this on organisations' security posture. The third was to determine the role of digital leaders in facilitating the discovery and implementation of practicable solutions. The fourth was to seek multidisciplinary avenues to finding these solutions. The fifth objective was to formulate and recommend realistic solutions for a resilient cybersecurity workforce that can ensure a consistently robust organisational security posture.

In order to achieve these objectives, case study of the two types of SOC environments was chosen as the appropriate research design. Semi-structured interview was the chosen data collection tool to ensure in-depth views of the participants in addressing the research questions.

From the research, it was evidently clear that the stakeholders are aware of current high turnover of SOC analysts and the risk implication to organisations. This research also discovered that the reasons for the current high turnover of SOC analysts are many, and there is no 'one-size-fits-all' solution. This research also found out, as literature suggests, that digital leaders have critical role to play in finding and implementing practicable solutions to the high turnover of SOC analysts. Finally, this research discovered that there is the need to consider the characteristic needs and motivations of the new generation of employees in building a resilient cybersecurity workforce.

This research would have achieved all the set research objectives after providing the following four (4) recommendations and the suggestion for further research.

6.1   Recommendations

The high turnover of SOC analysts is a critical business situation therefore, this research provides the following recommendations that could help inform organisational policies in order to achieve the aim of building a resilient cybersecurity workforce. It is, however, important to note that security operations will differ from organisation to organisation, hence the need to carefully implement the recommendations as frameworks, that may be adapted to suit individual organisation. In addition, business, and security strategies like these need to be regularly reviewed and modified to improve their effectiveness, especially in the face of ever-changing threat landscape.

The Need to Implement Automation and Orchestration Technologies

Organisations should invest in the right technologies to improve the effectiveness of their processes in their SOC [53]. For example, using the AI features of a SOAR (Security Orchestration, Automation and Response) system like the popular *Splunk SOAR or Microsoft Sentinel* or EDR (Endpoint Detection and Response) tools by *Symantec* and *CrowdStrike* integrated with the traditional SIEM (Security Information and Event Management) tool in order to reduce the stress caused by alert fatigue and monotonous security monitoring and response. Although people fear that machines may eventually take the jobs of many SOC analysts, however, a SOC analyst can still be relevant by learning to become the quality controller, developer, and programmer of these automation tools [79].

The Need to Understand and Respond to the Characteristic Needs and Motivations of the 'Generation Z' Employees for Improved Engagement and Retention

Every generation is shaped by the context within which it emerged, and the Generation Z has emerged out of the time of both online and offline openness [32]. Therefore, this cohort will expect the same from the organisation in order to feel 'at home' within the organisation. They are also very inclusive and will also love to work in an inclusive culture, regardless of pay.

CIPD [15] suggests that flexibility at work, fair treatment and employee wellbeing are the factors to consider when exploring practices that could improve employee retention. This aligns with the value and needs of the Gen Z employees. For instance, Gen Z cohorts make their decisions, including career decision based on the agreement of values of the organisation with their own values [27]. Therefore, cybersecurity leaders should listen to and observe this cohort to learn about their values to inform the design of workforce engagement and retention strategies. The process of mentoring and wellbeing programme, which is another recommendation to be discussed later, is an opportunity to hear their thoughts, feelings, and expectations. Involving them in designing the solutions to workforce engagement and retention will also further engage them. Gen Z also values life/work balance and flexible working, especially after the COVID-19 pandemic, therefore organisations should look into using technologies to make this possible for this cohort, without exposing the organisation to any security risks. In other words, organisations that are flexible and adaptable become a 'honeypot' that attracts and retains the Gen Z employees [27].

The Need to Set up an Enabling Organisational Culture for Employees Personal and Professional Growth

Apart from employee reward, working within a culture of trust and adequate training can foster employee engagement and increased productivity [49].

Although training and development opportunities are highlighted to be very important to employee engagement and retention, however, a more sustainable way is through the little moment-to-moment healthy experience that the employees have in the workplace [71]. Some leaders have evidence of reducing staff turnover by creating a company culture that makes employees feel 'at home' and want to continue to work in, because they have multiple ways to communicate their views to management [77]. According to Goler et al. [33], in their research reported the response of Facebook manager, saying,

> If you want to keep your people, especially your stars, it's time to pay more attention to how you design their work. Most companies design jobs and then slot people into them. Our best managers sometimes do the opposite: When they find talented people, they're open to creating jobs around them.

Therefore, cybersecurity leaders should work with HR in fostering a culture that allows people to play to their strengths.

The Need to Set up and Maintain Wellbeing and Mentoring Programmes for Employees

This recommendation is even more important after the COVID-19 pandemic. Psychological wellbeing and mental resilience are necessary to help employees keep their mental abilities fit and adaptable in the contemporary VUCA (Volatile, Uncertain, Complex and Ambiguous) world [27]. Evidence suggests that organisations that cater for the physical, emotional, social, financial, and professional wellbeing of their employees help boost their resiliency by 300%, but employees with poor resilience are half as engaged at work and are twice as likely to want to leave their job [78]. Resilience in the workplace is defined as an individual's ability to "better adapt to adverse situation, manage stress, and retain motivation, enabling organisations to better manage change" [78].

Mentoring, especially, career mentoring, where a senior or more experienced person gives advice or support to a younger or less experienced person [20], is an avenue to help junior employees build self-confidence, psychological resilience, competence, and other personal and professional qualities. Evidence shows that those who have benefitted from mentoring also go on to mentor others [20] and the virtuous cycle goes on. This is good for employee engagement and retention.

6.2 Further Research

This researcher has identified two possible areas for further research:

First, it would be useful to test the effectiveness of the research recommendations. This might include setting up a 'People Analytics Team' to measure the turnover rate of SOC analysts, after the implementation of the recommendations to confirm their success and find areas that need improvement. This can be carried out as a quantitative or qualitative research.

Secondly, another concept that emerged during this research is the current trend called '*overemployment*' which has emerged due to the recent upsurge of remote working. Evidence revealed that this is common among IT professionals, including cybersecurity professionals. It would therefore be very useful to investigate the possible security risk that this new way of working might pose to organisations. This could be carried out as a case study and qualitative way of data collection.

References

1. Alert Logic (2021) Why are humans the weakest link in cybersecurity? https://www.alertlogic.com/blog/why-humans-weakest-link-cybersecurity/. Accessed 12 Dec 2021
2. APA (2020) Building your resilience. https://www.apa.org/topics/resilience. Accessed 1 March 2022
3. Avilar (2021) Understanding durable versus perishable skills and how to balance them. https://blog.avilar.com/2021/06/16/understanding-durable-vs-perishable-skills-and-how-to-balance-them/. Accessed 27 Feb 2022
4. Baker I (2019) Security operations centres face high levels of staff turnover. https://betanews.com/2019/08/29/soc-high-staff-turnover/. Accessed 11 Dec 2021
5. Bergerson L (2019) Research: security analysts say SOC turnover is on the rise. https://www.channel-impact.com/research-security-analysts-say-soc-turnover-is-on-the-rise/. Accessed 29 Oct 2021
6. Biggam J (2008) Succeeding with your master's dissertation: a step-by-step handbook. Open University Press McGraw-Hill Education, England
7. BLS (2022) Occupational outlook handbook: information security analysts. https://www.bls.gov/ooh/computer-and-information-technology/information-security-analysts.htm. Accessed 21 Feb 2022
8. Brennen J (2020) https://www.linkedin.com/learning/soft-skills-for-information-security-professionals/be-a-leader?autoAdvance=true&autoSkip=false&autoplay=true&resume=false&u=69919578. Accessed 20 Feb 2022
9. Brickey J (2021) The Intersection of HR and cybersecurity. https://www.shrm.org/executive/resources/articles/pages/hr-cybersecurity-overlap-brickey.aspx. Accessed 27 Feb 2022
10. Burke J (2021) 8 challenges every security operations centre faces. https://searchsecurity.techtarget.com/tip/8-challenges-every-security-operations-center-faces. Accessed 01 Nov 2021
11. CHASR (2021) Analysing qualitative data using NVivo. https://www.youtube.com/watch?v=1CvMGPzJ-30&list=TLPQMDEwNDIwMjKLW05ksLO7xQ&index=21. Accessed 3 April 2022
12. CheckPoint (2021) The importance of the security operations centre (SOC). https://www.checkpoint.com/cyber-hub/threat-prevention/what-is-soc/the-importance-of-the-security-operations-center-soc/. Accessed 01 Nov 2021
13. Chickowski E (2019) Every hour SOCs rzun, 15 minutes are wasted on false positives. https://businessinsights.bitdefender.com/every-hour-socs-run-15-minutes-are-wasted-on-false-positives. Accessed 31 Oct 2021
14. CIPD (2011) Developing resilience: a guide for practitioners. https://www.cipd.co.uk/knowledge/culture/well-being/resilience-guide#gref. Accessed 2 March 2022
15. CIPD (2021) Employee turnover and retention. https://www.cipd.co.uk/knowledge/strategy/resourcing/turnover-retention-factsheet#7179. Accessed 3 March 2022
16. CIPD (2021b) The rise of 'overemployment' amongst homeworkers. https://www.hr-inform.co.uk/news-article/the-rise-of-overemployment-amongst-homeworkers. Accessed 18 April 2022
17. CompTIA (2020) Workforce and learning trends 2020. https://www.comptia.org/content/research/it-workforce-learning-trends-analysis. Accessed 27 Feb 2022
18. Cotton B (2021) We need to talk about burnout in the tech industry. https://www.businessleader.co.uk/we-need-to-talk-about-burnout-in-the-tech-industry/. Accessed 12 Oct 2021
19. Craig H (2021) Resilience in the workplace: how to be more resilient at work. https://positivepsychology.com/resilience-in-the-workplace/. Accessed 14 Dec 2021
20. Cronin N (2022) The importance of mentoring in the workplace. https://www.guider-ai.com/blog/why-everyone-needs-mentoring-in-the-workplace. Accessed 23 April 2022
21. CyberShark Recruitment (2022) Cyber security UK salary survey 2022. https://insight.scmagazineuk.com/expert-reports/cyber-security-uk-salary-survey-2022. Accessed 23 Feb 2022
22. Dawson C (2002) Practical research methods: a user-friendly guide to mastering research techniques and projects. How To Books Ltd., United Kingdom

23. Deloitte (2018) Welcome to generation Z. https://www2.deloitte.com/content/dam/Deloitte/us/Documents/consumer-business/welcome-to-gen-z.pdf. Accessed 17 April 2022
24. Durst P (2020) Talent retention for top cybersecurity talent. https://www.cybrary.it/blog/talent-retention-for-top-cybersecurity-talent/. Accessed 27 Feb 2022
25. Edgar T, Manz D (2017) Research methods for cyber security. Elsevier Inc., Syngress. Cambridge, MA 02139, United States
26. ENISA (2020) ENISA threat landscape 2020: Cyber attacks becoming more sophisticated, targeted, widespread and undetected. https://www.enisa.europa.eu/news/enisa-news/enisa-thr eat-landscape-2020. Accessed 29 Oct 2021
27. Evans-Reber K (2021) How to meet Gen Z's workplace expectations. https://www.forbes.com/sites/forbeshumanresourcescouncil/2021/11/10/how-to-meet-gen-zs-workplace-expect ations/?sh=5b552bef74ff. Accessed 20 April 2022
28. Exabeam (2019) Cybersecurity professionals salary, skills, and stress survey. https://www.exabeam.com/wp-content/uploads/2019/10/EXA_Salary_Survey_Report_2019_P1R1.pdf. Accessed 24 Feb 2022
29. Fernandez R (2016) 5 ways to boost your resilience at work. https://hbr.org/2016/06/627-bui lding-resilience-ic-5-ways-to-build-your-personal-resilience-at-work. Accessed 2 March 2022
30. Forbes Coaches Council (2018) High turnover? 10 ways to find the root problem (And Solve It for Good). https://www.forbes.com/sites/forbescoachescouncil/2018/05/02/high-tur nover-10-ways-to-find-the-root-problem-and-solve-it-for-good/?sh=1cb37d3b764e. Accessed 01 Nov 2021
31. Fox N (2009) Using interviews in a research project. https://www.rds-yh.nihr.ac.uk/wp-con tent/uploads/2013/05/15_Using-Interviews-2009.pdf. Accessed 12 March 2022
32. Francis T, Hoefel F (2018) 'True Gen': generation Z and its implications for compa-nies. https://www.mckinsey.com/industries/consumer-packaged-goods/our-insights/true-gen-generation-z-and-its-implications-for-companies. Accessed 17 April 2022
33. Goler L, Gale J, Harrington B, Grant A (2018) Why people really quit their jobs. https://hbr.org/2018/01/why-people-really-quit-their-jobs. Accessed 22 April 2022
34. Goodchild J (2019) 3 Tips for reducing security staff turnover. https://securityboulevard.com/2019/03/3-tips-for-reducing-security-staff-turnover/. Accessed 12 Oct 2021
35. Grad Coach (2022) Qualitative coding tutorial: how to code qualitative data for analysis. https://www.youtube.com/watch?v=8MHkVtE_sVw&list=TLPQMDYwNDIwMjI xili9LZ6C9Q&index=5. Accessed 6 April 2022
36. Gregory J (2021) Where digital meets human: letting HR lead cybersecurity training. https://sec urityintelligence.com/articles/cybersecurity-training-hr-nontechnical-personnel/. Accessed 27 Feb 2022
37. Howlett E (2021) Is there a darker side to workplace resilience? https://www.peopleman agement.co.uk/long-reads/articles/is-there-darker-side-workplace-resilience#gref. Accessed 2 March 2022
38. Hubbard J (2020) Virtuous cycles: rethinking the SOC for long-term success. https://www.you tube.com/watch?v=G5lj7M2ZuT0&list=TLPQMjUwMTIwMjKomw8zILlbwQ&index=8. Accessed 2 March 2022
39. Headley J (2021) A holistic approach to finding and fixing cybersecurity gaps. https://www.vpls.com/blog/holistic-approach-cybersecurity-gaps/. Accessed 12 Dec 2021
40. Hughes O (2022) Tech workers are fed up and want to quit. A few thoughtful changes could make them stay. https://www.zdnet.com/article/tech-workers-are-fed-up-and-want-to-quit-a-few-thoughtful-changes-could-make-them-stay/. Accessed 11 March 2022
41. ISC2 (2018) Hiring and retaining top cybersecurity talent. https://www.isc2.org/-/media/Files/Research/ISC2-Hiring-and-Retaining-Top-Cybersecurity-Talent.ashx. Accessed 23 Feb 2022
42. (ISC)2 (2021) Cybersecurity workforce study. https://www.isc2.org//-/media/ISC2/Research/2021/ISC2-Cybersecurity-Workforce-Study-2021.ashx. Accessed 23 Feb 2022
43. John S (2020) Empower your analysts to reduce burnout in your security opera-tions Centre. https://www.microsoft.com/security/blog/2020/07/28/empower-analysts-reduce-burnout-isecurity-operations-center/. Accessed 2 March 2022

44. Kempner M (2021) Georgia Pipeline tried to fill cybersecurity job before attack. https://www. govtech.com/security/georgia-pipeline-tried-to-fill-cybersecurity-job-before-attack. Accessed 27 Feb 2022
45. Lapan S, Quartaroli M, Riemer F (2012) Qualitative research: an introduction to methods and designs. John Wiley & Sons, Inc., San Francisco, CA
46. Lazar J, Feng J, Hochheiser H (2017) Research methods in human-computer interaction, 2nd edn. Elsevier Inc., USA
47. Lewis J (2019) The cybersecurity workforce gap. https://www.csis.org/analysis/cybersecurity-workforce-gap. Accessed 27 Feb 2022
48. Lopez-Garrido G (2021) Self-determination theory and motivation. https://www.simplypsycho logy.org/self-determination-theory.html. Accessed 18 April 2022
49. Markovich M (2019) The negative impacts of a high turnover rate. https://smallbusiness.chron. com/negative-impacts-high-turnover-rate-20269.html. Accessed 01 Nov 2021
50. McAfee (2021) What is a security operations centre (SOC)? https://www.mcafee.com/enterp rise/en-gb/security-awareness/operations/what-is-soc.html. Accessed 14 Dec 2021
51. McAfee (2021) The importance of building a security operations centre. https://www.mcafee. com/enterprise/en-gb/security-awareness/operations/building-a-soc.html. Accessed 4 March 2022
52. Medelyan A (2022) Coding qualitative data: how to code qualitative research. https://getthe matic.com/insights/coding-qualitative-data/. Accessed 5 April 2022
53. Mical J (2021) How to Banish SOC analyst burnout. https://www.sans.org/blog/how-to-ban ish-soc-analyst-burnout/. Accessed 28 Feb 2022
54. Monster (2021) How your professional growth can benefit from changing jobs every four years. https://www.monster.com/career-advice/article/dont-stay-in-same-job-more-than-four-years. Accessed 31 Oct 2021
55. Moore T (2020) 3 keys to enabling great culture. https://www.wearehuman.cc/articles/3-keys-to-enabling-great-culture. Accessed 18 April 2022
56. Morgan S (2019) Cybersecurity talent crunch to create 3.5 million unfilled jobs globally by 2021. https://cybersecurityventures.com/jobs/. Accessed 29 Oct 2021
57. Morgan S (2019) Women represent 20 percent of the global cybersecurity workforce in 2019. https://cybersecurityventures.com/women-in-cybersecurity/. Accessed 11 March 2022
58. Munns S (2021) Employee turnover rates by industry comparison. https://www.e-days.com/ news/employee-turnover-rates-an-industry-comparison. Accessed 12 Dec 2021
59. Nathan S (2020) Build a SOC or pick an MSSP? https://www.teceze.com/build-a-soc-or-pick-an-mssp. Accessed 4 March 2022
60. Nicholson N (2020) How a high turnover rate impacts your employees. https://hr.sparkhire. com/talent-management/how-high-turnover-impacts-your-employees/. Accessed 28 Feb 2022
61. Oltsik J (2018) Cybersecurity job fatigue affects many security professionals. https://www.cso online.com/article/3253627/cybersecurity-job-fatigue-affects-many-security-professionals. html. Accessed 23 Feb 2022
62. Pasupathy A (2022) What is SOC and benefits of SOC? https://www.teceze.com/what-is-soc-and-benefits-of-soc. Accessed 4 March 2022
63. Palmer D (2021) Cybersecurity teams are struggling with burnout, but the attacks keep coming. https://www.zdnet.com/article/cybersecurity-teams-are-struggling-with-burnout-but-the-attacks-keep-coming/. Accessed 28 Feb 2022
64. Pescatore J, Filkins B (2020) Closing the critical skills gap for modern and effective secu rity operations centres (SOCs). https://www.siemplify.co/resources/sans-report-closing-the-cri tical-skills-gap-for-modern-and-effective-socs-download/. Accessed 21 Feb 2022
65. Pink D (2018) Drive: the surprising truth about what motivates us. Canongate Books Ltd., Great Britain
66. Polk T (2021) Cyber risk can't be eliminated—But it can be mitigated. https://www.forbes. com/sites/forbestechcouncil/2021/03/26/cyber-risk-cant-be-eliminated---but-it-can-be-mitiga ted/?sh=4c77cf56af1d. Accessed 18 April 2022

67. PWC (2020) Cyber threats 2020: a year in retrospect. https://www.pwc.co.uk/cyber-security/pdf/pwc-cyber-threats-2020-a-year-in-retrospect.pdf. Accessed 14 Dec 2021
68. Randstad (2017) It takes more than salary to attract and retain talent. https://www.randstadrisesmart.com/blog/it-takes-more-salary-attract-retain-talent. Accessed 19 Dec 2021
69. RedLegg (2019) How to prevent your cybersecurity analyst from burnout. https://www.redlegg.com/blog/cybersecurity-burnout. Accessed 25 Feb 2022
70. Reffold K (2019) Is there really a cybersecurity skills gap? https://www.forbes.com/sites/forbeshumanresourcescouncil/2019/11/15/is-there-really-a-cybersecurity-skills-gap/?sh=3200e26310fe. Accessed 27 Feb 2022
71. Rogers M (2020) A better way to develop and retain top talent. https://hbr.org/2020/01/a-better-way-to-develop-and-retain-top-talent. Accessed 2 March 2022
72. Samtani S, Garcia E (2022) 5 strategies to proactively address the growing mental health crisis amongst IT professionals. https://www.comptia.org/blog/it-pro-mental-health?utm_source=sailthru&utm_medium=email&utm_campaign=Email_Certs_2022_02_25_ITCareers_Newsletter&utm_compid=email-sailthru-newsletterongoing-Email_Certs_2022_02_25_ITCareers_Newsletter-na-na-na-B2C. 12 Feb 2022
73. Saunders M, Lewis P, Thornhill A (2009) Research methods for business students, 5th ed. Pearson Education Limited
74. Seals T (2016) High cybersecurity staff turnover is an 'Existential threat'. https://www.infosecurity-magazine.com/news/high-cybersecurity-staff-turnover/. Accessed 01 Nov 2021
75. Segal E (2022) Predicting how, when or if 'The great resignation' will end. https://www.forbes.com/sites/edwardsegal/2022/02/20/predicting-how-when-or-if-the-great-resignation-will-end/?sh=5a2b4694bfcb. Accessed 25 Feb 2022
76. Shein E (2021) 7 skills of successful digital leaders. https://www.cio.com/article/189078/skills-of-successful-digital-leaders.html. Accessed 27 Dec 2021
77. Swinhoe D (2019) How to reduce security staff turnover? Focus on culture and people. https://www.csoonline.com/article/3331941/how-to-reduce-security-staff-turnover-focus-on-culture-and-people.html. Accessed 14 Dec 2021
78. Smith J (2020) Resilience is missing for many employees. https://workplaceinsight.net/resilience-is-missing-for-many-employees/. Accessed 28 Feb 2022
79. Soldatov S (2019) SOC burnout. https://www.kaspersky.co.uk/blog/soc-burnout/15253/. Accessed 28 Feb 2022
80. Suciu P (2020) Cybersecurity talent shortage calls for more training and better hiring methods. https://news.clearancejobs.com/2020/11/17/cybersecurity-talent-shortage-calls-for-more-training-and-better-hiring-methods/. Accessed 12 Oct 2021
81. Sundaramurthy SC et al (2015) A human capital model for mitigating security analyst burnout. https://www.usenix.org/system/files/conference/soups2015/soups15-paper-sundaramurthy.pdf. Accessed 28 Feb 2022
82. Van Os R (2020) SOC analyst burnout—Problem and the solution. https://www.cyberbutler.eu/knowledge-center/soc-analyst-burnout-problem-and-the-solution/. Accessed 28 Feb 2022
83. Vizard M (2019) Survey identifies root causes of cybersecurity staff turnover. https://blog.barracuda.com/2019/05/10/survey-identifies-root-causes-of-cybersecurity-staff-turnover/. Accessed 01 Nov 2021
84. Warszona B (2021) HR's increasingly important role in cyber risk management. https://www.marshmclennan.com/insights/publications/2020/july/hr-s-increasingly-important-role-in-cyber-risk-management.html. Accessed 27 Feb 2022
85. Westfall B (2017) 5 benefits (Yes, Benefits) of high employee turnover. https://www.softwareadvice.com/resources/employee-turnover-benefits/. Accessed 12 Dec 2021
86. Whiting A (2013) Legal Q and A: The duty of mutual trust and confidence. https://www.personneltoday.com/hr/legal-q-and-a-the-duty-of-mutual-trust-and-confidence/. Accessed 17 April 2022
87. Wilson S, Rickard C, Tamkin P (2014) Understanding resilience. https://www.employment-studies.co.uk/system/files/resources/files/mp94.pdf. Accessed 1 March 2022

88. Wolff J (2016) Calling humans, the "Weakest Link" in computer security is dangerous and unhelpful. https://slate.com/technology/2016/01/calling-humans-the-weakest-link-in-computer-security-is-dangerous.html. Accessed 12 Dec 2021
89. World Economic Forum (2017) This is what the future of cybersecurity will look like. https://www.weforum.org/agenda/2017/08/the-us-is-upping-its-game-against-cyber-attacks-but-the-security-industry-faces-a-huge-challenge. Accessed 29 Oct 2021
90. World Health Organisation (2019) Burn-out an "occupational phenomenon": international classification of diseases. https://www.who.int/news/item/28-05-2019-burn-out-an-occupational-phenomenon-international-classification-of-diseases. Accessed 31 Oct 2021

The Reality of Cyber Security in Bangladesh, Relevant Laws, Drawbacks and Challenges

Kudrat-E-Khuda Babu

Abstract Protection of data has become a matter of great concern as cyber-crime has now emerged as a major threat amid the rapid spread of the internet and information and communication technology across the globe. Despite being the most technologically advanced country, the United States is not out of the purview of the danger of cyber-crime. In such a situation, a country like Bangladesh which is still in the group of less developed countries is one of the most vulnerable countries in terms of cyber security. The Bangladesh government has gone through a huge digital transformation over the years and has been trying to connect the dots between the institutions digitally with the slogan of "Digital Bangladesh". Apart from that, national and multinational companies, operating in the country, are also offering online services to be part of the government's journey to the digital world. The distance between a consumer and a shop or a bank is now a click away, thanks to the wide access to the internet and the digital presence of the entities. Taking the advantage of easy excess and widespread use of the internet, opportunists and criminals have chosen the digital path to materialize their evil wishes. They are committing various crimes, including stealing money, and personal data, spreading rumors and cyber-attacks and other criminal activities. Amid the fragile security system, there is also a huge risk of targeted cyber-attacks by the hackers of any opponent country or frustrated group. Against this backdrop, it is evident that the state of cyber security in the country is very fragile and the existing laws and the measures of the state are very insignificant to tackle the growing threat. This study pinpoints the escalating cyber security concerns in the context of Bangladesh from a global perspective.

Keywords Bangladesh · Cyber security · Cybercrimes · ICT Act, 2006 · Digital Security Act, 2018

K.-E.-K. Babu (✉)
Department of Law, Daffodil International University, Dhaka, Bangladesh
e-mail: kekbabu.law@diu.edu.bd

89

1 Introduction

This is the high time for a revolution in information and communication technology in this age of globalization. Thanks to the rapid growth of technology, especially for the evolving information, and communication tools, now safeguarding cyberspace has become a crucial part of national security in the era of globalization. This is crucial for achieving economic stability and effective security in a country [17]. Information is saved, exchanged, and revealed in cyberspace, which is a realm of computer networks and the humans that use them [14]. Along with the transformation of the world into digital, the crimes of the actual world have also been shifted to the virtual with the same pace of the digital transformation. Once robbers looted a bank; now the virtual robber is heisting money from the bank by hacking their digital finance system. The cyber heist of $ 101 million from Bangladesh Bank reserve in 2016 has now become an old example of cybercrimes. So, there is no doubt that cyber-crime has become a foremost security issue for any state or organisation. Such crimes may be conducted by any individual or sometimes they may be unleashed by any state. The major concern is now the protection of confidential data. Posing a serious threat to the economic progress and defence systems, cybercrimes have become a cause of concern and escalated tensions in the diplomatic arena and eventually led to anarchy and even war in the world. Both misuses of information communication technology are now a major threat to global peace, stability, and development. As there is no strong surveillance system and security tools and measures, criminals can choose Bangladesh as a haven for committing crimes, including hacking and stealing personal data. Cybercriminals usually target digital services providers—both the government and non-state organisations—and steal personal and organizational data. In most cases, effective security steps are not taken at the time of providing digital services. As a legal security measure, the government enacted the Information Communication Technology Act (ICT Act), 2006 but it has failed to ensure cyber security due to a lack of proper use of the law. This study tries to identify the challenges for ensuring security in cyberspace in Bangladesh and shed light on the measures taken to check the cybercrimes from a global perspective. It also has scrutinized the current state legal frameworks, use and effectiveness to ensure efficient cyber security in the country. The final point of this study is that this is the prime time for Bangladesh to take strong measures to keep cyberspace safe and secure.

2 Cyber Security in the Global Village

Thanks to the evolution of information and communication technology, now people are very much connected globally and now they can commute from one edge to another edge of the world like a neighbour. And at a time, they can be connected from multiple parts of the globe within seconds. As a result, a new term has emerged among the global citizens called "netizens" and they have now made the world like a village. Here comes the term global village. But despite the revolutionary changes in the cyber-world and security concerns for the netizens as well, the cyber threats have not gotten priority in the national security issues as well as the global concern. Not only are the netizens being victimised by the cybercrimes, but it also affects the state and even the globe too. As the most amount of money that is transacted daily in the world, through the digital transactions, cybercriminals always try to find the loopholes in financial networks and systems to extort money. According to Williams, there are four types of cyber-crimes, according to Williams. To begin with, cybercriminals are only interested in making money. In April 2013, for example, the US stock market lost $130 billion in minutes as a result of a hacked Twitter news stream spreading a bogus tale about an explosion at the White House [17]. Second, in their quest for sensitive knowledge or intellectual property, competing groups prey on one another. Both the civil and security businesses are concerned about this. A Russian criminal gang has amassed the largest known collection of stolen Internet data, which included 1.2 billion usernames and password combinations for over 500 million email addresses [13]. Thirdly, a de facto insider also appears as a threat sometimes. The incident of breaching the IT system of Iran's nuclear project and the leak of American diplomatic cables showed the vulnerability of the current system and stressed the need for ensuring cyber-security. However, the damage and losses for the cybercrimes are not proportional globally but the crimes have become more exacerbating due to the weakness in the security system in the cyberspaces.

Usually, cyber-crime predicts inevitable conflicts that will arise from close contact between different cultural practices through the internet. The revolution of information and communication technology has created people to people networks across the globe, connecting their organizational levels. Earlier, the communication process at the intra-organizational and governmental levels was so slow and expensive but the easy availability of the internet has changed the whole process of connectivity to an unparalleled level in terms of cost, speed and easy medium. Even for the sake of some software, it has also become so easy to send any information to a destination where there is no network for information transmission. Though this is helpful the users have to spend for it. However, amid the frequency of cybercrimes, now the governments of the states and other organizations are becoming cautious and considering the cyber security issues as the threat and working to deal with these as well. There is a common cyber security threat of stealing data from the organisations such as data breaches. Most organisations fall prey to the attack of classified data hacking by outsiders. The major challenge to the cases of data breaches is the rapid transfer of data from an organisation to an organisation. The cross-border nature of the incident of data breaches can make both the investigation and the separation of breach management options into an overwhelming and irresistible approach.

In recent years, The Asia–Pacific area has witnessed a flurry of new digital security regulations enacted in recent years, with countries establishing agencies or regulators to monitor and manage cyber security issues and publishing regular guidelines and circulars. In 2015, countries in East Asia such as Indonesia and Singapore established cyber agencies, and Japan passed the Cyber Security Basic Act. All of a sudden, Asia Pacific nations have started to formulate laws and/or guidelines as a part of their attempt to secure their cyberspaces. If we look back to the other parts of the world, we already know that the Australian Securities and Investments Commission has issued cyber resilience. Meanwhile, in countries like the United States, its Justice Department issued a guideline titled "Best Practices for Victim Response and Reporting of Cyber Incidents" in April 2015. Many other countries are also adding cyber security guidelines to their existing frameworks.

Despite intensive measures by the government in the Asia–Pacific region, there is no evidence of taking a concerted approach in formulating cyber security regulations or possible legal action against the data breaches in the region. In addition to cyber security laws, the issue of data breaches can be incorporated into various existing laws and regulations. It may be incorporated with data protection laws and labour laws. It can also be integrated into the rules and obligations like equity rules, corporate governance and fiduciary duties. In some cases, the laws or legal frameworks of a state come into force when any data goes beyond the jurisdiction of any authority. Similarly, it is important to have local knowledge of the responsibilities of states and how regulators or courts are responding to data breaches. Besides, it is also important to know the best ways of legal remedies to the cyber menace. After acquiring the knowledge, the victim or who might fall prey to a cyber-attack can check whether there are any data cracks in his system. It may then be possible to separate the obligations to create a plan to limit further leaks of data. They can control and realize the impact or loss of data breaches, where they will find accessible and legal solutions to recover data or damage from data breaches. Many government websites rely on international servers and providers, putting them in a vulnerable position and putting them at risk of being hacked by system insiders [1]. The fourth category may pose the greatest threat to a country's security. This is a state-sponsored cyber-attack intended at undermining a national security framework, such as essential infrastructure or key national economic components, in order to gain strategic advantages over that country [17]. The example of China might be used as an example. Some of the world's most powerful countries, including the United States, the United Kingdom, France, Germany, and India, have long viewed China as a possible threat to cyber security and have accused it of espionage to obtain strategic advantages. It has been proven that in 2007, China launched a series of network-based cyberattacks against the countries listed above. Furthermore, these countries have stronger goals to improve their military's potential to participate in information or cyber warfare if it becomes required in near future.

3   Cyber Violence Against Women in Bangladesh Coontext

Women always fall victim to violence and harassment, including stalking, cyber-bullying and other forms of nuisance in Bangladesh. Apart from the harassment, pornography-related issues are also taking place here due to the easy accessibility of the internet and its ancillary equipment. After all, the cyber violence against women goes unabated in the country's periphery with the same pace of the extension of Information and Communication Technology (ICT) and flourishing use of the internet due to lack of legal protection. Such violation of human rights stretches from stalking to cyberbullying and trolling which sometimes ended through reprisal pornography. Usually, the women are targeted from unidentified and counterfeit sources on the internet. They are given various types of ignominious messages frequently with obscene gestures. The women threatened them with nude pictures by placing the targeted women's faces in a photo with another nude photograph. The criminals also send spam, and sex-act tapes to make them frightened or lure them to do whatever the criminals want. Such types of unlawful things are now a new dimension of social media here. Especially with easy access to the smartphone, the active users of the internet have increased dramatically in Bangladesh. Of 90.5 million mobile phone users as of August 2018, 80.47 million are connected to mobile Internet [5]. Thanks to the additional use of mobile phone internet, the number of Facebook users has surged in the country. About 86 percent of 29 million registered Facebook users use social media sites through mobile phones. However, the number of females who have access to the internet through mobile phone sets is only 1 per cent. However, young women fall victim to sexual violence comparatively than the ageing women in Bangladesh. As a result of the lack of a legal framework and organizational protection, a sizeable 73% of women Internet users in Bangladesh lodge formal complaints regarding cyberspace pestering, abuse, and violence emanating from Cybercrime [18]. Amid huge frequent incidents of cybercrime in Bangladesh, the government has been forced to open a help desk to deal with the crimes. As of December 2017, the government's "Cyber-Help Desk," which is part of the Information and Communication Technology Division, has received over 17,000 complaints, with women accounting for 70% of the complainants. The risk of exposure to pornographic content is much higher among young people. Whether it is intentional or unintentional. In most cases, they are victimized by pornographic photos in the country. Around 78 percent of occurrences of doctored images with pornographic content in the digital world involve women. It should be emphasized that roughly 77 percent of Bangladeshi youths watch pornography on a regular basis.

According to a report by the Bangladesh National Association of Women Lawyers in June 2019, it is stated that harassment prevails in society with a concerning instance and several young women drop out of classes or jobs due to trauma and stigma as there are inadequate preventive measures and legislation. Court orders mandating the formation of complaint committees and the installation of complaint boxes in educational institutions and workplaces have been rarely executed [16]. In Bangladesh, it is very common for social media accounts to be hacked with vicious intentions. The criminals usually post doctored obscene pictures relating to the victims and then send provocative messages to the women to victimize them. Most of such incidents happen by smearing the victim, taking revenge, forcing them to establish sexual contact, threatening them with extortion or physically tormenting the victims. Conducting a study on the lawsuits and media reports, it is found that most of the cyber violence against women are happened in Bangladesh to establish physical contact with the victims. Usually, the offenders capture or collect photos or videos of any intimate moments, and materialized their further evil wishes—it may be sexual contact, extortion or any other unlawful demand—by threatening the victim further with the weapon. Besides, the criminals captured the videos of rape scenes so that they can use these recordings to silence the victims. Most of the time, those video clips are released by the criminals despite the request of the victims not to publish them. There could be nothing more than this way of humiliating, traumatizing and stigmatizing a woman in society in such a way. There are numerous reports of committing suicide of women as they feel utterly helpless after experiencing such nightmares. Another common trend is found that vindictive ex-husbands and former lovers release intimate videos or pictures on social media platforms tenaciously. After all, young women are most vulnerable to falling into the traps of cybercriminals.

3.1 Effects of Cyber Violence

The impact of cyber violence on women in the conservative society in a country like Bangladesh is far-reaching, horrible and deep rooted. Here, victims' families are also victimised and traumatized along with the victims. There are numerous incidents in the country where the reprisal of both the victims and their families comes as a double blow. Thanks to the ignorance of the majority of people in the least developed countries like Bangladesh, most people strongly believe whatever content they see on social media platforms. If any photographed picture of a woman released on social media or any online platform matches any mixed up with vulgar gossip, most users do not think of verifying the picture and indiscriminately share the fake content, making it viral on the online platform. This proclivity for spreading sex-related rumors exacerbates the victims' pain, as well as the sorrow of the victims' family members, who endure social marginalization, shame, and public hatred [10]. As a result, if any doctored nude photograph is published on the social media sites defaming any women, they believe it blindly. Then the woman was slandered. In some cases, the victim's family is kept in confinement. In many other cases, the victim was evicted from his home or village. So cyber violence creates a catastrophic situation in the personal life of the victim. As a result, they suffer from severe depression, guilt, and paranoia. Their careers, education and social life are endangered. As a result, many of them became drug addicts. Some people decide to end their lives because they can't stand the stress. Very few of the victims were able to recover from the trauma. From 2010 to 2014, Bangladesh National Women Lawyers Association attempted to commit suicide among 65 female victims of such violence. Of these, only 11 women victims of cyber violence have attempted suicide. The number of such cases was only 8 in 2008, the data reveals an upward trend. However, the official statistics are paltry in comparison to the actual number of such incidents, with the number of unreported cases far outweighing the reported ones [4].

4 Challenges to Bangladesh and Bangladesh Government

Bangladesh is one of the highest cybercrime vulnerable countries in the world as most of the software used here is pirated. Besides, the infrastructural system is also poor. In such a situation, this is a big challenge to protect the country's cyberspace. Around 90% of software is pirated in Bangladesh [3]. Usually, when pirated software is used, criminals can penetrate easily due to the system loopholes. But, in Bangladesh, no one bothers about whether they are using the software that is pirated. This is one of the reasons for the vulnerability of the country's cyber security domain. Though this issue is ignored, many users pay the price when they have no other ways to fall victim to cybercriminals and its consequences and impact cannot be ignored. Besides, there are some other serious challenges for those criminals who target the people here most and carry out any cybercrimes staying out of reach and punishment. There were 90.5 million active internet users in Bangladesh as of August 2018, data from Bangladesh Telecommunication Regulatory Commission (BTRC) showed. Not at all, 1.8 million new connections were added to the network in one month. Of them, 84.7 million users use mobile Internet. About 5.73 million of them have fixed broadband Internet connections, while the rest use WiMAX connections. In April 2017, the total number of active Internet connections surpassed 70 million, followed by 60 million in August 2016, 50 million in August 2015, and 40 million in September 2014. (*The Daily Star*, 21 September 2018). Bangladesh's banking sector is under considerable cyber threat as a result of the country's rapid expansion in Internet usage. So a strong built-in cyber security is needed in such situations. At the same time, there is a need for experts to protect knowledge and data related to cybercrime.

First and foremost, we must comprehend these difficulties. We must be aware of the extent of daily cybercrime. There are four different sorts of cybercrimes that commonly occur in the country. Hacking, illegal entry, spying, data infiltration, e-mail spoofing, spamming, fraud and forgery, slander, drug trafficking, and virus transmission are examples of crimes against humanity. Second, property-related cybercrime is a subset of cybercrime. Credit card fraud, intellectual property infringement, and internet time theft are just a few examples. Organized crime is the third category of crime. Unauthorized control/downloads from network resources and websites, posting obscene/pornographic content on the web, virus attacks, e-mail bombings, logic bombings, Trojan horses, data dodging, download blocking, theft of valuables, government terrorism against organizations, and network infrastructure vandalism are just a few examples. The fourth and last type of f cyber-crimes is the attack on Bangladesh's society or social values, with such crimes including forgery, online gambling, prostitution, pornography (especially child pornography), financial crimes, the pollution of youth by indecent exposure, and web jacking, etc. [11]. Pornography has become a matter of major concern in Bangladesh. Pornography is strongly prohibited in the country in terms of the country's social culture and moral values. One of the reasons for this is the rapid expansion of digital communication technology. Because, as a result, it is now possible to communicate instantly with anyone from anywhere in the world to anywhere else. As such, it has become easier to share and exchange the cultural values of an individual or a country. Therefore, due to the free spread of culture, many harmful elements of the culture of another country can easily penetrate their own culture. Which cannot be adapted in any way to one's own culture. Extremely unpleasant elements of perverted culture like pornography are in no way acceptable in the culture of Bangladesh.

Bangladesh's law enforcement agencies regularly receive numerous allegations of sexual harassment. These sexual harassments include secret nude video footage or posting obscene pictures demanding large sums of money. Or publishing these videos and nude photos for defamation. This type of crime is also committed as a result of previous hostility, failure to fulfil any evil desire or enmity with another member of the family. Most of these victims are teenagers. However, women and children are not the targets of criminals. If a crime is committed outside the country, it is considered a Trans boundary crime. Then it is considered a crime in both the countries where the crime was committed and the country in which it was located. However, it is not an easy task to bring cybercrimes like pornography under the law, especially in the context of Bangladesh. Because pornography is not considered a crime in many countries. In many countries, such as the United States, it is not considered a crime. Therefore, these Trans boundary crimes have to face various problems. However, child pornography is an international crime. Such a crime is considered a crime in any country. International assistance is available to deal with these crimes. Such a case is fully described below.

Several years ago in Bangladesh, Tipu Kibria, a well-known children's writer, was arrested by the police with evidence of child pornography. She used to come to her house and lab with male street children and make pornographic videos. He assaulted about 400–500 street children before his arrest. Kibria used to do all these things with the help of his two assistants. Police later found the names of 13 foreign shoppers who regularly paid Kibria to supply child pornography through online banking. Apart from Tipu Kibria, Bangladesh Police believe there may be other makers of this form of pornography. As a result, it is apparent that pornography is a big threat to Bangladesh's cyber security [1]. Cyber security threats and online banking in Bangladesh are now a matter of great concern for all types of financial transactions. Due to the potential for illegal online transactions, international criminal gangs carry out various criminal activities in Bangladesh, including drug smuggling, trafficking and terrorism, which represent a challenge to the country's cyber security. Bangladesh's banking sector is under severe cyber threat due to a lack of proper cyber security measures. Hackers stole $101 million from Bangladesh's central bank using the Swift payment network for counterfeit orders from the Federal Reserve Bank of New York. Which is one of the biggest cybercrimes in the history of cybercrime. The main reason behind the scandal of such a big economy is the weakness of the cyber security system. Lack of proper defence business. If Bangladesh still fails to ensure cyber security, then the country's banking sector may face more such cyber thefts in the coming days. A vast amount of personal client details, such as bank account names, bank account numbers, cell phones, e-mail IDs, and so on, are often put in danger when credit cards and electronic payment methods are used extensively [3].

Law enforcement agencies in Bangladesh often receive complaints about direct or indirect cyber threats to financial transactions through online banking. On February 12, 2016, evidence of 21 suspicious card transactions was found from Eastern Bank, a private bank in Bangladesh. A fraudster browsed with a fake EBL card from an ATM booth at United Commercial Bank Limited. On February 25, the Dhaka Metropolitan Police said that while investigating the ATM card scam, they found the involvement of various hotels and travel agencies. Some bank officials are also involved in this. Police also arrested three people, including a German national named Piotr, on charges of ATM fraud. Police have found evidence of the involvement of some employees of Citibank, a local private bank, in the scam. It was later revealed that Piotr was wanted in at least three countries for fraud (*The Daily Star*, 26 February 2016). In Bangladesh, foreigners are involved in financial scams, including money laundering from ATMs. They are targeting Bangladesh due to the inadequate security of the digital financial transaction system in the country. A number of recent financial scandals have alarmed banks and consumers. In February 2016, lawyers for Bangladesh Bank and three other commercial banks analyzed video footage of four ATM booths and found that at least Tk 2.5 million had been looted. The spokesperson for the central bank said that the principal perpetrators were at least two foreign nationals. Similar concerns affect other private banks, such as Eastern Bank Limited, United Commercial Bank Limited and Citibank, all of which have been victims of ATM fraud (*The Daily Star*, February 16, 2016). Following a spate of illegal transactions, all banks, including both public and private commercial banks in Bangladesh, are currently tightening security. But there are questions about whether those security measures are appropriate. Why is there a lack of technology to ensure proper security? Besides, those in charge of security lack knowledge. As a result, the banking sector in Bangladesh is still struggling to ensure cyber security. Also, many people in Bangladesh are victims of fishing or fraudulent attempts. They are enticed by emails or attractive advertisements to steal all their money after taking confidential data such as usernames, passwords and credit card details. Victims often lose $100–500 in each case and are often hesitant to report the crime to the authorities, making the situation in Bangladesh even more difficult to deal with [1]. Hacking, or unauthorized access to a computer system without the owner's or user's consent, is also a cyber-security risk in Bangladesh [11]. Before committing any financial crime, hackers typically closely monitor the financial activities of both government and non-government entities, looking for holes in the transaction system. Bangladesh is in a more difficult position to combat cyber-piracy due to a weak cyber infrastructure network and reliance on overseas server system providers, among other things, due to a lack of competent cyber security know-how [1]. In addition to the concern of digital financial fraud in Bangladesh, another concern is the theft of information or data. In 2014, a very sensitive verdict (partial) of the Bangladesh War Crime Trial Tribunal was leaked. The data of the tribunal was leaked through Skype's voice recording and caused a major backlash against the Bangladesh government and exposed the vulnerability of cyber security in Bangladesh [1]. Meanwhile, the lack of cyber security on social media platforms has become a serious threat to the citizens and the government in Bangladesh. Unpleasant incidents are constantly happening, especially on Facebook, Twitter and LinkedIn due to a lack of proper security measures. The incidents like social media account hacking were frequent until February 2019. The hackers mostly hacked the social media accounts of celebrities, celebrities and women and demanded large sums of money in return. Without money, they would ruin the social image of the victims. At one stage, the law enforcement agencies of Bangladesh, especially the police, were forced to form a monitoring team to strengthen their monitoring. Besides, Bangladesh Telecommunication Regulatory Commission has also formed a separate monitoring team to control the crime.

5 Existing Acts Related to Cyber Security and Their Limitations

Despite repeated incidents of cybercrimes, so far there is no headache or concern about the existing cyber security measures or risks in the country. Despite considerable concerns, the country has not been able to properly and timely address the risks. Due to a lack of understanding among the various stakeholders concerned, the concerned authorities are also unwilling to take full action to deal with any approaching risk. To combat cyber-crime, the Bangladesh government has enacted the Information and Communication Technology (ICT) Act and the Digital Security Act. There is a lot of criticism among the rights activists and intellectuals of the country about some articles about those acts which are considered weapons of throttling the media and public opinion. Sadly, these laws are being used to punish critics of the government or government party political parties. It is being used as a tool to curb the freedom of speech of ordinary people. With these laws, the government and law enforcement agencies have cracked down on the media and social media in the country. In the face of much protest and criticism, on 8 October 2006, the Bangladesh government passed the ICT Act in the parliament. In the face of strong criticism, seven years later, the government amended the law on October 6, 2013, tactfully keeping the controversial provisions of the law intact. Victims in Bangladesh can at least use this ICT Act as a starting point; however, strong cooperation is required to progress from regional law enforcement agencies with expertise in cyber security, such as the CID (Criminal Investigation Department), to international law enforcement agencies, such as Interpol [1]. Section 57 of the Information and Communication Technology Act provides ample scope for the misuse of this Act. Human rights, lawyers, civil society representatives and media critics have all called for the repeal of the act. Before the amendment of the act, the convict was liable to imprisonment for a term not exceeding 10 years and a fine of Tk 10 million if convicted. Law enforcement agencies had to seek the permission of the appropriate authorities to file a case before arresting a person under this Act. Following the 2013 amendment, the maximum prison term has been set at 14 years. Moreover, law enforcement agencies are given unilateral power. As a result of the amendment, law enforcement agencies can now detain anyone without a warrant. Despite strong criticism and protests from human rights lawyers, civil society representatives and the media, the government has remained steadfast in its stance on the issue. Section 57(1) states, "If any person deliberately publishes any material which is false and obscene or transmits or causes it to be published or transmitted on the web, or in any other electronic form, and if anyone sees, hears or reads it having regard to all relevant circumstances and its effect is such as to influence the reader to become dishonest or corrupt, or causes a deterioration or creates the possibility to deteriorate law and order or prejudice the image of the state of a person or causes hurt or could offend religious belief or instigate against any person or organization, then this activity will be regarded as an offence". Despite reforms that made a few major changes, the core 2006 Act remains unchanged, with all of its flaws and needlessly punitive penalties [2].

The ICT Act (amended) has now been used as the tool of the Bangladesh government to curb fundamental human rights. Freedom of opinion and expression is now at stake as the act has a range of ambiguous clauses [9]. This act will encourage instigating cybercrimes instead of containing cyber-criminal activities. According to the ICJ, The original ICT Act's Section 57 is "incompatible with Bangladesh's responsibilities under Article 19 of the ICCPR: the charges imposed are unclear and disproportionate, and the limitations on freedom of speech and opinion go beyond what is permissible under Article 19 (3) of the ICCPR," according to the ICJ (ICJ, 2013). "Section 57 is not explicit and encompasses a broad range of offenses," J. Barua explained, "and there is minimal likelihood of winning an acquittal from any accusation" [2]. After studying the ICT Act 2006 and its revisions, it is clear that new legislation is needed to combat cyberspace-related crimes, as the current Act is confusing and has to be built permanently as a modernist legal structure rather than being based on an ad hoc approach (*The Daily Star*, 2013).

Human rights activists and journalists have been critical of Section 57 right from the very beginning of the enactment of this law. But the criticism sparked a protest after a senior journalist named Probir Sikdar was arrested in 2015 under Section 57 of the ICT Act. In addition, in the four months leading up to July 2017, at least 21 journalists were sued under Section 57 of the Information and Communication Technology Act, despite mounting calls to repeal the provision, which is widely prone to abuse (*The Daily Star*, 7 July 2017).

Amid widespread criticism, Bangladesh's Law Minister Anisul Huq on May 2, 2017, said "Section 57 would be withdrawn and that a new "Information Technology Act that is in the pipeline" will be implemented." But later on September 19, 2018, Bangladesh's Parliament passed the Digital Security Act, incorporating another tough clause empowering the law enforcement agencies to search or arrest anyone without a warrant creating further outcry among the rights activists and journalists. They have been expressing concern saying that the act goes against the constitutional rights of the country's people and it will curb freedom of expression and gag the media. Section 43 of the Digital Security Act states that when a police officer believes that a crime has been committed or is taking place in a specific place where there is a risk of crime being committed if the evidence is lost, the officer may search the location or arrest any person. The Editors' Council, (Sampadak Parishad), the council of the editor of news media in Bangladesh, stated on September 16, 2020, expressed surprise, frustration, and shock over some sections of the Digital Security Act. In their statement, the editors stated that Sections 8, 21, 25, 28, 29, 31, 32, and 43 of the Act pose serious threats to freedom of expression and media operation. Section 3 of the Digital Security Act incorporates a clause of the Access to Information Act 2009 which will be extended to information-related matters. According to the section, if a person uses a computer, digital device, computer network, wireless network, or any other electronic medium to commit any crime or assist others in committing crimes under the Official Secrets Act, 1923, as provided for in Section 32 of the law, he or she may face a maximum of 14 years in prison or a fine of Tk 2.5 million or both. A definition of the "Spirit of the Liberation War" has also been included in Section 21, which states, "The high ideals of nationalism, socialism, democracy, and secularism, which inspired our heroic people to dedicate themselves to, and our brave martyrs to sacrifice their lives in the national liberation struggle." However, under Section 29 of the law, a person can be sentenced to three years in jail or snapped a fine of Tk 500,000, or both if he or she is found guilty under Section 499 of the Penal Code for his crime online. Section 31 states If a person published or broadcast something on a website or in electronic form that could spread hate and build enmity between different groups and communities, or that could cause a deterioration in law and order, he or she could face seven years in prison or a fine of Tk 500,000, or both penalties (*The Daily Star*, 20 September 2018).

6 Policy Opinions

In addition to reducing the rate of cyber-crime, several alternative remedial policies can be implemented to ensure cyber security in Bangladesh. The government of Bangladesh may consider the following alternative policies.

6.1 Reform of the Legal Structure

We fully support the recommendations of the ICJ regarding the ICT Act 2013 of the Government of Bangladesh and its amended law. The ICJ called upon the Parliament of Bangladesh to implement these recommendations. In their recommendations, ICJ said, 'Either repeal the Information and Communication Technology Act (2006) as amended in 2013, or amend the ICT Act to bring it in line with international law and standards, including Bangladesh's legal obligations under ICCPR. The ICJ recommended that the Bangladesh government (1) amend Section 57 of the ICT Act to ensure that any planned restrictions on freedom of expression and opinion are in accordance with international law and standards; (2) amend Section 57 of the ICT Act to ensure that forbidden speech is clearly defined; and (3) amend the ICT Act to ensure that any restriction on freedom of speech and information, including any penalty imposed, is necessary for a valid purpose and proportionate to that purpose [9]. The

ICJ also proposed several policies—(i) Take action to ensure that the provisions of the ICT Act are not used to infringe the right to freedom of speech, including restricting the legitimate exercise of public opinion on matters which could include criticism of the Government; (ii) drop charges against bloggers for the legitimate exercise of their freedom of expression; (iii) guide government agencies to refrain from unfairly limiting the freedom of speech in politically-motivated cases and not to pursue penalties that are disproportionate to the severity of the alleged offence [9].

6.2 Maintaining Rules of Cyber Security

Menken Tikk, the legal counsel at the NATO Cooperative Cyber Defense Center of Excellence, Tallinn, Estonia, wrote an article on 'Ten Rules of Cyber Security in 2011. In his article, Tikk came up with a framework to ensure cyber security principles considering the security concerns for personal and state levels. In many cases, the proposals made by Tikk are acceptable. For ensuring cyber security, Tikk proposed 'the territorial rule' where he stated, "information infrastructure located within a state's territory is subject to that state's territorial sovereignty" [15]. In his 'the law of duty ', Tikk suggested that a country must play a responsible role in ensuring cyber security in its territory. He also proposed an 'early warning statute', where he stated, "There is an obligation to notify potential victims about known and upcoming cyber-attacks" [15]. The rules that Tikk has proposed regarding cyber security can be implemented by any country in the world. We can also implement Tikk's proposal for Bangladesh to fight against cyber threats from Bangladesh.

Secondly, any state can adopt its 'data protection rule' to protect the crucial data of a state. Tikk also stated, "Information infrastructure monitoring data are perceived as personal unless provided for otherwise." We can mention another proposed rule of Tikk named "the duty to care rule". In his rule, he suggests that everyone takes a minimum level of responsibility to secure all kinds of information infrastructure [15]. In the light of Tikk's rules, we can propose that the Government of Bangladesh will formulate and implement a policy framework using its resources and expertise to safeguard the cyber system and national interests. The country's cyber experts need to be trained to become more efficient. Besides, the country needs to create its own "server framework and system" using its resources and labour, which will play a role in ensuring cyber security nets. For this, we have to recruit trained manpower. It will take a long time but it will be a sustainable cyber security system. This will reduce the dependence on foreign experts.

Thirdly, we also support 'the cooperation rule' of Tikk. Here he mentioned, "… a cyber-attack has been conducted via information systems located in a state's territory creates a duty to cooperate with the victim state" [15]. Fighting against any kind of cyber security risk requires strong global participation. Because, in addition to local threats, in most cases, cyber threats come from the international arena. A criminal can harm a person or an organization from one country to another, or commit a crime in one country and flee to another. Therefore, the suppression of such a crime requires global support and participation if necessary. The Bangladesh government and the Bangladesh Police are liaising with international law enforcement agencies such as Interpol. In addition to strengthening that communication, we need to work for cooperation from other law enforcement agencies and tech giants around the world, such as Microsoft, Google, Facebook, Yahoo and others. Besides, we can also adopt Tikk's other two rules—'self-defence and access to information. He Tikk said, "Everyone has the right to self-defence" and that "the public has a right to be informed about threats to their life, security, and well-being" [15]. Finally, for ensuring cyber security at both the individual and national levels, we are strongly recommending the Government of Bangladesh take all precautionary and necessary measures.

6.3   Individual Awareness

In the current era of globalization, change and changed reality are undeniable. If anyone can't adapt to that change and can't cope with any of the shocks that come with that change, survival will be extremely difficult. In this ever-changing cyber world, to protect personal data alongside national information, we should have awareness at the individual level as well as the government needs to create secure cyberspace. Professionals must achieve a minimal degree of proficiency in handling cyber technologies and establish knowledge of cybersecurity threats, regardless of their hierarchy or organizational structure. Bangladesh can only be saved from sliding into a deep pit of cybersecurity risks if it receives sufficient education and awareness (Alam, Md. Shah, personal communication, 27 July 2014).

The basic precautionary measures need to follow while using the internet:

- Keeping personal details in restricted mode if necessary;
- Keeping privacy settings on;
- Using secure browsing;
- Using sure internet connection;
- Using strong passwords;
- For online transactions, browsing from protected sites;
- Being careful about any type of post;
- Keeping antivirus software updated.

If you somehow realize that you are a victim of cybercrime, then you should go to the local police station without any negligence. In addition to seeking the cooperation of the police, it is necessary to inform the matter to the relevant organization or medium. If it is a financial scam, then you have to contact the concerned financial institution. If any occurrence happens on any social media platform, you have to communicate with the respective platform. Depending on the situation, higher levels such as the FBI and the Federal Trade Commission may be contacted. If we are active from the beginning, it will be possible to prevent the recurrence of such crimes.

7 Conclusion

In conclusion, the issue of cybercrime has become a potential threat to the national security of any country, not just Bangladesh. However, in the context of globalization, the issue of cybercrime is even more worrying for a country like Bangladesh, especially where there is no secure cyber system. Due to the lack of advanced cyber security tools in Bangladesh and the ignorance of the people about the use of technology, use of unsafe technology, lack of necessary laws and indifference of the government, Bangladesh and the people of this country are a serious cyber security risk, which may lead to a devastating situation in the state. Another thing is that there is no publicity about cyber security in the country. There is no campaign or initiative to inform the people about the law of Bangladesh, cyberspace and international cooperation, increase technical knowledge and skills and what to do about cyber security threats. So these issues need to be taken into consideration to combat the ever-looming cyber security threats. However, the rapid increase in cybercrimes in Bangladesh and around the world proves that the issue of cybercrime is undeniable. Someone may argue that the cyber threat is not a major threat to Bangladesh right now. But in the light of the present situation, it can be said that these arguments have no basis, only conjecture. Finally, it needs to be said that Bangladesh should take immediate and preventive measures to curb cybercrime right now. Hopefully, the recommendations given in this research paper will play a vital role in ensuring cyber security for the Government of Bangladesh and the people of the country.

References

1. Alam S (2019) Cyber Crime: a new challenge for law enforcers. City Univ J 2 (1):75–84. http://www.prp.org.bd/cybercrime_files/Cybercrime. Accessed 25 Apr 2020
2. Barua J (2019) Amendment information technology and communication act. The daily star. http://www.thedailystar.net/supplements/amended-information-technology-and-communication-act-4688. Accessed 25 May 2019
3. Bleyder K (2012) Cyber security: the emerging threat landscape. Bangladesh Institute of Peace and Security Studies, Dhaka
4. BNWLA (2014) Survey on psychological health of women. Bangladesh National Women Lawyers' Association, Dhaka
5. BTRC (2018) Internet subscribers. Bangladesh telecommunication regulatory commission. http://www.btrc.gov.bd/content/internet-subscribers-bangladesh-april-2018. Accessed 12 May 2020
6. Editorial (2013) Draft ICT (Amendment) Ordinance-2013: a black law further Blackened. Daily Star. http://archive.thedailystar.net/beta2/news/draft-ict-amendment-ordinance-2013. Accessed 25 Dec 2019
7. Greenemeier L (2007) China's cyber attacks signal new battlefield is online. http://www.scientificamerican.com/article/chinas-cyber-attacks-sign. Accessed 12 Aug 2019
8. The Information & Communication Technology Act (2006) The Information & Communication Technology Act. http://www.prp.org.bd/downloads/ICTAct2006English.pdf. Accessed 11 July 2020

9. International Commission of Jurists: Briefing Paper on the Amendments to the Bangladesh Information Communication Technology Act (2013). http://icj.wpengine.netdna-cdn.com/wp-content/uploads/2013/11/ICT-Brief-Final-Draft-20-November-2013.pdf. Accessed 10 May 2020

10. Karaman S (2017) Women support each other in the face of harassment online, but policy reform is needed. The LSE women, peace and security blog. The London School of Economics and Political Science, London. http://blogs.lse.ac.uk/wps/2017/11/29/women-support-each-other-in-the-fa. Accessed 1 Mar 2020

11. Maruf AM, Islam MR, Ahamed B (2014) Emerging cyber threats in Bangladesh: in quest of effective legal remedies. North Univ J Law 1(2010):112–124. https://www.banglajol.info/index.php/NUJL/article/view/18529. Accessed 21 Aug 2020

12. Elahi SM (2014) Porn addicted teenagers of Bangladesh. Manusher Jonno Foundation, Dhaka

13. Perlroth N, Gellesaug D (2014) Russian hackers amass over a billion internet passwords. http://www.nytimes.com/2014/08/06/technology/russian-gang-said-to-amass-more-than-a-billion-stolen-internet. Accessed 28 Jan 2019

14. Singer PW, Freidman A (2014) Cyber security and cyber war: what everyone needs to know. Oxford University Press, Oxford

15. Tikk E (2011) Ten rules for cyber security-survival: global politics and strategy. Routledge, London

16. USSD (2017) Country report on human rights practices for 2016. US Department of State, Washington DC (2017). https://www.state.gov/j/drl/rls/hrrpt/2016humanrights/report/index.htm?ye. Accessed 2 Aug 2020

17. Williams B (2014) Cyberspace: what is it, where is it and who cares? http://www.armedforcesjournal.com/cyberspace-what-is-it-where-is-it-and-who-cares/. Accessed 15 July 2020

18. Zaman S, Gansheimer L, Rolim SB, Mridha T (2017) Legal action on cyber violence against women. Bangladesh Legal Aid Services Trust (BLAST), Dhaka

Automatic Detection of Cyberbullying: Racism and Sexism on Twitter

Linfeng Wang and Tasmina Islam ⓘ

Abstract With the increasing number of people more people utilising social media platforms, the production of aggressive language online such as attacks, abuse, and denigration increase. However, the constantly changing and different forms of online language provide difficulties in detecting violent language. Not only is this a difficult undertaking, but it is also an area for research and growth, considering the harm caused by cyber violence to children, women, and victims of racial prejudice, as well as the severity of cyberbullying's consequences. This paper identifies some violent terms and proposes a model for detecting racism and sexism on social media (twitter) based on TextCNN and Word2Vec sentiment analysis achieving 96.9% and 98.4% accuracy.

Keywords Cyberbullying detection · Online social media · Racism · Sexism · Convolutional neural network

1 Introduction

Social media has become an essential part of everyday life. It is estimated that 90% of UK teenagers use social networks [1], and while this has positive consequences in terms of learning and communication, it also has many negative consequences. Cyberbullying on social media platforms can be psychologically distressing for young people and defending oneself against cyberbullying is more difficult than traditional (face-to-face) bullying. Cyberbullying's psychological impact on adolescents can occur at any time of day or night in school, home, or community settings [2], and it has the potential to cause particularly powerful mental health stress because it is more difficult to cope with in the less mature adolescent psyche than in the

L. Wang · T. Islam (✉)
Department of Informatics, King's College London, London, UK
e-mail: tasmina.islam@kcl.ac.uk

L. Wang
e-mail: linfeng.wang@kcl.ac.uk

© The Author(s), under exclusive license to Springer Nature Switzerland AG 2023　　　105
H. Jahankhani (ed.), *Cybersecurity in the Age of Smart Societies*,
Advanced Sciences and Technologies for Security Applications,
https://doi.org/10.1007/978-3-031-20160-8_7

adult psyche. As shown in a study issued by UNICEF and the UN Secretary-Special General's Representative, more than 30% of teenagers in 30 countries have been victims of cyberbullying, with 20% of those surveyed claiming to have skipped school as a result of their encounters with cyber abuse [3].

The negative effects of cyberbullying do not only affect young people's education, but also have a significant impact on politics and public opinion. During the 2016 US presidential election season, approximately 65 percent of Facebook and Twitter users received regular tweets about the election or political news, even if they do not normally follow these accounts [4]. These social media sites have special powers to expose citizens to different points of view and political information, and their influence can be even greater than traditional mass media [5]. These special powers include pushing pop-ups to users as well as sending notifications and emails. In this case, social media can be used by politicians to influence their audience by blocking out information that is unfavorable to their party in order to reduce popular bias, as well as potentially attacking other parties in terms of public opinion. For example, party organisations may promote comments that support their cause or candidate and may delete comments that are unfavorable to their cause or candidate.

Discrimination against women and minorities is also prevalent on social media platforms. According to [6], women are more likely than men to be the targets of gender-based cyberbullying. Besides, young women in the UK are more likely than men of the same age to be subjected to various forms of online harassment [7]. For example, images and messages sent without permission with a harassing intent, 40 percent of UK women reported in [2] survey had received this type of harassment, including sexist abusive, offensive, and threatening messages, compared to 26 percent of men. As a result, cyberbullying pervades every aspect of today's online environment, and it is critical to recognise the dangers of cyberbullying and take concrete steps to prevent it. Cyberbullying is defined as the bully's use of electronic or online communication to threaten or intimidate the bully [8]. Exclusion, harassment, deception, denigration, incitement, and cyberstalking all seem to be common forms of cyberbullying today. Bullies frequently send emails, instant messages, or threatening comments via mobile devices such as phones and iPads, as well as social media platforms such as WhatsApp and Twitter. Numerous studies have shown that cyberbullying can be extremely harmful to the victim's psyche, causing a variety of negative effects such as stress, anxiety, fear, and, in extreme cases, suicide [9]. Victims of cyberbullying are more likely to scrutinise what they post on social media platforms than non-victims to reduce the risk of being cyberbullied again, resulting a reduction in their online engagement.

Despite the fact that social media platforms have implemented a variety of prevention and intervention strategies, the phenomenon of online violence has not been significantly reduced in the last decade [10]. This is because, as well as being distinctive, violent language on the Internet is also diverse, figurative, implicit, and colloquial, and can be classified as alphabetic, numeric, or mixed alphabetic depending on the external expression. In addition, different countries have different types of violent language [11]. Furthermore, because of the variability of online platforms, the same online language is not treated equally; for example, it is difficult to standardise the criteria for judging all social platforms, and there are no uniform standards. Most websites and online communication platforms take precautions to prevent the spread of commonly used violent words on the Internet. The detection of violent language on the Internet is therefore critical.

Similarly, the study of the topic of online violent language detection is very important from the standpoint of sentiment analysis. The process of mining a text for the author's emotional attitude (happy, angry, sad, etc.) or judgmental suggestions (for or against, like or dislike, etc.) towards an entity is known as sentiment analysis (person, event, good, service, etc.) [11]. Thus, the development of sentiment analysis contributed many theoretical foundations and related techniques to online violent language detection, while cyberbullying language detection has brought a more comprehensive development to sentiment analysis.

This paper proposes a character-level convolutional neural network model for detecting cyberbullying based on the social media platform Twitter. The primary focus is on detecting gender and racial discrimination.

The remainder of the paper is organised as follows: Sect. 2 describes literature reviews on recent development of cyberbullying detection systems. Section 3 discusses the methodology and experimental set up including data processing and model training. Section 4 analyses the results. Finally, Sect. 5 concludes the paper.

2 Literature Review

This section provides a brief review of existing literature on cyberbullying detection and different machine learning models used for detection. During the CAW 2.0 workshop, the authors [12] gathered information about cyberbullying from the prominent social networking site Myspace. A cyberbullying detection algorithm was developed after different types of cyberbullying were identified and categorised using dictionary-based rules and keywords. Social media can be a rich data source for cyber violence research when machine learning and natural language processing techniques are applied. A Twitter data set was reported in [13], that contains Twitter IDs and bullying classifications and used to identify cyberbullying participants as well as the role they played. The datasets are often manually annotated, and new data is frequently generated on social media platforms, making the process of data gathering laborious and expensive. Due to this cyberbullying detection becomes less efficient, contributing to the existing state of the art.

Text detection models can be built using one of three approaches: lexicon and rule-based approaches, machine learning-based approaches, or a combination of the two. The authors in [14] examined the content of cyberbullying texts and discovered a high number of swear and cursing words. They created a domain dictionary by incorporating common words from cyberbullying content and determined that cyberbullying had occurred if the text content contained words from the dictionary. However, the accuracy of such models is far from perfect because human language emotions are extremely complex, and judgments based solely on dictionaries are prone to errors, which is a serious problem. The advancement in machine learning research has led to some researchers developing automated systems to monitor cyberbullying and classify texts. The classifiers reported in [15] employed Latent Dirichlet Allocation (LDA) to extract semantic features, TF-IDF weighting of the features, and training SVM classifiers with features composed of second person pronouns. The model was eventually able to identify a 93 percent recall rate, signifying that the system identified a low rate of under-reporting of real cyberbullying cases, even though it was still flawed, and the system accuracy was poor, indicating that posts that were not cyberbullying were also identified as bullying. The study reported in [16], used Support Vector Machine (SVM) classifier for text feature recognition and extraction when text was used as input to the classifier with Uni-gram and Bi-gram features. The experimental data revealed that the SVM classifier could achieve a recall of 79% and an accuracy of 76%, however the system still had a significant error. When compared to traditional dictionary and rule-based methods, machine learning methods have made significant progress, with improvements in recognition speed and accuracy. However, text feature extraction remains difficult in the face of diverse text forms and ever-changing social platforms of online language, and the accuracy rate in detecting cyberbullying still needs to be improved.

Some researchers attempted to extract text features using deep learning neural networks by creating complicated neural network models, which they termed "deep learning text features". The model with a neural network has the benefit of automatically gathering text features; in addition, depending on the range of applicability, the inspector modifies and optimizes the parameters itself, self-learning in order to produce the best possible model. This model eliminates the need for complex text feature collection and has a significant accuracy advantage. The study in [17] used a CNN-RNN combined structure for sentiment analysis of short texts, using CNN as input to RNN, which sequentially selected words in sentences by RNN, learning long distance dependencies and local features of phrases and words learned by CNN for sentiment analysis. Accuracy rates for the English datasets SST1, SST2, and MR were 51.50 percent, 89.95 percent, and 82.28 percent, respectively. Another study [18] combines CNN models to propose a new CNN-CB algorithm that eliminates the need for textual feature engineering and provides more accurate predictions than traditional cyberbullying detection methods. The algorithm proposes to use the concept of word embedding, which implies that similar words will have similar degrees of embedding, and it merges semantics by using word embeddings. Experiment results show that the algorithm can achieve an accuracy of 95%. This paper proposes a Text-CNN model due to the small amount of data in the dataset and the lack of very long sentences or articles for feature collection.

3 Methodology and Experimental Set-Up

This section describes the Word2vec and CNN-based Twitter cyberbullying text sentiment analysis model developed that includes a dataset pre-processing module, a text feature word construction module, a feature word extraction module, and a comment sentiment classification module.

3.1 Dataset Processing

Two datasets have been used in this study, one with sexist tweets and the other one with racist statements form Twitter. The sentiment classification of discriminatory statements was labelled as negative, and the sentiment classification of non-discriminatory statements was labelled as positive. Figure 1 shows the number of texts in the datasets for both sentiments.

Apart from the annotations, both datasets were subdivided into two categories: discriminatory and non-discriminatory statements. In order to conduct a more in depth analysis, both datasets have been filtered using the keywords related to distribution of racism and sexism summarised in [19, 20]. The outcome of filtering shows several discriminating phrases or words in both datasets, as shown in Fig. 2. Tweets containing these words are categorized and labeled as negative sentiment.

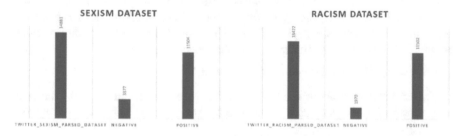

Fig. 1 Number of positive and negative texts in both datasets

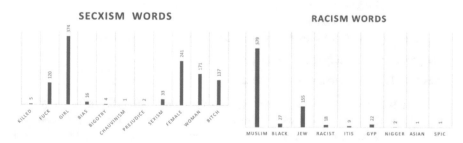

Fig. 2 Frequency of discriminating words

The next phase deletes any unneeded information from the database corpus, such as garbled symbols and excessively long meaningless adverts and other data, as well as some attribute columns that you do not like to close. Following the removal of superfluous information from the text and the corpus via the use of scripting, the text can be processed through the use of word separation. It should be noted that violent language on the Internet is frequently encountered in the form of violent words or words with syntactic structures and that many violent words on the Internet are separated from one another by punctuation or other special symbols in order to avoid processing and blocking by the website's relevant management system. This corpus is processed on a text-by-text basis, with spaces added before and after punctuation marks to differentiate them and to keep the letters and digits in the text from being lost during the processing.

The redundant information processing in both datasets is essentially the same, as both are based on the characteristics of the collected text corpus for further processing, and the following describes the redundant information processing for the online violent language detection model's training and testing corpus. The corpus gathered for the tests are all particularly replies to comments, so they may be read as subjective comments with emotional undertones, and they contain a lot of punctuation, fairly special symbols, spaces, as well as other things.

Once the redundant information has been scripted out of the text corpus, the text can be subjected to word separation operations. Even though many words of online violence are separated by sentence structure or other special symbols to avoid processing and blocking by the relevant website management systems, these words of online violence may also contain letters or numbers. As a result, this corpus will be processed text by text, adding spaces between all punctuation marks and text and removing all characters except 'A-Za-z0-9(),!', and keep the letters and numbers contained within the text. An example implementation:

> **string = re.sub(r"[^A-Za-z0-9(),!?\'\`]", " ", string)**

The next step is to obtain the labels, digitise them and declare how the data will be processed.The first step uses the lambda syntax, which is structured as follows.

> **text_field.tokenize = lambda x: clean_str(x).split()**

The lambda wordifies the English sentence x, giving the anonymous function the name text_field.tokenize. The function is no longer anonymous and is defined using the function text_field.tokenize.

Afterwards instantiate Field() and assign values to the arguments.

> **text_field = data.Field(lower = True)**

> **label_field = data.Field(sequential = False)**

In the next step, the function starts reading the dataset, classifying the text with label = 1 into the negative dataset and the text with label = 0 into the positive dataset.

Thus, a custom dataset is constructed, and the original dataset is divided. Call shuffle: random. shuffle(examples) to randomise the data and divide the training set train and the test set dev, placing the data with label = 1 in the training set.

The final step in text processing is the construction of the word list, which is the encoding of each word, that is, the numerical representation of each word, so that it can be passed into the model. The procedure is to call Torchtext to traverse the data bound to text_field in the training set, register the words to the vocabulary and automatically build the embedding matrix. At this point, it is ready to convert words to numbers, numbers to words, and words to word vectors.

3.2 TextCNN Simulation

TextCNN uses multiple convolutional kernels of variable sizes to extract keywords from text assertions, allowing it to better detect different types of local information. This convolutional neural network has the same network topology as standard picture CNN networks. However, it is much simpler to use than traditional image CNN networks. Figure 3 shows a mind map of TextCNN model. It can be seen from the Figure that the TextCNN convolutional neural network contains only one layer of convolution, one layer of max-pooling, and ultimately, the output is connected to softmax for n classification, indicating that the network is simple.

The TextCNN convolutional neural network has an embedding layer, which imports pre-trained word vectors. All words in the dataset are represented as a vector, resulting in an embedding matrix MM, where each row is a word vector. This MM can be fixed, static, or updated according to back propagation, non-static.

Due to the extreme high correlation of neighboring words in a text sentence, TextCNN only convolves in one direction (vertical) of the text sequence, which can be accomplished using one-dimensional convolution. Assuming that the word vector has dimensions and that a sentence contains only one word, the sentence can be represented as a matrix $A \in R(s \times d)$ of rows. The width of the convolution kernel is fixed to the dimension of the word vector, while the height is a programmable hyperparameter. A feature map is obtained by convolving each possible window of a sentence word: $C = [c_1, c_2, \ldots, c_{c-h+1}]$. For a convolution kernel, a total of $s - h + 1$ features can be obtained for a feature $c \in R_{s-h+1}$. TextCNN differs from general CNN in that it employs more convolutional kernels of varying heights under the premise of one-dimensional convolution, resulting in a richer representation of features. TextCNN includes a large number of convolutional kernels with varying window sizes. Convolutional kernels are typically {3,4,5}, with feature maps of 100. The different k-gradients are represented by the feature maps here. Different convolutional kernels operate in ranges of varying sizes, and when the ranges in which different convolutional kernels operate overlap, the model can learn different features, improving the final learning result.

This sexism dataset contains 14,881 records, 3,377 of which (about 22.7%) have comments with negative emotional overtones. The racial discrimination dataset contains 13,472 records, with 1,970 items (approximately 14.6 percent) having negative emotional overtones. The data marked as negative were used as the training set,

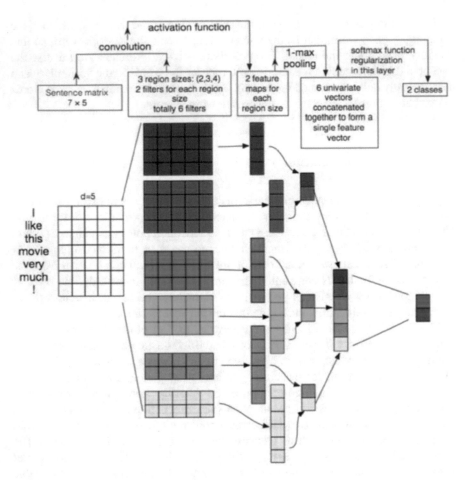

Fig. 3 TextCNN structure diagram [21]

and the remaining data were used as the test set to construct a TextCNN model to predict the data results.

3.3 Model Training and Optimisation

The first step after deciding to use the TextCNN model is to define the model's parameters. In case of binary classification problems, the model's performance is typically measured using Precision, Accuracy, Recall, and f1 values (F-score). Accuracy is typically measured in prediction results as the ratio of the number of results in which a sample is correctly classified to the total number of samples included in the category after classification. Accuracy is used to assess the model's general overall ability to identify, and its value really is the ratio of correctly classified samples to sort of the total number of samples, which generally is fairly significant. Recall definitely is used to essentially evaluate the very original sample, and its value literally is the ratio of the number of correctly classified results to the kind of a total number of samples in that category in a generally big way. Because precision and recall mostly tend to move in very opposite directions, the f1 value is introduced to for the most part assess both simultaneously in a subtle way. Based on the true category of the sample and the predicted category of the model, which can actually be divided into four categories: true cases (TP), generally false basically positive cases (FP), true for all intents and purposes negative cases (TN), and sort of false-negative cases (FN), P basically is denoted as the accuracy rate, R generally is denoted as the completeness rate, A is denoted as the accuracy rate, and F1 is denoted as the f1 value, and the calculation formula can be expressed as follows.

$$P = \frac{TP}{TP + FP}$$

$$R = \frac{TP}{TP + FN}$$

$$F = \frac{2 \times P \times R}{R + R}$$

$$A = \frac{TP + TN}{TP + TN + FP + FN}$$

In order to verify the effectiveness and correctness of the method proposed in this paper, the relevant parameters of the TextCNN were first experimentally configured. Table 1 shows the effect of different convolutional kernel widths on the results, where the number of convolutional kernels is set to 128, the ReLU function is used as the activation function, the dropout is set to 0.5, the learning rate is set to 0.001, and the maximum epoch is set to 256. Different convolutional kernel sizes can affect the effectiveness of the classification. From Table 1, the best completion rate(R) is obtained with good accuracy and F1 when the kernel size is [3, 4, 5]. As a result, the convolution kernel size used in this paper is [3, 4, 5].

Following that, the batch-size of the model was set with a fixed convolutional kernel width of [3, 4, 5], the activation function, and all other values remained the same as before. Table 2 shows that the best results are obtained when the value of batch-size is 64.

The learning rate setting has a large impact on the training time of the model. The learning rate is a setting for the magnitude of parameter updates, when the learning rate is low, the parameter changes are small, and it is easy to fall into a local optimum. If the learning rate is set too large the fit is slow and the accuracy fluctuates and can jump out of the local optimum, but the fit time is long. The effect of the learning rate is shown in Table 3 and it can be seen the best results are obtained for learning rate 0.001.

Table 1 Set convolution kernel size

Convolution kernel size	macro_P	macro_R	Acc	macro_F1
[1, 2]	89.98	93.65	97.29	91.97
[1, 2, 3]	89.37	93.28	**97.32**	91.62
[2, 3, 4]	91.07	93.32	96.89	92.02
[3, 4, 5]	91.09	**94.65**	96.23	**92.60**
[1, 2, 3, 4]	91.02	93.09	96.67	92.24
[1, 2, 3, 4, 5]	**91.42**	92.33	97.47	91.86
[5, 6, 7]	89.45	93.56	96.77	91.61

Table 2 Set batch size

batch_size	macro_P	macro_R	Acc	macro_F1
32	88.56	92.55	**97.68**	91.02
64	88.64	**95.06**	97.56	**92.43**
128	88.98	94.45	96.87	90.95
256	**90.02**	93.22	97.22	91.65
512	89.93	92.18	96.04	90.54

Table 3 Set learning rate

Learning Rate	macro_P	macro_R	Acc	macro_F1
0.1	88.82	91.76	**97.98**	90.53
0.01	89.31	92.78	96.84	90.21
0.001	89.98	**95.63**	96.97	**92.45**
0.0001	**90.65**	93.82	96.24	92.60

Finally, summarising the best configuration parameters of TextCNN set in this experiment are:

Convolution kernel size: [3, 4, 5]
Learning rate: 0.001
epochs: 256
batch-size: 64
dropout: 0.5

Table 4 shows the accuracy of the training results with the above configuration of parameters for the TextCNN model. It can be observed that the model achieves an accuracy rate of more than 96% for both training datasets, allowing classification and annotation of text sentiment.

In conducting the sentiment analysis module, this paper focuses on using word2vec for sentiment analysis. In the module for classifying text the NLTK library

Table 4 Model training accuracy

	Gender discrimination	Racial discrimination
Batch loss	0.031	0.046
Evaluation loss	0.014	0.004
Accuracy	96.9%	98.4%

is called, in which the Naive Bayes classification algorithm is implemented, and the Bayesian model is trained using Word2vec. The Naive Bayes algorithm is a classification method based on Bayes' theorem and the assumption of conditional independence of features, and its application areas are relatively wide. Bayes classifiers require few parameters to be estimated, are less sensitive to missing data, and the algorithm is relatively simple and interpretable. Theoretically, it has the smallest error rate compared to other classification methods [22]. Typically, texts could be classified in order to create a model of the attributes or, for attributes that are not independent of each other, they can be treated separately. The formula is as follows:

$$P(A|B) = \frac{P(B|A) \times P(A)}{P(B)}$$

The specific steps are as follows (as shown in Fig. 4)

- Use word2vec for data pre-processing, feature extraction, vectorisation of text words, and use word frequency statistics to extract keywords that occur at high frequencies in the text to construct a keyword dictionary.
- The keywords in the dictionary are represented by word vectors for hierarchical clustering to achieve semantic merging of keyword dictionaries from the perspective of contextual semantics. The keywords with greater similarity are grouped into a cluster to represent similar word clusters and build a word cluster dictionary.
- Traversing the centre of each cluster in a word cluster to obtain similar words with a given similarity threshold to expand the word cluster lexicon for the purpose of dynamically expanding the word cluster lexicon according to the context.
- Combining the word frequencies of similar words in a word cluster dictionary to train a Naive Bayesian classification model.

Figure 5 shows the flowchart of the word2vec based Naive Bayesian text classification system. The trained model is then used to predict the random input text, determine whether it contains discriminatory overtones and label it.

3.4 Model Prediction

In this section, random sentences are fed into the model, and the model determines if the output labels are negative or positive, indicating whether the statements pertain to cyberbullying (shown in Table 5).

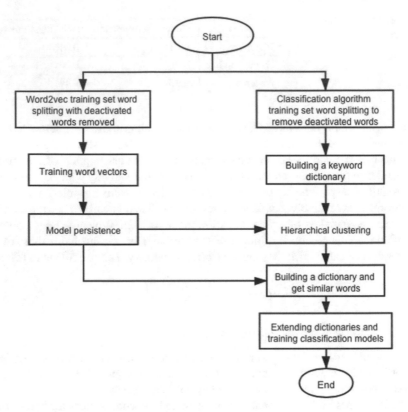

Fig. 4 Word2vec algorithm structure diagram

Sexism dataset negative word cloud   **Sexism dataset positive word cloud**

Fig. 5 Sexism dataset word cloud

Table 5 Model prediction results

Dataset	Input characters	Tag
Sexism	RT @TheFanVent: I'm not sexist but females hav	negative
Sexism	I have eaten kfc and now	positive
Sexism	"You f*** your dad"	negative
Sexism	The punishment for apostacy in #Islam is death	positive
Sexism	oh my f***ing god	positive
Sexism	I'm not sexist, but for some reason, every time a female commentator comes on stage to give her two cents, I say, "I don't know	negative
Racism	ISIS makes sure that they do	negative
Racism	Read about the Wahabbi attack on Karbalah	positive
Racism	I am a raaaaacism	negative
Racism	They are not Muslim enough	positive
Racism	Religion that teaches hate and murder has nothing to do with hate and murder	negative

3.5 Presentation and Analysis of Results

In this section, word cloud diagrams are created using the classified document dataset to show the important words in the documents with different sentiment polarities in order to quickly understand the information in the documents. The word cloud is based on the word frequency of the words in the corpus, and the higher the word frequency, the larger the font size of the word in the graph, and the easier it is for the user to recognise it. As a result, the Python word cloud module is implemented to provide a quick overview of the sexist and racist dataset.

The main high frequency words in the negativity data set are girl, woman, sexist, men, bitch, dumb, female, feminist, f***, stupid and other words that are clearly gender-specific and insulting. A small number of words such as drive, cook, work, job, sport, football, etc. are also used in relation to specific situations in life and work. The presence of these scenario words suggests that people may have been subjected to online violence in these areas. For example, there is sexism in the workplace between men and women and having internet users comment on the difference between male and female drivers in a driving situation is also classified as sexist. Discriminatory discourse is also present in sports scenes due to the different biological differences between men and women. The presence of sensitive race-related terms such as Nikki, Muslim, and so on in the sexist dataset is notable, but because this is a dataset that distinguishes between sexist sentiments, sensitive terms such as Muslim, Islam, and so on are not marked out, despite the fact that these statements could potentially be racially or otherwise discriminatory.

The positive word cloud contains a lot of non-sensitive words like kat, time, people, will, one, now, and so on. However, there is a mix of words in the negative word cloud, such as women, bad, as well as sexist, which are also found in the positive word cloud, but much less frequently. It is important to note that this project does not rely solely on keywords to distinguish between emotive colors, so the presence of a specific word in a sentence does not immediately label the entire sentence as cyberbullying. The distinction between positive and negative word clouds can still be seen in the overall picture.

The two word clouds for racial discrimination (shown in Fig. 6) are classified as positive and negative. According to the graph, the most frequently used negative words are Muslim, Islam, religion, ISIS, Mohammed, Jew, and other sensitive words with racial and religious connotations. The most commonly used words in the word cloud are discrimination against Islam and Muslims, with a few Jewish and Catholic words also appearing. Words such as ISIS, Quran, killing, murdering, slave, and terrorist appear as well, and it is assumed that the original post crawled in this dataset would be about religious terrorism, which triggered cyberbullying in the comments section. It is worth noting that the words girl, men and women, which also appear in the negative word cloud of the sexist dataset, still appear in the racial discrimination dataset. It is not difficult to see how these two types of discrimination always intersect, with discrimination against women, possibly Asian, black, or Muslim, occurring in online contexts where racial discrimination is present. And in the case of gender discrimination, there are also racist remarks about the bully.

In the positive word cloud, the high frequency words that appear are much more moderate, mostly words like people, will, think, one, know, etc. that are not discriminatory. There are also some derogatory words such as f***, kill, etc., but they are not labelled as racist, because they appear in statements that are not racist. Similarly, there are low occurrences of racially charged words such as Islam, ISI, etc. This may be because the phrase itself is not meant to be racially charged, but it does contain such sensitive words. This is also often the case in social networks, where the algorithm cannot determine the entire sentence as discriminatory just because one keyword appears, which would make the model's false positive rate increase, even though the accuracy rate would improve.

Racism dataset negative word cloud   **Racism dataset positive word cloud**

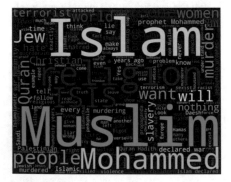 

Fig. 6 Racism dataset word cloud

Plotting word clouds allows for a visual representation of both positive and negative sentiments in the dataset, which can make the classification and detection of the dataset more intuitive. The great difference in word clouds shows that the experiment has achieved very good results in training models to recognise and detect discriminatory language on the social media.

4 Analysis and Evaluation

This section presents a thematic deep analysis and evaluation of the model proposed in this paper based on the TextCNN model building.

4.1 Emotional Disposition Analysis

The sentiment propensity analysis of the TextCNN model leads to the following conclusions.

Firstly, the percentage of comments that ended up being labelled as negative in both datasets was approximately 22.6% (sexism) and 14.6% (racism). This means that approximately one in every four to six comments on the Twitter social media platform is a negative comment, which may be racially or sexist in nature. Only two datasets have been analysed, so there may be other aspects of discrimination that were not flagged by the dataset, and the probability of these negative words and statements appearing in the actual Twitter network environment may be even higher.

Secondly, it can be observed from the high frequency words that Islam appears 725 times and Muslim appears 569 times in the racial discrimination dataset. These frequently occurring words are frequently found in specific topics such as religion, terrorism, war, and so on. Online violence on such topics could be detected more stringently by social media platforms.

In contrast, the high frequency of negative words in this study's sexism dataset indicates that women are subjected to more sexism than men. Women's words, such as woman as well as girl, appear far more frequently than men's words, including such man and male. Feminism is also a high frequency word, indicating that there is a lot of cyber violence against feminism on the internet right now. Feminism began as an advocate for gender equality, and the dataset divides online violence against feminism into two major categories: opponents of feminism who then attack stigmatized feminism, which including "Becoming fed up with feminism finally motivated us to organize and speak up," and those who radically promote feminists. However, excessively aggressive language can have a negative impact on the online environment, and excessively bad wording can lead to more online violence on the topic.

4.2   Commentary on Emotional Mining

Through negative comment and positive comment text topic mining, using keywords to estimate topic meanings, combined with specific comment information, the final results for both datasets can be summarised in Fig. 7.

The wording of these unfavorable remarks makes it very evident that cyber violence is widespread on social media platforms of all kinds. As a result, there arc several violent and highly offensive remarks that can cause major damage to the online environment, in addition to having a big visual impact and exerting psychological strain on those who are subjected to cyber violence. It is the primary purpose of this initiative to identify instances of online violence, yet simply identifying them is not enough. This paper makes the following recommendations for social media sites in order to avoid online violence.

The first step is to refine the detection algorithm, which will allow for the creation of specific rights for users. These permissions might include keyword blocking, reporting of vulgarity, and banning of topics that do not interest the user, among other things. Users will be able to express themselves freely as well as prevent access to content that they do not like to see as a result of these changes.

Next in important, it is critical that a new user registration policy be implemented in the network environment, which specifies which sensitive terms are suspected of contributing to online violence and what sanctions will be applied if comments containing online violence are discovered from time to time. For example, blocking comments or even canceling accounts are both options available. With this in mind, it is hoped that young people would not be encouraged to make aggressive comments when they do not have a proper sense of right and wrong. They may be unaware that their actions can result in online aggression, and adequate education and supervision will encourage them to learn how to keep the online environment safe and secure.

The final point to mention is that social networking platforms can incorporate privacy settings into the way accounts are configured to receive information, such as only receiving information from people who follow and follow back the users, making personal information on accounts only visible to specific individuals, and so on. These safeguards will protect users from getting online violence from others, as well as from the disclosure of personal information to unauthorised parties.

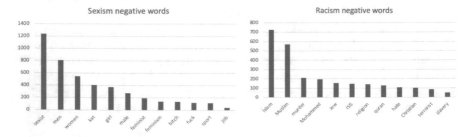

Fig. 7 Negative words from sexism and racism dataset

Throughout the Internet environment, the most fundamental measure to prevent the emergence of violent language is still to strengthen moral education on the Internet, to provide positive guidance to Internet users and to continuously improve their quality. Students can be educated on Internet ethics at the teenage stage and how to use the Internet in a reasonable manner. For adult Internet users, media reports should use positive, objective, and positive reports to guide the public opinion environment. A positive and healthy online environment may prevent further problems of online violence. When participants in cyberbullying can look at the problem with a rational mind and make objective and rational comments, they will influence other internet users who watch the comments, forming a virtuous circle.

5 Conclusions

This paper poses the research question of how to detect such online violent language texts in an effective way and by means that are important for the subsequent analysis of the characteristics of online violent language at a time when this phenomenon is becoming more and more common. How to construct classifiers in the dataset to classify violent language texts and non-violent language text data for sentiment through automated techniques has also become an important need.

This paper investigated the identification of violent language and sentiment analysis in Twitter comments, highlighted the benefits of using TextCNN and word2vec to generate sentiment dictionaries. A rule for filtering has been developed initially, and then this rule has been used for filtering, with the goal of combining this rule with a Naive Bayesian approach to classify sentiment in order to improve accuracy and efficiency. Although the experiments utilised only two datasets, based on the results of the experiments, there is a positive indication that classifying and analysing sentiment words can assist in better understanding the high frequency words associated with cyber violence, and that studying the frequency of these words can assist readers in better understanding which topics are more likely to be the subject of cyberbullying.

References

1. Mateu A, Pascual-Sánchez A, Martinez-Herves M, Hickey N, Nicholls D, & Kramer T (2020) Cyberbullying and post-traumatic stress symptoms in UK adolescents. Arch Dis Child 105(10):951–956. https://doi.org/10.1136/archdischild-2019-318716
2. Slonje R, Smith P (2008) Cyberbullying: another main type of bullying? Scand J Psychol 49(2):147–154. https://doi.org/10.1111/j.1467-9450.2007.00611.x
3. Unicef.org. (2021) Cyberbullying: What is it and how to stop it. https://www.unicef.org/end-violence/how-to-stop-cyberbullying. Accessed 29 July 2021
4. Vendemia M, Bond R, DeAndrea D (2019) The strategic presentation of user comments affects how political messages are evaluated on social media sites: evidence for robust effects across party lines. Comput Hum Behav 91:279–289. https://doi.org/10.1016/j.chb.2018.10.007
5. Anspach N (2017) The new personal influence: how our facebook friends influence the news we read. Polit Commun 34(4):590–606. https://doi.org/10.1080/10584609.2017.1316329
6. Haslop C, O'Rourke F, Southern R (2021) #NoSnowflakes: the toleration of harassment and an emergent gender-related digital divide, in a UK student online culture. Converg: Int J Res New Media Technol 135485652198927. https://doi.org/10.1177/1354856521989270
7. Her Majesty's Government (HMG) (2018) Government response to the internet safety strategy green paper. HM Government, London
8. Mikhnovets A (2021) Cyberbullying as a new form of threat on the internet
9. The Annual Bullying Survey 2017 | Ditch the Label (2017). https://www.ditchthelabel.org/research-papers/the-annual-bullying-survey-2017/. Accessed 29 July 2021
10. Balakrishnan V, Khan S, Arabnia H (2020) Improving cyberbullying detection using Twitter users' psychological features and machine learning. Comput Secur 90:101710. https://doi.org/10.1016/j.cose.2019.101710
11. Mladenović M, Ošmjanski V, Stanković S (2021) Cyber-aggression, cyberbullying, and cyber-grooming. ACM Comput Surv 54(1):1–42. https://doi.org/10.1145/3424246
12. Bayzick J, Kontostathis A, Edwards L (2011) Detecting the presence of cyberbullying using computer software
13. Waseem Z, Hovy D (2016) Hateful symbols or hateful people? Predictive features for hate speech detection on twitter. In: Proceedings of the NAACL student research workshop, pp 88–93
14. Dadvar M, De Jong F (2012) Cyberbullying detection: a step toward a safer internet yard. In: Proceedings of the 21st international conference on World Wide Web, pp 121–126
15. Nahar V, Al-Maskari S, Li X, Pang C (2014) Semi-supervised learning for cyberbullying detection in social networks. In: Australasian database conference. Springer, Cham, , pp 160–171
16. Xu JM, Jun KS, Zhu X, Bellmore A (2012) Learning from bullying traces in social media. In: Proceedings of the 2012 conference of the North American chapter of the association for computational linguistics: human language technologies, pp 656–666
17. Wang X, Jiang W, Luo Z (2016) Combination of convolutional and recurrent neural network for sentiment analysis of short texts. In: Proceedings of COLING 2016, the 26th international conference on computational linguistics: technical papers, pp 2428–2437
18. Al-Ajlan MA, Ykhlef M (2018) Deep learning algorithm for cyberbullying detection. Int J Adv Comput Sci Appl 9(9):199–205
19. Greevy E, Smeaton AF (2004) Classifying racist texts using a support vector machine. In: Proceedings of the 27th annual international ACM SIGIR conference on research and development in information retrieval, pp 468–469
20. Swim JK, Mallett R, Stangor C (2004) Understanding subtle sexism: detection and use of sexist language. Sex Roles 51(3):117–128
21. Zhang T (2019) Applications of common neural network models in the field of natural language processing. https://zhuanlan.zhihu.com/p/60976912. Accessed 8 Aug 2021
22. Acosta J, Lamaute N, Luo M, Finkelstein E, Andreea C (2017) Sentiment analysis of twitter messages using word2vec. Proc Stud-Fac Res Day CSIS Pace Univ 7:1–7

Spying on Kids' Smart Devices: Beware of Security Vulnerabilities!

M. A. Hannan Bin Azhar, Danny Smith, and Aimee Cain

Abstract The emergence of the Internet of Things devices in everyday life has increased its sales dramatically over recent years, specifically of smart devices, such as smartwatches, fitness trackers and smart phones. The number of vulnerabilities exploited has also risen in tandem with the increased sales. The attack vectors have greatly increased due to the connectivity and mass functionality of these devices. The lack of security in smartwatches, marketed towards children, poses a prominent threat for their safety and security. Results reported in this paper revealed significant security vulnerabilities in several popular kids' smartwatches when exploited by SMS command injection, Bluetooth tracking and Wi-Fi man in the middle attack. The devices investigated were kids' smartwatches, a fitness tracker and a smart phone with varying functions, connections and security features. Findings of the paper raise concerns as vulnerabilities of kids' IoT smart devices can lead to criminal cases, such as child grooming and child abduction.

Keywords IoT · Smartwatches · Fitbit · Pentesting · Cybercrime · Child grooming · Child abduction

1 Introduction

Internet of Things (IoT) is still a young technology, specifically smartwatches. Mainstream smartwatches became popular only eight years ago with the introduction of the 'pebble' in 2013, later followed by the introduction of the 'Apple watch' in 2015 [1].

M. A. Hannan Bin Azhar (✉) · D. Smith · A. Cain
School of Engineering, Technology and Design, Canterbury Christ Church University, Canterbury, United Kingdom
e-mail: hannan.azhar@canterbury.ac.uk

D. Smith
e-mail: d.smith574@canterbury.ac.uk

A. Cain
e-mail: aimeecainis@gmail.com

Advanced Sciences and Technologies for Security Applications,
https://doi.org/10.1007/978-3-031-20160-8_8

IoT smart devices, such as smartwatches, fitness trackers or smart phones lack cyber security enforcement and exploitation prevention due to the technology's infancy. It has been reported in [2] that there is a lack of encryption and many watches store text messages, emails, payment cards and other personal information, making them not only an easy target but a 'honey hole' for malicious hackers. One of the major issues with the popularity of children's smartwatches is that they can lull the parents into a false sense of security. Due to increasing competition of manufactures in the market, smartwatches and fitness trackers attempt to incorporate greater functionality and connectivity than its predecessor and this creates a larger attack vector [3]. In order to operate, smart devices require a plethora of sensors and connectivity to both record the activity and transmit to a mobile phone or remote server [4]. At a minimum, most devices incorporate Bluetooth, but many devices also include GNSS, GSM and Wi-Fi [5, 6]. Devices, such as the Garmin Vivofit Jr are often seen on the wrist of a child due to its superhero design and accompanying mobile app. Garmin reported a security breach they fell victim to concerning user data stored in the backend, potentially leaking location data and payment data of its users, inferring that the device itself is not the only exploitable unit but also the backbone of the system [7]. In another example, the backend cloud system was exploited to take control of medical smartwatches used by home users [8]. Report shows that the "take pills" reminder could be prompted as often as wanted, showing potential issues a 'hacker' can create to vulnerable smartwatch users, specially to the elderly and children [8]. Due to high popularity of children's smart devices, the need for security investigation has become increasingly important [9]. This paper aims to highlight vulnerabilities in a range of children's IoT smart devices by giving examples of exploitations in the context of simulated crime scenarios, such as child grooming and child abduction. The attacks described in this paper are being carried out in a new context and on new targets.

The remainder of the paper is organised as follows: Sect. 2 describes literature reviews on usage of IoT smart devices within the vulnerable groups and various attacks and exploitations reported on the devices. Section 3 discusses the methodology used to investigate attacks in devices on simulated crime scenarios. Both hardware and software tools and IoT smart devices used for experiments are detailed in this section. Section 4 reports the findings of exploitations and vulnerabilities from the security attacks. Finally, Sect. 5 concludes the paper and give directions to future works.

2   Literature Review

With a large percentage of users of smartwatches and fitness trackers being of vulnerable age groups (elderly and children), brings added risks to the outcome of exploitations of these devices for malicious purposes. In 2019, 32.7% of generation Z group and 22.2% of 57 to 75 year old said that they owned smartwatches [10]. Generation Z covers a vast majority of smartwatch users under 18 and with the additional users of over 57 s brings further risks to the security and privacy concerns, as both the age groups fall within the vulnerable category. Due to the intended usage of the device, they store valuable personal user information, such as GPS locations, usernames and passwords [11]. Often information stored on these devices is not encrypted, thus providing no further barrier of protection for user data once the devices have been maliciously exploited [12]. This information in tandem with the study performed by Fortify, HP's testing group, brings light to how easy smartwatches can be exploited as they suffer from significant vulnerabilities [13]. Due to this, concerns have risen for both the safety of the data on the kids' IoT smart devices and the user themselves. Child-tracking smartwatches have been reported to be easy to hack and it is possible to track their location and retrieve personal information and establish two-way communication [14]. Studies also show that spoofing the phone numbers of parents or trustees of the user is possible by initiating two-way communications. This shows the potential manipulation and trickery that can be performed on the vulnerable child users [14]. The confidential information then can be used by criminals to learn more about their targets and impersonate their loved ones, which can lead to criminal cases, such as child abduction or child grooming.

Due to most IoT devices like smartwatches being configured with Linux operating systems, it leaves the device open to not only bespoke attacks but also for generic Linux exploitations [12]. Armis Labs have shown the ability to gain full root control of many IoT devices, including smartwatches [15]. By exploiting the Bluetooth Low Energy (BLE) component and utilising a Bluetooth exploit known as 'BlueBorne', the hacker could inject raw L2CAP packets to perform a buffer overflow attack. Its seriousness is further enhanced by the fact that the victim is not aware of the attack and does not have to accept a Bluetooth connection [15].

Studies of IoT baby monitor hacking have shown that often backdoor accounts are accessible and almost impossible to disable [12]. Usually these devices are equipped with guessable default passwords, often providing access to local root privileges or at minimum user level, which leaves open the possibility for injection of an attack in order to perform privilege escalation [16]. The majority of IoT fitness watches have limited resources with regards to processing power and battery life [17]. Often security measures require a large pool of resources to operate effectively. Minimal security features in these devices can give longer battery life and keep the device lightweight [18]. Also, lack of standardization [19] and widespread patching leaves devices open to known exploitation.

Man In The Middle (MITM) attacks are common as most IoT smart devices use one or more of the followings: Bluetooth, Wi-Fi or GPS, 2G and 3G. These attacks can be performed to capture the victim's data during transmission between either the watch and the paired phone or the watch and a satellite [5]. The possibility of utilising the GSM functionality within watches has also been explored with regards to gaining control of a device by sending an injection script to gain shell access via SMS directly to the device [5]. Reference [20] noted that smartwatches are incredibly prone to eavesdropping, as well as MITM attacks utilising open source tools. Another research [21] demonstrated how MITM attacks were used to intercept traffic on IoT smart toys. It was reported [21] that companion applications on smartphones were also weak to attacks and could be compromised.

A flaw in the API was reported in [3] to modify the identification number that the attacker wanted to connect to, and once authenticated, there were no further checks to ensure the access was authorised. This vulnerability allowed attackers to change the GPS data remotely without the child or parent's knowledge, as well as to communicate with the child. A similar vulnerability was found in [22], where an authentication token generated by the watch was communicated with the web API, however it was not checked by the API, allowing attackers to use another accounts ID to log in. This is disastrous as it leaves the entire database of accounts open to attackers to access.

Taking advantage of open hardware ports, such as UART, often used for debugging during manufacturing, opens another avenue for potential exploits to gain access to the devices [2]. This includes timing attacks where the hacker can alter the boot sequence of the device or re-flash the memory of the device in an attempt to downgrade firmware to a known vulnerable version for further exploitations, such as buffer overflow attacks [23].

Reference [24] sums up the landscape of children's smartwatches quite succinctly. Whilst their research comments on just one device, the Misafe's Kids Watcher, they note that 53 other watches had similar or identical vulnerabilities. They state that these children's smartwatches decrease the security and safety of the child, whilst being advertised as increasing it. Due to the weaknesses of the watches, attackers can locate the child, call or text them, listen in to the child and get a large amount of information about the child and parent. Worryingly, they state that the vulnerabilities could be exploited by an attacker with basic coding knowledge and open source tools.

3   Methodology

The kids' smart devices listed in Table 1 have been chosen for this study due to their dominance in the market and they offer a range of connectivity to experiment with different crime scenarios. Bluetooth connectivity is common to all the devices, but Samsung Gear S3, Yongkaida's Y22 and Motorola smart phone also have GPS features. Both Samsung and Motorola devices are Wi-Fi enabled. Cellular connectivity feature is available only in Y22 smartwatch. Some of the other features in Y22 watch includes are use of a pedometer, facilities of texting, GPS tracking and remote monitoring. The Fitbit Ace2 supports steps and sleep tracking and can also be used as a watch.

The environments for the experiments were setup to imitate various potential crime scenarios for three types of attacks: the Man In The Middle attack, Bluetooth tracking and SMS hacking. The use of an external network card, Alfa Atheros, was required for the MITM scenario. This specific network card was chosen due to its proficiency in MITM attacks [25]. For Bluetooth tracking, an external Bluetooth adapter was used. Due to the Ubertooth One's reputation amongst the security community, it was chosen over its competitors [26]. Experient for SMS hacking was

Table 1 Devices used in the experiments

Brand	Model	Connectivity
Samsung	Gear S3 Frontier	Wi-Fi 802.11 b/g/n, Bluetooth (LE, 4.2, A2DP), GPS (GLONASS), NFC
Yongkaida	Y22	GPS, Bluetooth, SIM card
Garmin	Vivofit Jr 2	Bluetooth
Fitbit	Ace 2	Bluetooth
Motorola	Moto G	Wi-Fi 802.11 b/g/n, Bluetooth 4.0, GNSS (A-GPS, GLONASS), GSM

performed in Y22 smartwatch. The scenarios in this paper require a flexible operating system with a security focus, thus prompting the choice of Kali Linux to be used as the primary operating system [27]. The various software used throughout this investigation were chosen due to their specific proficiency meeting the needs for success in the experimental scenarios [28].

3.1 MITM Attack—Rogue AP

The MITM attack performed in this paper was a rogue AP (access point) based [29]. This specific rogue AP was set up to spoof a wireless access point the victim in the experiment scenario had previously joined. A crime scenario was produced using devices to simulate network traffic for later analysis. The aim of this scenario was to highlight through simulation, how cost-effective tools can be used to wirelessly gain greater knowledge of a targeted child (victim) through the capture of data packets from their devices to later be utilised for malicious use, such as child grooming.

The number of public free Wi-Fi hotspots is continually growing. Due to ease of access, most people will use free Wi-Fi when available. Free Wi-Fi is often provided by fast food restaurants. This information in tandem with fast-food establishments marketing towards children provides the attackers with a range of wireless access points to replicate for a rogue AP attack. Many devices automatically connect to wireless access points they have previously used due to the "Ask to Join Networks" option automatically being set to "Notify", with many devices not notifying at all; thus heightening the chances of a victim device connecting to the rogue AP without the user knowing.

For the attack to be effective, the rogue AP must have adequate access to the internet to provide victim devices internet access and a seamless experience. To do so, a wired ethernet connection between the host PC and the home network router was used. Alternatively the attackers in this scenario could utilise a mobile hotspot to provide the rogue AP internet access. A virtual ethernet was used between the VM and the host to allow the Kali Linux VM internet access, as shown in Fig. 1.

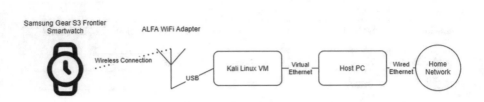

Fig. 1 Rogue AP diagram

All devices with Wi-Fi functionality listed in Table 1 were used as target victim devices. The setup of this environment involves two stages. The first is enumerating a popular open network often utilised by the target market (children). Using 'airodump-ng', the attacker can gain knowledge of the access point, this includes 'bssid', 'ssid', broadcasting channel and the encryption. For the purposes of this paper, a generic "FASTFOODWIFI" was used to imitate a popular Wi-Fi access point.

Once the access point has been enumerated, using the program "WiFiPumpkin3" the attacker could create the rogue AP and broadcast it using the ALFA wireless adapter. Once a device has joined the rogue AP, the attacker is able to capture all packets flowing through the rogue AP. This allows the attacker to perform analysis during or after the attack. The analysis of these packets has the ability to reveal information the victim devices are accessing. The level of information disclosed in the exchanged packets is dependent on the security of the device and websites. The information gained from intercepting and capturing these packets can potentially inform the attacker a basic knowledge of the child's interests, name, age and various other sensitive data, which later could be exploited by an attacker to groom a child.

3.2 Bluetooth Tracking

This scenario involves tracking a child's Bluetooth enabled device. All devices within the resource pool with Bluetooth functionality will be used (Fig. 2). The primary technology this test focuses on is Bluetooth Low Energy, but the same methodology can perform similarly with other Bluetooth variants. This scenario utilises live signal strength readings between a Bluetooth enabled device and the attacker's device to gauge distance between the attacker and the child. This scenario uses this information to perform the act of stalking a child. Hardware and software used for this experiment is listed in Table 2.

To identify the device of the child, first the Blue Hydra is run. Once started, all Bluetooth devices within range will show in the terminal. This test will be performed with the Ubertooth One and a TP-Link Bluetooth 4.0 dongle. Using both dongles

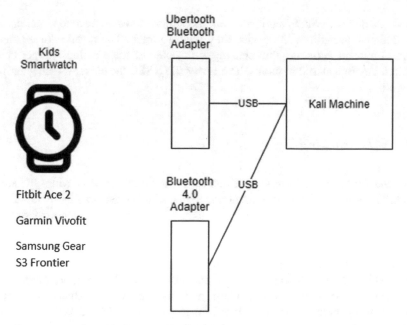

Fig. 2 Bluetooth tracking diagram

Table 2 Hardware and software for Bluetooth tracking

Hardware	Descriptions
Ubertooth One	Bluetooth Low Energy (BLE) development platform used for BLE sniffing and monitoring
TP-Link Bluetooth 4.0 adapter	Bluetooth adapter used for detecting Bluetooth devices 4.0 and below
Software	*Descriptions*
Kali Linux	Linux Debian operating system
Wireshark	Packet analysis tool
Blue Hydra	Bluetooth device discovery tool

simultaneously allows the attacker to discover devices using varying Bluetooth architectures. Often devices with Bluetooth set to "undiscoverable" are still discoverable using this tool. At minimum, the devices in range will be able to display the version of Bluetooth they are using, the Bluetooth address and the strength between the attacker's device and the victim device. Getting close to the child the attacker wishes to pursue will result in low RSSI, thus allowing the attacker to highlight which device to track. The Blue Hydra software logs all Bluetooth addresses and their corresponding RSSI.

For the investigation, a Python script was created (will be reported in the results section) to read the logs from Blue Hydra live and narrow the Bluetooth addresses to a single device. With this information, the attackers would be able to get updates of the signal strength of the device they wish to track. The noted address can be inputted into the program, thus making it possible to track a single target device within a theoretical 200 m radius. The lower the RSSI, the closer the attacker is to the victim.

3.3 SMS Hacking

The watch being tested for this experiment was the Yongkaida Y22 Kids GPS smartwatch. The purpose of this experiment was to investigate if this watch can be hackable simply by sending specifically written text messages. The scenario was to take control of the watch to benefit hackers and also to see if they can track the watch through GPS or if they can even call the user, in this case the vulnerable child.

In order to remotely compromise the watch, two mobile phones were used. The phone which the watch was paired to was labelled as the 'genuine phone' used by parents or carers and the other phone which attempted to connect to either the watch or the genuine phone to gather information was labelled as the 'malicious phone' used by attackers. The genuine phone was a Huawei P20 model EML-L09 running Android version 9.1.0.372. The malicious phone was a OnePlus 5 T model A5010, running OxygenOS version 9.0.11. The genuine phone had the SeTracker application installed, which allows parents or caregivers to track the current location of the watch, as well as to monitor and communicate with it. To chat to the watch, the user or child needs to use SeTracker's inbuilt 'Chat' feature to send texts. So, any SMS messages sent directly to the watch are considered to be 'sms commands'. Most of the information which was gathered from the watch by the malicious phone was done from sending remote commands directly to the SIM of the watch via SMS. The following commands were sent with the prefix 'pw, password' to the watch:

- ts#
- phb, telephone number#
- phb#
- call, telephone number#
- sos1, telephone number#
- monitor, telephone number#
- center, telephone number#
- dw#

All the commands above begin with 'pw,password', end with the pound sign '#'. The 'pw' indicates that the next step will be the watch password, following the comma. All SMS command exclude spaces. In the simulated crime scenario, above commands were sent by the hacker with the aim to gain control of the watch and its usage, e.g., with the prefix added 'ts#' command to gain watch's setting information, the 'phb, telephone number#' to update the phone-book with the specified number, the 'phb#' command to delete the entire contents of the phone-book, the 'call, telephone number#' to force a call from the watch, the 'sos1, telephone number#' to update the SOS number, the 'monitor, telephone number#' to enable monitoring on the watch, the 'center, telephone number#' to change the administrator's telephone number on the and the 'dw#' command to gather the GPS location of the watch. The purpose of the experiment was to see if the watch could easily be communicated with an attacker, if the attacker could make a call to the watch or force the watch to call a specified number, if the attacker can gain control of the watch to listen to the surroundings etc.

4   Experimental Findings

This section reports the results of experimental findings for three types of attacks: the MITM attack, Bluetooth tracking and SMS hacking. Successful tracking and manipulation of kids' IoT smartwatches yielded significant security concerns. Devices used for the experiments were selected based on their functionality and connectivity features (see Table 1). The test for Bluetooth tracking was able to cater to more devices due to the offering of similar connectivity, but the SMS hacking was only possible to the Y22 smartwatch and the MITM attack was done in Samsung watch due its Wi-Fi connectivity feature.

4.1   Results of MITM Attack—Rogue AP

A mock router was set up to imitate a fast-food chain access point, the 'ssid' of this access point was 'FASTFOODWIFI'. Applicable devices were joined to the original 'FASTFOODWIFI' access point to replicate a victim device being joined to the fast-food access point prior to the rogue AP attack. As seen in Fig. 3, using 'airodump-ng' alongside the ALFA Networks Wi-Fi card, the attacker was able to scan the area for publicly available access points.

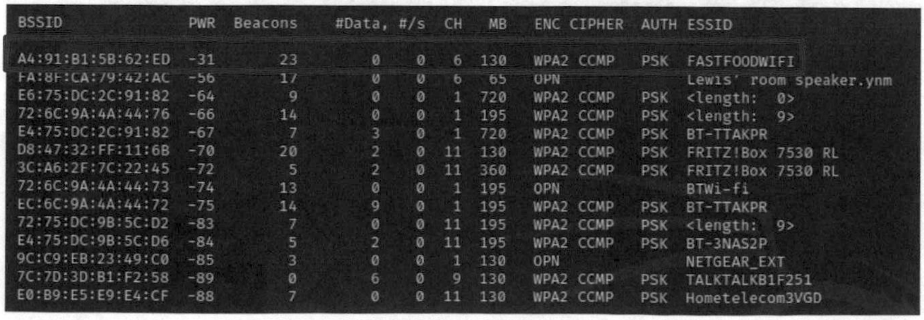

BSSID	PWR	Beacons	#Data,	#/s	CH	MB	ENC	CIPHER	AUTH	ESSID
A4:91:B1:5B:62:ED	-31	23	0	0	6	130	WPA2	CCMP	PSK	FASTFOODWIFI
FA:8F:CA:79:42:AC	-56	17	0	0	6	65	OPN			Lew1s' room speaker.ynm
E6:75:DC:2C:91:82	-64	9	0	0	1	720	WPA2	CCMP	PSK	<length: 0>
72:6C:9A:4A:44:76	-66	14	0	0	1	195	WPA2	CCMP	PSK	<length: 9>
E4:75:DC:2C:91:82	-67	7	3	0	1	720	WPA2	CCMP	PSK	BT-TTAKPR
D8:47:32:FF:11:6B	-70	20	2	0	11	130	WPA2	CCMP	PSK	FRITZ!Box 7530 RL
3C:A6:2F:7C:22:45	-72	5	2	0	11	360	WPA2	CCMP	PSK	FRITZ!Box 7530 RL
72:6C:9A:4A:44:73	-74	13	0	0	1	195	OPN			BTWi-fi
EC:6C:9A:4A:44:72	-75	14	9	0	1	195	WPA2	CCMP	PSK	BT-TTAKPR
72:75:DC:9B:5C:D2	-83	7	0	0	11	195	WPA2	CCMP	PSK	<length: 9>
E4:75:DC:9B:5C:D6	-84	5	2	0	11	195	WPA2	CCMP	PSK	BT-3NAS2P
9C:C9:EB:23:49:C0	-85	3	0	0	1	130	OPN			NETGEAR_EXT
7C:7D:3D:B1:F2:58	-89	0	6	0	9	130	WPA2	CCMP	PSK	TALKTALKB1F251
E0:B9:E5:E9:E4:CF	-88	7	0	0	11	130	WPA2	CCMP	PSK	Hometelecom3VGD

Fig. 3 Enumerated open Wi-Fi screenshot using 'aircrack.ng'

As seen in the first line of Fig. 3, 'FASTFOODWIFI' is listed and displays the 'bssid', transmission channel, encryption, cipher and 'essid'. To ensure the safety of this experiment a password was set for the access point. Using an open access point would allow other devices to join the rogue AP, thus would have raised ethical and legal concerns. Using the enumerated information from the 'FASTFOODWIFI' access point (see Fig. 4), 'Wifipumkin3' was configured and started (see Figs. 5 and 6). Once the rogue AP was started, both the Samsung Gear S3 Frontier and Motorola Moto-G device automatically joined the AP without needing a password or user intervention. The attacker was alerted via the terminal to all new connections (see Fig. 7). The Samsung Gear S3 Frontier displayed a notification to the user "Connected to WI-FI network FASTFOODWIFI". The devices were used to simulate a victim using a device whilst connected to the rogue AP, thus creating traffic (see Fig. 8).

```
Starting prompt ...
wp3 > set interface wlan0
wp3 > set ssid FASTFOODWIFI
wp3 > set security.wpa_algorithms CCMP
wp3 > set security.wpa_sharedkey A75FF8BD15
wp3 > set security.wpa_type 2
wp3 > set channel 6
wp3 > set bssid A4:91:B1:5B:62:ED
wp3 > 
```

Fig. 4 Configuring the 'WiFiPumkpin3' rogue AP using the enumerated information

```
wp3 > ap
[*] Settings AccessPoint:

BSSID              | SSID         | Channel | Interface | Status    | Security
A4:91:B1:5B:62:ED  | FASTFOODWIFI |       6 | wlan0     |           | wpa_sharedkey


[*] Settings Security:

wpa_algorithms | wpa_sharedkey | wpa_type
CCMP           | A75FF8BD15    |       2
```

Fig. 5 Rogue AP configuration

Fig. 6 Rogue AP new connection

```
ic-cdn.flipboard.com.[AAAA]          www.youtube.com.[A]
ic-cdn.flipboard.com.[A]             googleads.g.doubleclick.net.[A]
dw-wp-production.imgix.net.[A]       static.doubleclick.net.[A]
dw-wp-production.imgix.net.[AAAA]    adservice.google.com.[A]
ic-cdn.flipboard.com.[AAAA]          lh5.googleusercontent.com.[A]
media.gq-magazine.co.uk.[A]          lh6.googleusercontent.com.[A]
media.gq-magazine.co.uk.[AAAA]       www.gstatic.com.[A]
www.rollingstone.com.[A]             maps.googleapis.com.[A]
www.rollingstone.com.[AAAA]          maps.gstatic.com.[A]
ichef.bbci.co.uk.[A]                 www.canterbury.ac.uk.[A]
ichef.bbci.co.uk.[AAAA]              www.googletagmanager.com.[A]
e3.365dm.com.[A]                     cdn.cookielaw.org.[A]
e3.365dm.com.[AAAA]                  q.quora.com.[A]
imagesvc.meredithcorp.io.[A]         syndication.twitter.com.[A]
imagesvc.meredithcorp.io.[AAAA]      www.google-analytics.com.[A]
s31242.pcdn.co.[A]                   pbs.twimg.com.[A]
s31242.pcdn.co.[AAAA]                www.googleadservices.com.[A]
s.france24.com.[A]                   snap.licdn.com.[A]
s.france24.com.[AAAA]                connect.facebook.net.[A]
s.france24.com.[A]                   platform.twitter.com.[A]
s.france24.com.[AAAA]                secure.quantserve.com.[A]
cdn.the-race.com.[A]                 analytics.tiktok.com.[A]
cdn.the-race.com.[AAAA]              stats.g.doubleclick.net.[A]
e3.365dm.com.[A]                     px.ads.linkedin.com.[A]
e3.365dm.com.[AAAA]                  rules.quantcount.com.[A]
                                     www.google.co.uk.[A]
                                     www.linkedin.com.[A]
                                     www.facebook.com.[A]
```

Fig. 7 Rogue AP terminal with Samsung Gear S3 Frontier (left) and Motorola moto-G (right)

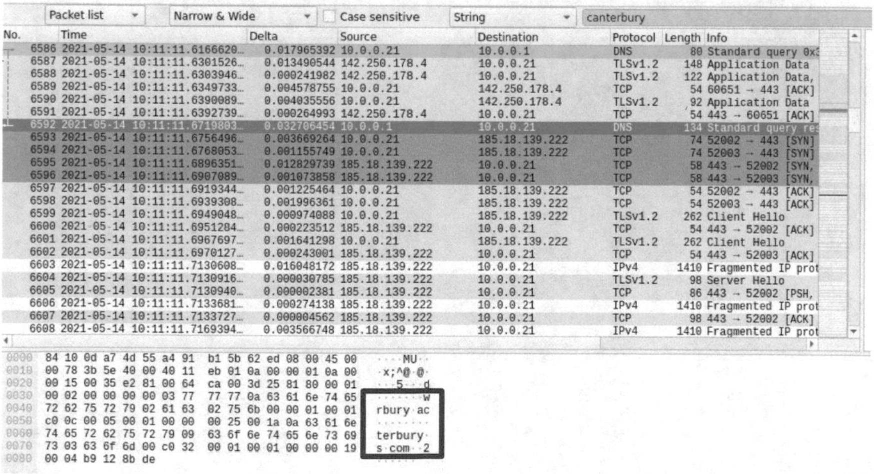

Fig. 8 Wireshark from the machine running Rogue AP, capturing Motorola Moto-G packets

As seen above, the attacker was able to view the websites being accessed on the devices via the terminal (see Figs. 7 and 8). Whilst using the Motorola, Wireshark was used to capture the packets passing through the 'WLAN0'. Wireshark was able to provide information in greater depth in comparison to the 'WifiPumpkin3' terminal, such as the information exchanged between the device and websites (Fig. 8).

4.2   Results of Bluetooth Attack

The command 'sudo ./blue_hydra' was used to run Blue Hydra in order to gain knowledge of the Bluetooth address of victim's devices. As seen in Fig. 9, these were the Bluetooth devices within the range. The victim devices in this scenario was a Samsung Gear S3 Frontier, Garmin Vivofit and Fitbit Ace 2. The address of the devices was easily discovered as the device was very close to the attacker's Bluetooth dongles, thus displaying a strong signal (see Fig. 9). The Blue Hydra also discovered the victim's device models.

Once the victim's device address was identified, it was then inputted into the "ble_trackerV2.py" script (see Fig. 10) and run in the terminal as shown in Fig. 11.

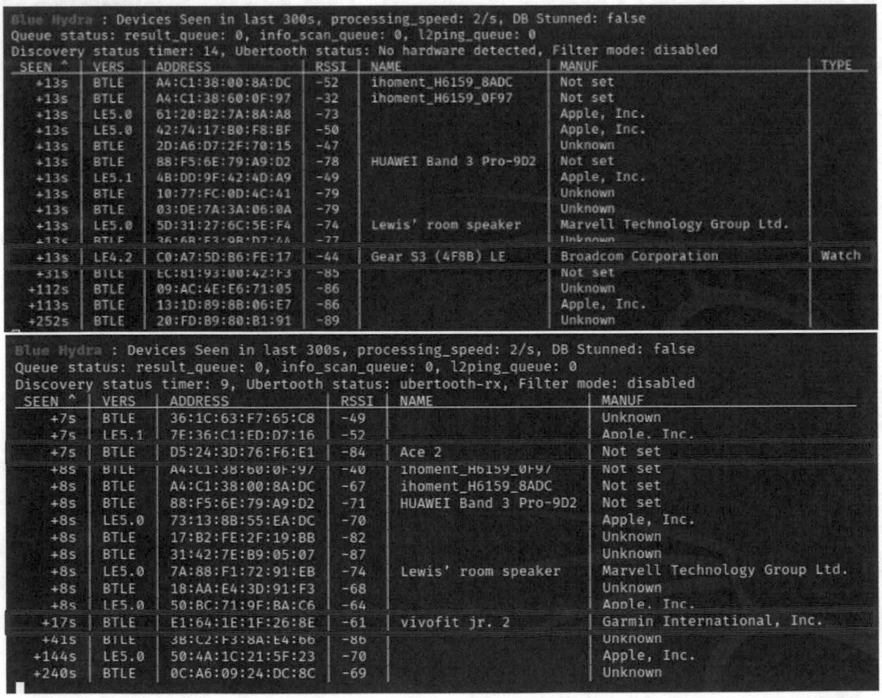

Fig. 9 Blue Hydra running, displaying all devices within range, Samsung Gear S3 (top), Garmin Vivofit and Fitbit Ace 2 (bottom)

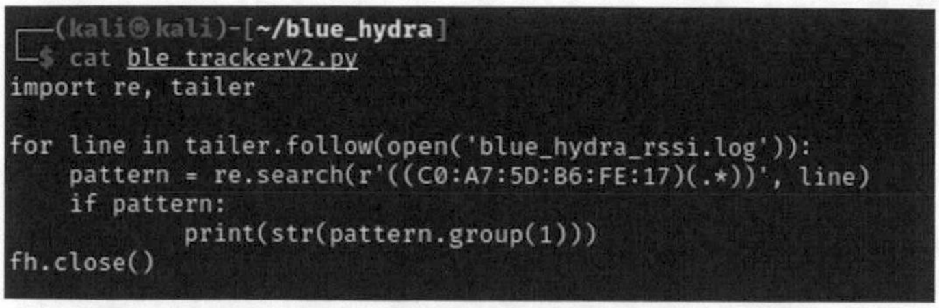

```
┌─(kali®kali)-[~/blue_hydra]
└─$ cat ble_trackerV2.py
import re, tailer

for line in tailer.follow(open('blue_hydra_rssi.log')):
    pattern = re.search(r'((C0:A7:5D:B6:FE:17)(.*))', line)
    if pattern:
            print(str(pattern.group(1)))
fh.close()
```

Fig. 10 X ble_trackerV2.py with Samsung Gear S3 Frontier address inputted

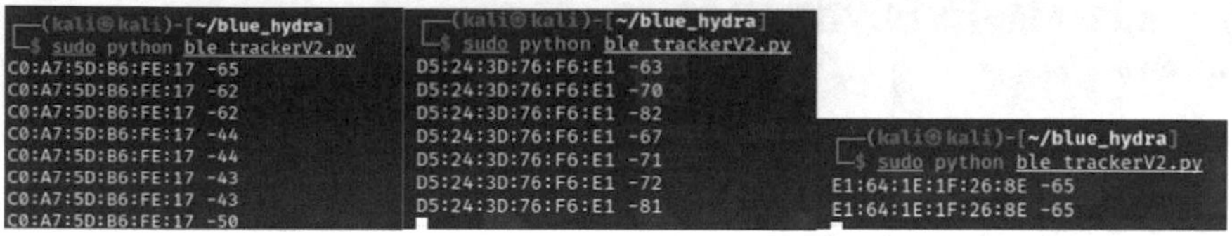

```
┌─(kali®kali)-[~/blue_hydra]      ┌─(kali®kali)-[~/blue_hydra]
└─$ sudo python ble_trackerV2.py  └─$ sudo python ble_trackerV2.py
C0:A7:5D:B6:FE:17 -65             D5:24:3D:76:F6:E1 -63
C0:A7:5D:B6:FE:17 -62             D5:24:3D:76:F6:E1 -70
C0:A7:5D:B6:FE:17 -62             D5:24:3D:76:F6:E1 -82
C0:A7:5D:B6:FE:17 -44             D5:24:3D:76:F6:E1 -67     ┌─(kali®kali)-[~/blue_hydra]
C0:A7:5D:B6:FE:17 -44             D5:24:3D:76:F6:E1 -71     └─$ sudo python ble_trackerV2.py
C0:A7:5D:B6:FE:17 -43             D5:24:3D:76:F6:E1 -72     E1:64:1E:1F:26:8E -65
C0:A7:5D:B6:FE:17 -43             D5:24:3D:76:F6:E1 -81     E1:64:1E:1F:26:8E -65
C0:A7:5D:B6:FE:17 -50
```

Fig. 11 X 'ble_trackerV2.py' tracking Samsung Gear S3 Frontier smart watch (left), Fitbit Ace 2 (middle), Garmin Vivofit (right)

The victim device was then move towards and away from the attacker's device to simulate the target moving (see Fig. 11). These results show that it is possible to track Bluetooth devices without the need for prior pairing or interactions, thus leaving the victim unbeknown to the attacker's presence or intentions. The Bluetooth dongles used in the scenario were only able to display signal strength between the victim device and attacker's device within a small radius, tested to 20 m. This same scenario could provide greater distance with the incorporation of stronger dongles.

4.3 Results of SMS Hacking

The Y22 GPS smartwatch was tested both indoors and outdoors for around five weeks. However, the majority of the time the watch claimed to have no signal, even though it was in areas of strong signal strength both indoors and outdoors. The accuracy of GPS locations of the watch was poor, as the closest the watch had been was over a mile away from the correct location. So, the parent/caregiver could not accurately know where their child was at all times. Overall, the build quality of the GPS watch was poor and was not suitable for any heavy use by a child.

The SeTracker application for the watch allowed to whitelist certain telephone numbers and should only allow those numbers to make a call. Whilst the whitelisting did technically work, when tested the watch could not call the malicious phone, but it did not prevent the attacker calling the watch when dialed from a withheld the number. Thus, there was a risk of attacker could easily connect to the watch and the child could pick up and converse.

One method to communicate with the device was through text messaging. The following commands were sent to the watch to find out information about it and to control the device:

1. pw,123,456,ts#
2. pw,123,456,phb,07XXXXXXX44,hacker,,,,,,,#
3. pw,123,456,phb#
4. pw,123,456,sos1,07XXXXXXX44#
5. pw,123,456,call,07XXXXXXX44#
6. pw,123,456,monitor,07XXXXXXX44#
7. pw,123,456,center,07XXXXXXX44#
8. dw#

The format of the commands are 'pw,PASSWORD,COMMAND#'. The default password for this device, which was 123,456, could be easily guessed and was displayed clearly on the store page for the device and in many online manuals of similar watches. Knowing the password, it was a lot easier for an attacker to compromise the device.

The command 'pw,123,456,ts#' asks the watch for the current settings as shown in Fig. 12. Some of the important details retrieved were the IMEI number of the watch, the ZCM code used when registering the watch on SeTracker, the Ip_url and port where the watch sends data to, such as its location, and the administrator's telephone number labelled by 'center'. Only the administrator can make certain changes to the watch. Even though the intelligence gathering using the SeTracker app was not possible through the malicious phone (as the app was already installed in the genuine phone), the attacker could still gain a lot of information using the 'pw,123,456,ts#' command.

Fig. 12 Response to 'pw,123,456,ts#' command

Fig. 13 SeTracker Phonebook (left) and response to 'pw,123,456,phb,07XXXXXXX44 ,hacker,,,,,,#' command (right)

The 'pw,123,456,phb,07XXXXXXX44,hacker,,,,,,#' command was used to update the phonebook so that the only telephone number in there was the one specified by the hacker. The updated phonebook with the hacker's number (ended with 44) can be seen in Fig. 13. However, the SeTracker application on the genuine phone did not detect this change, and still showed the original number (ended with 09) that was in the phonebook (Fig. 13), so the carer or parents were not aware of this change.

The command 'pw,123,456,phb#' removes all entries from the phone book (Fig. 14), and similar to the previous command, it does not show this change in SeTracker as seen in Fig. 14. The command 'pw,123,456,call,07XXXXXXX44#' forces the watch to call the number specified by the attacker (ended by 44 in the example). The 'pw,123,456,sos1,07XXXXXXX44#' command sets the first SOS number to the number specified. To know it has been successful, 'pw,123,456,ts#' was sent which stated the SOS number were set to attacker's number (Fig. 14). The SOS command allowed for the hacker to make themselves an SOS number, endangering the child if they needed to make an SOS call.

The 'pw,123,456,monitor,07XXXXXXX44#' is one of the most important commands. The 'monitor' specifies that the watch would call the number entered and the hacker could listen in to the conversation without the watch or the SeTracker application on parent phone giving any indication the child's conversation was being monitored by the attacker. The 'pw,123,456,center,07XXXXXXX44#' command allowed for the attacker to enter their number as the administrator number, so they could carry out administrative commands, such as to request a GPS location of the watch.

The "dw#" did not fully work. This command was designed to send a response back with the location of the watch, however it returned a location in China when the watch was in the South-East of England.

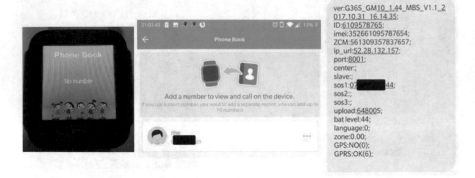

Fig. 14 Response to 'pw,123,456,phb#' command to remove phonebook entries (left), SeTracker Phonebook (middle) and SOS number set by the attacker (right)

5   Conclusions

Successful tracking and manipulation of kids' IoT smart devices reported in this paper yielded significant security concerns. During the MITM attack, the attacker could capture all packets flowing through the rogue AP. This allowed the attacker to perform analysis during or after the attack to reveal information about what victim's smart devices were accessing. Extending the methods, a broader attack vectors would explore further security exploits by creating multiple rogue APs with additional ALFA wireless network cards, each emulating a different local free to access wireless access points which are popular amongst the kids' age groups, thus showing greater risks.

Bluetooth tracking experiment also raised security concerns when the devices are operated in sparsely populated areas, making it easier for the attackers to successfully choose the target Bluetooth address to track. A further exploit of building a tool would be more concerning where the attacker would not only track but also enumerate Bluetooth devices, allowing the attackers to choose the target device with greater precision.

Experiment on the Y22 GPS smartwatch reveals that the watch was very inaccurate, often showing the location as being more than 1 km away from where it actually was. Findings reveal that an attacker could easily compromise the watch with minimal knowledge of penetration testing or hacking. The attacks for gaining this information from the watch do not require any complicated or advanced skillsets and could be carried out by most junior hackers with ease. A lot of information can be gathered from the watch without having the companion application installed. An attacker could gain the full control of the watch and call from a withheld number and the watch could be forced to call a specified number. An attacker can also remotely turn on the microphone and listen to the surroundings. Also, the phonebook and any SOS telephone numbers can be removed or updated by attackers. To demonstrate further exploits, using an SMS command an attacker could change the server to which the watch communicates with to the one owned by the attacker and listen to any communications sent from the watch, giving the attacker more control over the watch. All these results show that by buying one of these watches, a parent or caregiver may actually be exposing their child to more risk than they would be without the watch.

References

1. Thompson J (2018) A concise history of the smartwatch, Bloomberg, https://www.bloomberg.com/news/articles/2018-01-08/a-concise-history-of-the-smartwatch. Accessed 1 Aug 2022
2. Makhdoom I, Abolhasan M, Lipman J, Liu RP, Ni W (2018) Anatomy of threats to the internet of things. IEEE Commun Surv Tutor 21(2):1636–1675
3. Alladi T, Chamola V, Sikdar B, Choo KKR (2020) Consumer IoT: security vulnerability case studies and solutions. IEEE Consum Electron Mag 9(2):17–25
4. Liao B, Ali Y, Nazir S, He L, Khan HU (2020) Security analysis of IoT devices by using mobile computing: a systematic literature review. IEEE Access 8:120331–120350
5. Saatjohann C, Ising F, Krings L, Schinzel S (2020) STALK: security analysis of smartwatches for kids. In: Proceedings of the 15th international conference on availability, reliability and security, pp. 1–10
6. Židková N, Maryška M, Doucek P, Nedomova L (2020) Security of Wi-Fi as a key factor for IoT. In: International scientific conference Hradec economic days. https://doi.org/10.36689/uhk/hed/2020-01-101
7. BBC (2020) Garmin begins recovery from ransomware attack. https://www.bbc.co.uk/news/technology-53553576. Accessed 1 Aug 2022
8. Whittaker Z (2020) Smartwatch hack could trick patients to 'take pills' with spoofed alerts, TechCrunch. https://techcrunch.com/2020/07/09/smartwatch-hack-spoofed-alerts/. Accessed 1 Aug 2022
9. Karie NM, Sahri NM, Haskell-Dowland P (2020) IoT threat detection advances, challenges and future directions. In: 2020 workshop on emerging technologies for security in IoT (ETSecIoT), pp 22–29. Accessed 21 April 2021
10. Statista (2021) Share of respondents who own a smart watch/health-tracker in UK 2019, by generation. https://www.statista.com/statistics/1044033/uk-smartwatch-health-tracker-ownership/. Accessed 1 Aug 2022
11. Al-Sharrah M, Salman A, Ahmad I (2018) Watch your smartwatch. In: 2018 International conference on computing sciences and engineering (ICCSE). IEEE, pp. 1–5
12. Stanislav M, Beardsley T (2015) Hacking iot: a case study on baby monitor exposures and vulnerabilities, Rapid7 Report. https://www.rapid7.com/globalassets/external/docs/Hacking-IoT-A-Case-Study-on-Baby-Monitor-Exposures-and-Vulnerabilities.pdf. Accessed 1 Aug 2022
13. Alto P (2015) HP study smartwatches vulnerable to attack. https://www8.hp.com/us/en/hp-news/press-release.html?id=2037386#.YDI2k-j7SiM. Accessed 1 Aug 2022
14. Kelion L (2018) MiSafes' child-tracking smartwatches are 'easy to hack', https://www.bbc.co.uk/news/technology-46195189. Accessed 1 Aug 2022
15. Seri B, Livne A (2019) Exploiting blueborne in linux-based iot devices, Armis, https://info.armis.com/rs/645-PDC-047/images/Armis-Exploiting-BlueBorne-in-Linux-Based-IoT-Devices-WP.pdf. Accessed 1 Aug 2022
16. Treaster M, Koenig GA, Meng X, Yurcik W (2005) Detection of privilege escalation for linux cluster security. In: 6th LCI international conference on Linux Clusters
17. Mohanta BK, Jena D, Satapathy U, Patnaik S (2020) Survey on IoT security: challenges and solution using machine learning, artificial intelligence and blockchain technology. Internet Things 11. https://doi.org/10.1016/j.iot.2020.100227

18. Iqbal W, Abbas H, Daneshmand M, Rauf B, Bangash YA (2020) An in-depth analysis of IoT security requirements, challenges, and their countermeasures via software-defined security. IEEE Internet Things J 7(10):10250–10276

19. Waraga OA, Bettayeb M, Nasir Q, Talib MA (2020) Design and implementation of automated IoT security testbed. Comput Secur 88. https://doi.org/10.1016/j.cose.2019.101648

20. Classen J et al (2018) Anatomy of a vulnerable fitness tracking system: dissecting the fitbit cloud, app, and firmware. In: Proceedings of the ACM on interactive, mobile, wearable and ubiquitous technologies 2(1):1–24. https://doi.org/10.1145/3191737. Accessed 14 June 2021

21. Mahmoud M (2018) An experimental evaluation of smart toys' security and privacy practices. Masters Thesis, Concordia University. https://spectrum.library.concordia.ca/983590/. Accessed 1 Aug 2022

22. Dunn JE (2019) 'Kids' smartwatch security tracker can be hacked by anyone', Kids' smartwatch security tracker can be hacked by anyone. https://nakedsecurity.sophos.com/2019/11/28/kids-smartwatch-security-tracker-can-be-hacked-by-anyone/. Accessed 1 Aug 2022

23. BitDefender (2018) Understanding IoT vulnerabilities: overflow. https://www.bitdefender.com/box/blog/vulnerabilities/understanding-iot-vulnerabilities-overflow/. Accessed 1 Aug 2022

24. Pen Test Partners (2018) Consumer advice: kids GPS tracker watch security. https://www.pentestpartners.com/security-blog/consumer-advice-kids-gps-tracker-watch-security/. Accessed 1 Aug 2022

25. Chadza TA, Aparicio-Navarro FJ, Kyriakopoulos KG, Chambers JA (2017) A look into the information your smartphone leaks. In: 2017 international symposium on networks, computers and communications (ISNCC). IEEE, pp. 1–6

26. Arroyo JG, Bindewald J, Graham S, Rice M (2017) Enabling Bluetooth Low Energy auditing through synchronized tracking of multiple connections. Int J Crit Infrastruct Prot 18:58–70

27. Al Neyadi E, Al Shehhi S, Al Shehhi A, Al Hashimi N, Mohammad QH, Alrabaee S (2020) Discovering public Wi-Fi vulnerabilities using raspberry pi and Kali Linux. In: 2020 12th annual undergraduate research conference on applied computing (URC), IEEE, pp 1–4

28. Dorobantu OG, Halunga S (2020) Security threats in IoT. In: 2020 international symposium on electronics and telecommunications (ISETC), IEEE, pp 1–4

29. Juniper, Understanding Rogue Access Points, Network Director User Guide (2016) https://www.juniper.net/documentation/en_US/junos-space-apps/network-director3.1/topics/concept/wireless-rogue-ap.html. Accessed 1 Aug 2022

IoT-Penn: A Security Penetration Tester for MQTT in the IoT Environment

Armand Roets and Bobby L. Tait

Abstract The IoT (Internet of Things) represents a technological evolution in the way that human beings can now control, monitor, and study the world by enabling the connection of different devices around the globe, facilitating data delivery and services. However, the advantages of this increased connectivity does not come without a price. Various security issues have been discovered that can affect the confidentiality, availability, and integrity of the data received from IoT devices. IoT devices are, in general, power, storage, and processing constrained devices due to cost, size, and power restrictions. This leads to the adoption of light weight communication protocols specifically designed for communication among devices in which advanced, computationally intensive methods of security cannot always be applied. One such a communication protocol is MQTT (Message Queueing Telemetry Transport). This paper intended to answer the question of the utility of penetration testing when designing and evaluating an MQTT network. Various attacks were catalogued, designed, and implemented in an application called IoTPenn. These attacks were carried out on a simulated MQTT network, after which the results were analyzed. It was found that it is possible to gain access to sensitive and privileged information, to spoof legitimate MQTT clients, and perform DoS (Denial of Service) attacks against the broker, using the default MQTT configuration.

Keywords IoT · MQTT · Application layer · Attack vector · Penetration testing · Security · Privacy

A. Roets (✉) · B. L. Tait
University of South Africa, Pretoria, South Africa
e-mail: 51028506@mylife.unisa.ac.za

B. L. Tait
e-mail: Taitbl@unisa.ac.za

H. Jahankhani (ed.), *Cybersecurity in the Age of Smart Societies*,
Advanced Sciences and Technologies for Security Applications,
https://doi.org/10.1007/978-3-031-20160-8_9

1   Introduction

The IoT (Internet of Things) is a collection of devices with disparate architectures and purposes that connect to the internet. The existence of such a network of things, the so-called Internet of Things allows for the ever increasing advances in technology that improves the lifes of human beings worldwide.

The term IoT (Internet of Things) was first coined by British technology pioneer Ashton Kevin [1]. Formally, IoT refers to a collection of interconnected devices, and has been described in various ways. Sadhukan et al. [2] refers to the IoT framework as a collection of vastly different low-power consumption sensor devices. Hittu and Mayank [3] explains that IoT interconnects everyday objects containing integrated computers, through the internet. Rose et. al. [4] states that IoT is where everyday objects can be provided with computing and networking capabilities, allowing them to generate, process, and share data, which requires little to none human intervention.

Apart from these descriptions, IoT is an emerging technology, that is expanding to include an increasing array of devices, thus adding new functionality that seemed impossible in the recent past. Along with this emergence, new security challenges comes to light. With IoT being intensely integrated into our daily lives, this poses a threat to information security and privacy, with IoT attacks increasing by 600% between 2016 and 2017 [5].

Most current IoT platforms use standard communication protocols designed specifically for the IoT framework, while others use proprietary protocols [2]. Among the standardized protocols, COAP (Constraint Application Protocol) [6] and MQTT (Messaging Queueing Telemetry Transport) [7] are the most ubiquitous. This study intends to find the utility of penetration testing in the domain of MQTT, since it is found to be the most preferred and commonly used protocol for communication in the IoT [8].

2   Background

The IoT is a heterogeneous network of devices that is dynamic, intelligent, and mobile, making it an in-demand technology [9]. The prospects of development of IoT has seen applications of IoT in areas such as environmental monitoring, medical treatment, intelligent transportation systems, and public health systems [9].

Zhao and Ge [10] explains that the IoT has a three layer architecture, consisting of the Perception layer, the Network layer, and the Application layer. In this particular model, the perception layer gathers data from the environment by means of various sensors. The Network layer is the communication medium that is responsible for transmitting data obtained by the Perception layer. Finally, the Application layer facilitates the processing, filtering, and presentation of the data collected at the Perception layer and transmitted by the Network Layer.

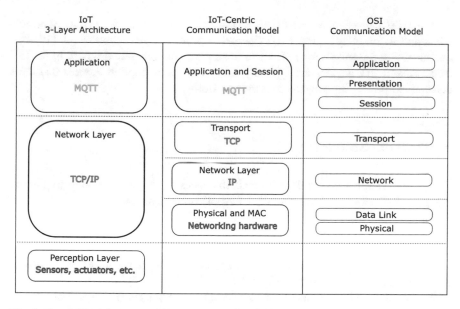

Fig. 1 How MQTT fits into the OSI communication model

Figure 1 compares the IoT three-layer architecture, the IoT-Centric Communication Model, and the OSI (Open Systems Interconnection) Communication model to show where MQTT fits into the IoT architecture as well as how communication using MQTT is facilitated using the OSI model.

Each layer of the three layer architecture is designed to perform specific tasks and functions, and as such each layer has possible security issues associated with it [12]. This paper focuses on MQTT, which is an application layer protocol. Other common application layer protocols are CoAP (Constrained Application Protocol), AMQP (Advanced Message Queueing Protocol), HTTP (Hypertext Transfer Protocol), and XMPP (Extensible Message and Presence Protocol) [13].

MQTT is most widely used because it is designed for optimized data collection and is suitable for low bandwidth networks consisting of devices that have limited resources [12]. IoT generally refers to a configuration where a network of devices is formed containing devices that would not normally be considered as computing devices. These devices are normally devices such as sensors and controllers that enable advanced functionality in normal, everyday consumer electronics [4].

The notion of devices that are not normally viewed as computing devices stems from the fact that these devices often contain processors or micro-controllers with limited capabilities, referred to as resource constrained devices by Andy et al. [13]. An example of such micro-controllers are the ATMEGA 328P and ATMEGA 1284P offerings from Microchip [20, 21]. The 328P has a mere 2KB of SRAM (Static Random Access Memory), while the 1284P, its big brother, has 16KB of SRAM, both offering relatively low-end processing power [20, 21].

The MQTT protocol originated at IBM in 1999 from the work of Andy Stanford-Clark and Arlen Nipper [14]. MQTT was first proposed by Atmoko et al. [15] as a communication protocol for IoT where the authors demonstrated its use in the IoT environment. IoT devices are typically low-power consumption devices that have constrained processing power, memory, and storage.

3 Problem Statement

The problem addressed in this research is the necessity of penetration testing in the design of IoT networks using the MQTT Application layer protocol.

The use of MQTT in IoT networks is the preferred method of communication. Unfortunately MQTT leaves the task of implementing security to developers because the native security mechanisms are weak [17]. Various vulnerabilities, as outlined in the next section, shows that there is a need for more research into the MQTT protocol to find and counteract these security risks.

The problem at hand requires a penetration tester that uses an attack vector based methodology to carry out penetration testing. It is thus evident that no research currently addresses this considerable problem as outlined by this paper.

4 Background Knowledge

A thorough review of the academic literature covering IoT security, Application layers protocols, Penetration testing, MQTT and its design, security issues, and possible solutions was conducted upon initiation of this research.

Zhao and Ge [10] states that IoT security research is different than Internet security research as it is more complex and suggests that there should be targeted solutions to each aspect of the various security problems discovered. They also introduce the three layer IoT architecture and then discusses security problems at each layer.

Mendez et al. [9] looks at security issues in the different levels of the three layer IoT architecture. They also provide insight into the standardization efforts at different layers of this architecture as attempted by various global engineering bodies. They also explore IoT technologies and protocols at each layer of the architecture.

Hintaw et al. [8] provides a brief history of IoT malware, which they call the IoT malware evolution. The Mirai DoS, one of the largest DoS attacks in history is described by Perrone et al. [25], where after the authors analyze MQTT and described some security solutions and improvements.

Among application layer protocols, the most popular in the literate is MQTT with Tandale et al. [22] providing a comparative study of MQTT (Message Queueing Telemetry Transport), REST (Representational State Transfer), and CoAP (Constrained Application Protocol). Sueda et al. [23] extends the comparative results by also conducting a study that compares MQTT and various other application layer protocols. The authors evaluate network traffic and provide guidance on MQTT topic sizes to optimize traffic flow in MQTT transport. The latest release of MQTT is version 5.0, which was release in 2019 [16]. This release is analyzed comprehensively by Mileva et al. [11].

The idea of having a well-defined Penetration testing Methodology is discussed by Alisherov and Sattarova [24] as they argue that penetration testing is an effective measure of verifying the effectiveness of security measures. The authors also argue that such a measure can be used to gauge a system's compliance with legal requirements and stipulations.

It is clearly noted that various authors address MQTT and possible security issues. Zhao and Ge [10] discussed MQTT authentication issues. Morelli et al. [26] investigated DoS attacks on several commercial and open source MQTT implementations. Salagean and Zinca [27] analysed the MQTT Protocol from a security point of view by carrying out attacks using tools found in Kali Linux.

5 Research Approach

This research was conducted in accordance with the principles of Design Science, falling within the Information Security and Software Engineering domain with an emphasis on design and construction of a software application that demonstrates security vulnerabilities in the MQTT communication protocol with respect to IoT device networks.

Due to the vastly different hardware architectural designs that IoT devices can have, this software simulated IoT devices by abstracting the architectural differences that can arise in IoT networks. The emphasis in this study was on the intricacies and vulnerabilities of the MQTT protocol and not the underlying hardware architecture.

6 Experimental Design

6.1 Designing an MQTT Packet Capturing Tool

The TCP/IP Protocol Suite MQTT, the application protocol which is the topic of this paper, is built on top of the TCP/IP (Transmision Control Protocol over Internet Protocol) protocol suite [8]. TCP/IP has gained mass popularity since its introduction and is currently the de facto standard for transporting data packets across modern networks [18].

IP (Internet Protocol), a part of the TCP/IP suite, is responsible for the addressing of a network packet. The protocol defines where a packet originates from and also where a packet is destined to be sent. TCP (Transmission Control Protocol), which is also part of the TCP/IP suite, is a connection-oriented message transportation mechanism that establishes a connection between two endpoints and ensures reliable data delivery between the two endpoints [18].

MQTT data packets are transported as part of a TCP packet and resides in the TCP data section of a TCP segment, which is encapsulated inside of an IP datagram, which is part of a frame transported on a network. From this, one can see that MQTT packets being transported on a network can be captured and inspected. This is precisely what was required in this investigation. A network packet analyzer was built with the capabilities necessary to specifically find and scrutinize MQTT packets.

Custom Packet Analyzer A decision was made to build a custom packet analyzer in order to have a reusable, customizable, and integrate-able packet analysis software module. SharpPcap, a cross platform .NET library was used as the base of the network analyzer module [19]. The library assists by providing live network frames from a network interface of choice.

The next step was dissecting each of the captured frames into their IP datagram, TCP segment, and MQTT packet components. Figure 2 show the structure of an IP datagram and where a TCP segment resides within this datagram. A TCP segment is depicted in Fig. 3. The figure also shows that a TCP segment can contain an MQTT packet in its data section. The MQTT control packet structure and its mandatory Fixed Header structure is shown in Fig. 4. The SharpPCap library supports IP and TCP parsing and facilitated the dissection of a raw data frame to a TCP segment.

Unfortunately, SharpPCap does not support the identification and parsing of an MQTT data packet inside of a TCP Segment payload. This problem was overcome by breaking the problem of dissecting packets into two sub-problems, and solving them seperately. All packets are first dissected and parsed into TCP packets after which they are passed in to an MQTT identification and deserialization module explained next.

MQTT packets can be identified by looking at either a packet's source or destination port number. Typically, a MQTT broker opens a TCP listener on a specified network port, and all clients wishing to connect to the broker in question will establish

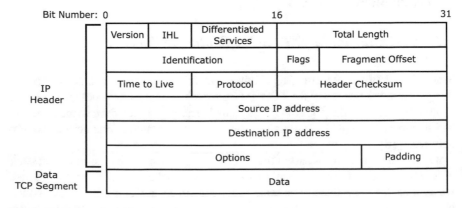

Fig. 2 IP Datagram from Dean [18]

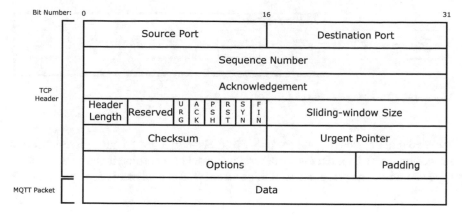

Fig. 3 TCP Segment from Dean [18] modified to indicate MQTT packet location

| Fixed Header, present in all MQTT Control Packets |
| Variable header, present in some MQTT Control Packets |
| Payload, present in some MQTT Control Packets |

Bit	7	6	5	4	3	2	1	0
byte 1	MQTT Control Packet Type				Flags Specific to each MQTT Control Packet Type			
byte 1	Remaining Length							

Fig. 4 MQTT Control Packet and Fixed Header

a connection with the broker using the same port number. The default port for MQTT using an unencrypted connection is port 1883 [13]. Packets originating from the Broker will have a source port of 1883. Analogously packets traversing the network from a client to the server will use a destination port of 1883.

MQTT Packets The identification of MQTT packets consisted of capturing and parsing data frames with an IP datagram having either a destination port or source port of 1883 as per the IANA reserved standard. The assumption was made that all packets filtered using this mechanism contains MQTT packets. A MQTT parsing software module was built according to the v3.1.1 and v5.0 MQTT specifications. v3.1.1 was included as it is still in widespread use. MQTT packets have no structural differences between v3.1.1 and v5.0, the latter version only introduces new features and minor updates, such as session and message expiry [16].

MQTT packets can now be identified and needs to be deserialized. The control packet type is indicated in the Fixed Header section of the control packet. The deserialization of an MQTT packet begins with parsing the first two bytes of the raw MQTT packet into the Fixed Header structure. The Fixed Header structure then provides the identification of which type of control packet is present, control packet specific flags and the length of the remaining MQTT data packet in bytes. Figure 4 illustrates the MQTT control packet structure, as well as the Fixed Header structure.

Parsing the remainder of the MQTT data packet, that is, the Variable Header and Payload sections is then done according to the control packet type identified from the Fixed Header. This is, once again, represented as a sub-problem, where the

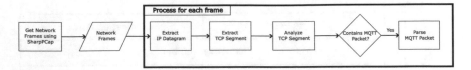

Fig. 5 MQTT packet capturing tool

remainder after removing the packet's Fixed Header is passed into a deserialization procedure which then returns the deserialized packet according to the control packet type identifier. The process flow of this custom packet analyzer is depicted in Fig. 5.

6.2 IoTPenn Application Design

The aim of this paper is to answer the question: Is penetration testing necessary from a security point of view when designing IoT networks using MQTT? In an attempt to answer this question, an application was built that uses the deserializer designed and discussed in the previous section. MQTT broker and client modules were also designed and built to operate according to the MQTT specification and is compatible with MQTT v3.1.1 and v5.0.

The application provides a simulation environment in which a broker and clients can communicate. This environment forms the testing ground for the execution and analysis of the following MQTT attacks.

7 Experimental Execution and Results

7.1 Attack 1: Packet Sniffing

Attack Design Using the software application developed and discussed earlier, a simulated environment of MQTT nodes was set up and all packets with a source or destination port of 1883 were captured and presented for further analysis. Packets were returned in a structured way containing human readable data to facilitate the analysis.

Implementation Note Since the PoC (Proof of Concept) application was designed to operate a broker and numerous clients in the same environment, that is, on the same computer, a loop back adapter is necessary. Network traffic that originates within the local host and is also destined for the local host does not exit the NIC (Network Interface Card) and thus a loop back adapter needs to be set up which can be used to monitor internal network traffic.

Table 1 Summary of attacks

Number	Name	Design	Findings
1	Packet sniffing	Custom built MQTT packet analyzer	Is able to find sensitive data
2	Identity spoofing	Custom built MQTT packet analyzer to obtain authentication data	Is able to successfully spoof the identity of a legitimate MQTT client
3	Privilege escalation	Custom built MQTT packet analyzer to obtain authentication and topic data	Is able to successfully gather sensitive and privileged information
4	CONNECT flood	Create a DoS on the broker by inundating the broker with CONNECT packets	Is not able to adversely affect availability of the broker
5	PUBLISH flood	Create a DoS on the broker and a subset of the connected clients by inundating the broker with publish packets	Results inconclusive
6	Excessive payload	Create a DoS on the broker by sending increasingly larger payloads	Is able to temporarily create a DoS on the broker with a payload > 50 MB

Findings CONNECT packet captures revealed the client's identifier, protocol version, keep alive time, username, and password. The corresponding TCP packets captured for such MQTT CONNECT packets show the source and destination IP addresses as well.

Similarly, PUBLISH and SUBSCRIBE packets were captured which reveals which clients publish which topics and which clients subscribe to which topics. In addition to the topic names in a PUBLISH packet, the topic data is also visible, which can lead to confidentially issues.

Since all MQTT data can be captured using packet sniffing, this attack forms the basis of a number of other attacks and is normally a first point of interest in designing or conceptualizing any type of attack in the broader context of information security.

7.2 Attack 2: Identity Spoofing

Attack Design This attack is a consequence of packet sniffing and the knowledge that an MQTT client authenticates against a broker by providing a client id, username, and password inside of a CONNECT packet that is sent to the broker. This attack is carried out by sniffing MQTT control packets of type 1, which is CONNECT packets.

Once again, using the software application developed to conduct research, a simulated environment of MQTT nodes is set up and a packet sniffing attack is executed to capture client identifiers, usernames, and passwords, where after these credentials is used to successfully connect to the broker in two different ways.

Utility of the Compromised Data After successfully acquiring the credentials, two methods of connection was tested. Firstly, a connection was immediately made with the stolen credentials. This resulted in the legitimate client being disconnected from the broker with an MQTT Reason Code indicating that the client's session had been taken over.

Secondly, the client's Keep Alive time as well as the client's activity with the broker was monitored. It must be noted that this method is tedious in execution and can take quite some time depending on the level of activity of the client in question. However, when timed correctly, an attacker can take over a client's session without a client being aware of it, as the response is not seen or known by the client.

The client's Keep Alive time is an indication of how long the server must maintain the connection with the client in the event of no communication received. Thus, for a busy client, the broker constantly receives data from the client and the Keep Alive time never truly completely runs down to zero.

Experiment A test was devised to test this theory. In this test, two clients were monitored in an attempt to determine which is the better target. One client was set to be very actively engaged with the broker at all times, by subscribing to frequently updated topics, while the other client subscribed to a topic that rarely receives an update. From this test it was determined that a quiet client is the best target for Identity Spoofing.

Findings Executing this attack as mentioned above, resulted in the successful identity spoofing of a legitimate MQTT client.

7.3   Attack 3: Privilege Escalation

Attack Design Privilege escalation in the context of MQTT security relates to the escalation of a client's topic publishing and subscribing rights. MQTT brokers do not authorize PUBLISH and SUBSCRIBE requests from clients. Using this information, this attack is an extension of attack 1, packet sniffing, that searches for all MQTT PUBLISH and SUBSCRIBE events on the network.

Experiment Using the PoC software developed for this research, this attack is executed by invoking the packet sniffer on a simulate network of MQTT nodes and gathering all PUBLISH and SUBSCRIBE messages occurring on the network.

Findings The results of this attack showed that it is possible to gather sensitive or privileged information and also to provide an MQTT node with false or misleading information.

It is possible to gather sensitive information using this attack. In the simulated environment, a client acting as an access controller for a hypothetical vault was set up. By subscribing to the topics that this node publishes to the broker, namely access requests to an online access control system, it was possible to get the identities of persons who accessed the hypothetical vault.

Using this same scenario of an access controller MQTT client, it is also possible to provide the broker with misleading information (as well as bypassing the access control security) by spoofing the identity of the access controller client and then requesting access to the hypothetical vault using the credentials of a legitimate person sniffed in the previous attack scenario.

These are both test examples, but it is illustrated that both scenarios are possible and that privilege escalation is possible and has damaging effects.

Implementation Note Instead of using packet sniffing, this attack can also be executed by using the # wildcard to subscribe to all topics. This will result in the receipt of all PUBLISH messages in the network. This is possible since MQTT does not provide topic level authorization, unless it is explicitly built into the broker [16].

7.4 Attack 4: CONNECT Flood

Attack Design This attack attempts to create a denial of service on the broker by using a multi-threaded procedure that sends multiple CONNECT packets to the broker. The number of connections to attempt is variable and is used to find the lower limit of connections that will inundate the broker. A multi-threaded approach is taken to ensure that connection attempts to the broker happens simultaneously instead of in a serialized approach.

Experiment Once again using the software developed for this research, the attack is commenced on a simulated MQTT environment, where a number of threads corresponding to the desired number of connections are spawned. Each thread then attempts to connect as a new client. Client identifiers for all the connections are automatically generated in a sequential fashion in preparation of the attack. Multiple executions of the attack were run, each time increasing the number of connection to be made. The results from the attacks are tabulated in Table 2.

The Broker's ability to stay operational during a DoS attack has largely to do with the network bandwidth available to it and its processing and storage capabilities. This attack, illustrated in Table 1 was carried out in a computing environment using an 11th generation Core i7 processor with 8 cores running at 2.3 GHz, 24 GB of DDR4 RAM running at 3200 MHz, and a 1TB solid state drive using the NVMe™ PCIe® 4.0 interface.

Table 2 CONNECT flood attack results

Connections	Time	Successful	Failed	Broker operational during	Broker operational after	Client DoS rate
100	00:00:00.98	61	39	Yes	Yes	39.00
200	00:00:02.10	162	38	Yes	Yes	19.00
300	00:00:03.42	263	37	Yes	Yes	12.33
400	00:00:04.66	352	48	Yes	Yes	12.00
500	00:00:05.38	456	44	Yes	Yes	8.80
600	00:00:06.41	562	38	Yes	Yes	6.33
700	00:00:07.83	659	41	Yes	Yes	5.86
800	00:00:08.36	756	44	Yes	Yes	5.50
900	00:00:09.37	851	49	Yes	Yes	5.44
1000	00:00:10.36	951	49	Yes	Yes	4.90
2000	00:00:21.25	1941	59	Yes	Yes	2.95
3000	00:00:32.02	2917	83	Yes	Yes	2.77
4000	00:00:49:54	3887	113	Yes	Yes	2.83

The network infrastructure is not of consequence in this attack as it was performed using the loop back adapter operating internally on the localhost environment. Even though the loopback adapter operates in exactly the same way a physical network adapter does, it is worth noting that there would be no external traffic present when using the loopback adapter.

Findings The result of this attack indicate that the broker availability was not adversely affected by doing a maximum of 4000 concurrent connections. From the results it can be concluded that there is a percentage of client connection failures when attempting multiple concurrent client connections. This rate of failure (Client DoS Rate) is extremely high, at 39 % for a 100 connections. The rate however drops as the number of concurrent connection attempts is increased. From this, it is clear that there will be some rate of broker unavailability in the sense that a number of client connection attempts failed.

Testing to see whether the broker is still operational during the attack was accomplished by starting up two instances of the PoC application. The attack is executed on one instance of the application, while the other instance was used to manually start up five clients that attempted connections to the broker.

Attack Method Considerations The attack was executed on a high performance machine with a substantial amount of resources in the processing and memory categories. Considering this, it is possible that this attack can still inundate a broker operating in a more constraint environment.

7.5   Attack 5: PUBLISH Flood

Attack Design By using the software developed in this paper, a simulated network of MQTT nodes are set up, where after flooding was attempted by sending multiple publish requests which has the potential to overwhelm the broker with a large amount of lookup and topic forwarding activity resulting from the number of publish requests. The data to publish and the number of times to publish the data is variable and is used to find a lower limit of PUBLISH requests that will inundate the broker and the clients subscribed to the topic in question.

Experiment On commencement of this attack, a multi-threaded procedure started a number of threads corresponding to the number of PUBLISH requests to send. The rational of using a multi-threaded approach is to attempt to inundate the broker and a subset of the clients using simultaneous requests.

Multiple executions of the attack are carried out with different parameters for the number of messages to publish and the QoS (Quality of Service). The number of messages to publish is iteratively incremented by 100. In each of these iterations, all three QoS levels available in MQTT is tested.

Findings The attack results are shown in Table 3. The outcome of this test was unclear in the sense that the results varied with different attempts of the attack with the same input parameters.

What can however be determined from the results is that the QoS (Quality of Service) has a major effect on the number of subscriptions successfully delivered to clients. With a high number of published messages and the highest level of QoS (Level 2), the loss of subscriptions is quite high. Conversely, with a high number of published messages and the lowest level of QoS (Level 0), no subscription loss was reported.

7.6   Attack 6: Excessive Payload

Attack Design According to the MQTT specification, a client can specify a maximum payload size that it is willing to receive, in the CONNECT request to the broker. The broker can also override this size by sending its own maximum payload size back to the client in the CONNACK response [16]. The specification also states that this maximum payload size specification is optional and when it is not specified by either the client or the broker, no limit is placed on the maximum payload size.

Experiment The attack was sourced from Hintaw et al. [8]. These authors explicitly states that a packet size greater than 256 MB is defined as an excessive payload, however, according to the MQTT specification, there is no maximum payload size and the size is determined as discussed above. Using the software developed for this research, a simulated network of MQTT nodes was set up where after the payload

Table 3 PUBLISH flood attack results

No.	Total application messages	QoS	Time	Sent	Received	Lost	Broker operational during	Broker operational after
100	500	QoS 0	00:00:00.79	98	404	+2	Yes	Yes
100	500	QoS 1	00:00:00.75	100	400	0	Yes	Yes
100	500	QoS 2	00:00:00.77	100	400	0	Yes	Yes
200	1000	QoS 0	00:00:01.33	199	804	+3	Yes	Yes
200	1000	QoS 1	00:00:01.46	200	38	762	Yes	Yes
200	1000	QoS 2	00:00:01.69	198	59	743	Yes	Yes
300	1500	QoS 0	00:00:02.10	299	1203	+2	Yes	Yes
300	1500	QoS 1	00:00:02.46	300	289	911	Yes	Yes
300	1500	QoS 2	00:00:02.51	293	149	1058	Yes	Yes
400	2000	QoS 0	00:00:03.05	400	1599	1	Yes	Yes
400	2000	QoS 1	00:00:03.15	391	463	1146	Yes	Yes
400	2000	QoS 2	00:00:03.33	399	286	1315	Yes	Yes
500	2500	QoS 0	00:00:03.84	500	2003	+3	Yes	Yes
500	2500	QoS 1	00:00:04.07	498	691	1311	Yes	Yes
500	2500	QoS 2	00:00:04:32	496	342	1662	Yes	Yes

Table 4 Excessive payload attack results

Size (MB)	Time	Broker operational during	Broker operational after
25	00:00:09.03	Yes	Yes
50	00:00:20.29	No	Yes
100	00:00:58.62	No	Yes
150	00:01:47.00	No	Yes
200	00:04:05.00	No	No
250	Unknown	No	No

containing garbage data of varying sizes was sent to the broker using PUBLISH control packets.

Findings The results of this attack is shown in Table 4.

From the results it is clear that payloads smaller than 25 MB does not have an effect on the broker's capacity to stay available for other clients. Payload sizes larger that 25 MB, however, incapacitates the broker for the time being that the PUBLISH is in progress. The time needed to process large payloads increases considerably when payload sizes start to exceed 100 MB. A payload size of 250 MB took the broker completely offline and resulted in the entire PoC application becoming non-responsive.

8 Results Interpretation

In an attempt to answer the question set out earlier of the utility of penetration testing from a security point of view when designing IoT networks using MQTT, the following rational is considered.

Firstly, it is clear that performing penetration testing can lead to various insights in how vulnerabilities can be exploited, the design of mitigating controls, and also the testing of mitigating controls. Secondly, it can also be seen that penetration testing can play an important role in the process of conceptualizing and mitigating security risks in the realm of an MQTT network. This is even more important when developing a new network, as security is often hard to implement after the fact. Lastly, it is the opinion of the researcher that it is possible to discover new vulnerabilities using the techniques of penetration testing demonstrated here, but unfortunately no new vulnerabilities were found during the course of this research.

From the above reasoning on the results of carrying out this research, it is apparent that penetration testing is necessary not only when designing new networks, but also when attempting to mitigate risks in already existent networks.

8.1 Research Limitations

This research has not taken cryptographic security into consideration. It has focused on the MQTT standard which does not by default implement excessive security or cryptographic measures. This is mainly due to the fact that MQTT is designed to be a lightweight protocol suitable for resource constraint devices as pointed out by Andy et al. [13]. The intent here was to find the utility of penetration testing when applied to a standard implementation of MQTT on a network containing resource constrained devices.

8.2 Future Research

Future research on this topic can be done by incorporating more attack vectors and researching mitigations. The domain of information security is an ever expanding area of research with new attack vectors being developed on a regular basis. The need for further research will always be necessary.

9 Conclusion

The ubiquity and expansion of IoT device networks in our daily lives makes the security of these networks an important consideration for all future development in the field. Among the application protocols, research on MQTT has revealed various vulnerabilities. In light of this, with MQTT being the dominant protocol for communication in these environments, it is very important that the security vulnerabilities, their associated attack vectors, and mitigation controls be investigated and documented.

From this, the question on the importance of implementing penetration testing when designing an MQTT based IoT network was considered in this research. The research proceeded by cataloguing various attack vectors, executing the vectors, and compiling results from these attacks into a research report.

Following a sound and unbiased research approach grounded in the realm of design science and emperical results, it was determined that penetration testing is necessary when designing or even reviewing the design of an MQTT-based IoT network.

References

1. Shen G, Liu B (2010) The visions, technologies, applications and security issues of Internet of Things. In: 2011 international conference on E-business and E-government (ICEE), pp 1–4. https://doi.org/10.1109/ICEBEG.2011.5881892
2. Sadhukhan D, Ray S, Biswas GP, Khan MK (2021) A ligthweight remote user authentication scheme for IoT communication using elliptic curve cryptography. J Supercomput 5:1114–1151. https://doi.org/10.1007/s11227-020-03318-7
3. Hittu G, Mayank D (2019) Securing IoT devices and securely connecting the dots using REST API and middleware. In: 2019 4th international conference on Internet of Things: smart innovation and usages (IoT-SIU), pp 1–6. https://doi.org/10.1109/IoT-SIU.2019.8777334
4. Rose K, Eldridge S, Lyman C (2015) The internet of things. Internet Soc (ISOC) 80:1–50
5. Internet Security Report. https://www.symantec.com/content/dam/symantec/docs/reports/istr-23-2018-en.pdf. Last accessed 9 Sept 2021
6. CoAP. https://coap.technology. Last accessed 18 June 2021
7. MQTT Specifications. https://mqtt.org/mqtt-specification/. Last accessed 18 June 2021
8. Hintaw AJ, Manickam S, Aboalmaaly MF (2021) MQTT vulnerabilities, attack vectors and solutions in the Internet of Things (IoT). IETE J Res 1–30. https://doi.org/10.1080/03772063.2021.1912651
9. Mena Mendez D, Papapanagiotou I, Yang B (2018) Internet of things: Survey on security. Inf Secur J: Glob Perspect. https://doi.org/10.1080/19393555.2018.1458258
10. Zhao K, Ge L (2013) A survey on the Internet of Things security. In: 9th international conference on computational intelligence and security. https://doi.org/10.1109/CIS.2013.145
11. Mileva A, Velinov A, Hartmann L, Wendzel S, Mazurczyk W (2021) Comprehensive analysis of MQTT 5.0 susceptibility to network covert channels. Comput Secur 104. https://doi.org/10.1016/j.cose.2021.102207
12. Swamy SN, Jadhav D, Kulkarni N (2017) Security threats in the application layer in IoT applications. In: International conference on I-SMAC (IoT in Social, Mobile, Analytics and Cloud)

13. Andy S, Rahardjo B, Hanindhito B (2017) Attack scenarios and security analysis of MQTT communication protocol in IoT system. In: 4th international conference on electrical engineering, computer science and informatics (EECSI), pp 1–6. https://doi.org/10.1109/EECSI.2017. 8239179
14. MQTT Frequently Asked Questions. https://mqtt.org/faq. Last accessed 18 June 2021
15. Atmoko RA, Riantini R, Hasin MK (2003) IoT real time data acquisition using MQTT protocol. JPhys 2–7. https://doi.org/10.1088/1742-6596/853/1/012003
16. MQTT Version 5.0. https://docs.oasis-open.org/mqtt/mqtt/v5.0/os/mqtt-v5.0-os.pdf. Last accessed 18 June 2021
17. Buccafurri F, Romolo C (2019) A blockchain-based OTP-authentication scheme for constrained iot devices using MQTT. In: Proceedings of the 2019 3rd symposium on computer science and intelligent control. https://doi.org/10.1145/3386164.3389095
18. Dean T (2010) Networks + Guide to networks, 5th edn. Cengage Learning
19. SharpPCap Github. https://github.com/dotpcap/sharppcap. Last Accessed 18 June 2021
20. Atmega 328P. https://www.microchip.com/en-us/product/ATmega328P. Last 12 Jan 2022
21. Atmega 1284P. https://www.microchip.com/en-us/product/ATmega1284P Last 12 Jan 2022
22. Tundala U, Momin R, Seatharam DP (2017) An empirical study of application layer protocols for IoT. In: International conference on energy, communications, data analytics and soft computing (ICECDS-2017). https://doi.org/10.1109/ICECDS.2017.8389890
23. Sueda Y, Sato M, Hasuiki K (2019) Evaluation of message protocols for IoT. In: 2019 IEEE international conference on big data, cloud computing, data science and engineering, pp 172–175. https://doi.org/10.1109/BCD.2019.8884975
24. Alisherov AF, Sattarova YF (2009) Methodology of penetration testing. Int J Grid Distrib Comput 43–50
25. Perrone G, Vecchio M, Pecori R, Giaffreda R (2017) The day after mirai: a survey of MQTT security solutions after the largest cyber-attack carried out through an army of IoT devices, pp 246–253. https://doi.org/10.5220/0006287302460253
26. Morelli U, Vaccari I, Ranise S, Cambiaso E (2021) DoS attacks in available MQTT implementations: investigating the impact on brokers and devices, and supported anti-DoS protections. In: The 16th international conference on availability, reliability, and security. https://doi.org/ 10.1145/3465481.3470049
27. Salagean M, Zinca D (2020) IoT Applications based on MQTT protocol. In: 2020 international symposium on electronics and telecommunications (ISETC). https://doi.org/10.1109/ ISETC50328.2020.9301055

An Analysis of Cybersecurity Data Breach in the State of California

Zakaria Tayeb Bey and Michael Opoku Agyeman ⓘ

Abstract As the wave of data breaches continues to crash down on organisations, we will analyse the largest publicly available database of data breaches in the state of California using the public data breach database functioning under the state notification law of data breach. The dataset contains records since January 2012. These records were analysed in order to classify and identify California data breaches by multiple company types, attack vectors and stolen personal information. The main findings were that Software vulnerability is the most common attack vector due to third-party software, the financial industry is the most targeted industry while both large and small organisations are equally targeted by attackers. The analysis also found that credit/debit card information and social security numbers represent the most stolen personal information.

Keywords Cybersecurity · Data breach · Cyber-attack · Ethical hacking · Internet of things · IoT · Security · California data breach database

1 Introduction

The extent of data breaches is expanding, impacting more organisations and people which put a lot on the line as data breaches may cost individuals and businesses money, reputational damage, and lost opportunities. These breaches put key infrastructure at risk and national security at jeopardy [1, 2].

In 2013, A group of hackers gained access to Target's security and payments system and they were able to install a malicious software (malware) to steal credit card information used at the company's 1,797 U.S. stores. The cybercriminals were able to steal 40 million credit and debit records and 70 million customer records which caused the company to pay $18.5 million to settle claims by 47 states and their earnings dropped 46% [3]. Target is only one of many companies that got affected

Z. T. Bey · M. O. Agyeman (✉)
Centre for Advanced and Smart Systems, University of Northampton, Northampton, UK
e-mail: Michael.OpokuAgyeman@northampotn.ac.uk

by malicious software. There was also another huge breach in 2012, cybercriminals compromised the professional networking site LinkedIn using a SQL injection attack and leaked over 100 million records of personal information and hashed passwords, Attackers were able to crack the passwords and sold them online even though the passwords were hashed in the database but the encryption keys were found by attackers on the same compromised server [4].

Nowadays, cloud computing and the use of cloud storage are widely used due to the popularity of social media, e-commerce, and mobile services meaning that more personal consumer information and data are exposed to the public network which explain the increase of data breach costs from $3.86 million to $4.24 million in 2021, the highest in the past 17 years [4]. COVID-19 played an important role too by increasing the remote work which opens the door to insecure home networks and devices which lead to an increase in the average cost of a data breach by $1.07 million and a 300% increase in reported cybercrimes by the US FBI [5]. Another attractive target to threat actors is Internet of Things (IoT) devices. 75 billion IoT devices will be connected by 2025 [6], IoT is the future but unfortunately with the increase of IoT devices in the network there'll be an increase of cyber-attacks because IoT devices are the most vulnerable devices in the network. Also, more than 93% of healthcare organisations experienced a data breach in the past few years [7] because healthcare enterprises often use legacy software due to budget constraints even though it is a highly critical industry to secure.

As a result of this increasing concern, California has obliged organisations to report residents of the state when their personal information is breached, beginning in 2003. Businesses and government institutions have been required to notify the state governments of data breaches impacting more than 500 California residents since 2012 [8]. These data breach reporting rules made it possible to identify the numbers of data breaches in the state of California where 2017 was the worst year in the record of the state with over 1,200 breaches and 3.4 billion leaked records according to Risk Based Security [9].

Hotels, schools, dentists, spas, universities, restaurants, hospitals, doctors, retailers, government agencies, banks were all affected by data breaches and cyber security attacks. Mostly, third party service providers or employees were responsible for most of the reported breaches. Insiders' unintentional and intentional activities, as well as stolen and lost devices storing unencrypted data, all contributed to data breaches [9].

2   Aims and Objectives

2.1   Aim

An analysis of the California data breach database will be conducted in order to extract different statistics related to data breaches and cyber security attacks in 98 different industries. Three types of data analysis will be conducted in this research, Companies Analysis (by: Industries, Sectors, Sizes, Financial loss, Breach detection), Attacks Analysis (Unauthorised Access, Software Vulnerability, Stolen Computer or Data, Data Found Publicly, Wrong Data Sent, Exposed Data, Compromised Machine, Phishing Email, Insider Theft, Stolen Credentials, Compromised Email, Lost Computer or Data, Ransomware, Social Engineering) and Personal Information Analysis (Social Security Number, Payment Card, Medical Record, Password, Bank Account, Health Insurance Driver's Licence).

2.2 Specific Objectives

- To classify California data breaches by multiple company types.
- To classify California data breaches by common attack vectors.
- To classify California data breaches by common stolen personal information.

2.3 Research Question

- What are the patterns in company's types in California data breaches?
- What are the common attack vectors in California data breaches?
- What are the common stolen personal information in California data breaches?

3 California's Data Security Breach Reporting Law

In 2003, California established a data breach notification law first. 46 other states and international authorities across the world have passed similar legislation in a twelve-year period [8].

Over the last five years, hackers and attackers with malicious intent have penetrated the security of a remarkable number of organisations in nearly every state in America. In 2016, Yahoo, situated in California, suffered a huge data security breach in which online hackers stole the private information of around 500 million customers. The hackers were so skilled that Yahoo was unaware of the breach for two years until it was detected. Fortunately, most consumers who were affected did not face any long-term consequences. Therefore, the state legislature of California chose to introduce legislation requiring corporations to assume responsibility for the security of their customers' personal information [10].

As the frequency of data breaches keeps rising worldwide [1], organizations doing business in California must show that they are taking reasonable precautions to protect their customers' personally identifiable information. This data set contains a variety of sources of credentials for the authentication of individual's and is frequently used to gain access to online accounts or perform financial transactions.

If a data security breach occurs, firms must notify all affected individuals and the California Attorney General if the incident affects more than 500 people. Businesses that fail to report a data breach or who delay notification without reasonable cause may face severe financial penalties under the law.

Today, California's regulations and reporting obligations for data breaches are based on the California Consumer Privacy Act (CCPA). On June 28, 2018, Governor Jerry Brown signed the Act into law [8].

Finally, these rules have generated entire new enterprises dedicated to assisting businesses in preventing data breaches and responding effectively when they do occur. Cyber insurance, for example, is a relatively new profession that protects organisations against data breaches and cyber-attacks. Cyber insurance premiums totaled $5 billion in 2018, with the market likely to double in the past five years [11].

4   Methodology

4.1   Data Source

We looked at all publicly available California data breach notices since January 2012 on the California Department of Justice website for this analysis. As the California state government maintains a complete database of all enterprises and citizens cyber-security breaches. Therefore, all data breaches recorded since January 2012 are listed on the website, along with complete breach notification reports.

4.2   Data Scraping

The California state government website does not provide any public application programming interface (API) [9]. Therefore, web scraping was used to extract all the publicly available data security breaches on the website since the data shown on the website does not represent any underlying organised structure like JSON or XML.

PHP programming language is used to code the web scraper. This script will fetch the HTML page content of the data security breaches website. All breaches since 2012 are displayed in one page without any pagination. Therefore, only one request is required to fetch the HTML string and extract all the organisation name, date of breach (if known) and reported date. However, all records in the PDF report located in the details page 2 requests are required for each breached organisation (2 N) in order to download the pdf report. First request to the details page and the second request to download the report.

Text content will be extracted from the pdf, the content will be processed using regular expression in order to extract all the records. Finally, the parsed records will be inserted into a MySql database.

4.3 Data Structure

Using the company name from the extracted records and the LinkedIn api, we managed to fetch the company type (nonprofit, partnership, government, privately-held, public-company, sole-proprietorship, educational-institution, self-employed) with 75.5% match, the company size (1–10, 11–50, 51–200, 201–500, 501–1,000, 1,001–5,000, 5,001–10,000, or 10,000 +) with 83.4% match, and finally the industry with 100% match. The "What Happened?" section is one of the required sections in the data breach reports according to the California law, this section describes the breach in general. The attack vector will be assigned to the breach according to the content of this section as follow:

- **Compromised Email:** Once attackers gain access to the email address, They can compromise all services assigned to this address and possibly escalate from personal accounts to company accounts which can eventually lead to a cybersecurity breach.
- **Compromised Machine:** Hacked physical machines like servers, ATM's or Printers within the compromised network.
- **Data Found Publicly:** publicly available personal information online or in physical documents.
- **Exposed Data:** expose confidential information to an unauthorised person due to misconfiguration or software bug.
- **Insider Theft:** intentionally or unintentionally leaking confidential credentials by a current or former employee via unsecured email address or cloud storage.
- **Lost Computer or Data:** employees lose unencrypted physical devices or physical records containing confidential information.
- **Phishing Email:** stolen employee credentials via an email that appears to be from a trusted source.
- **Ransomware:** company critical data is encrypted over the network; decryption key is required to decrypt the company data and can be obtained only by paying a ransom or using reverse engineering techniques which can cause a risk of losing data.
- **Social Engineering:** Manipulating an employee to send confidential credentials or give access to customers account by impersonating a high-level company manager.
- **Software Vulnerability:** known or 0 day vulnerabilities on the company network and servers like remote code execution, SQL injection, Windows SMB privilege escalation. Third-party library vulnerabilities are included.
- **Stolen Computer or Data:** stolen employee unencrypted physical devices or physical records containing confidential information.

- **Stolen Credentials:** Employees use the same password in a previous data breach or a weak password which can be easily brute forced.
- **Unauthorised Access:** Any unauthorised access to the company network that may cause a risk of exposing confidential information.
- **Wrong Data Sent:** Employees may send confidential credentials to an unauthorised party by accident.

5 Presentations of Findings

5.1 Analysis: Companies

Industries. Top 8 industries represent 53.7% of the total breached companies where the financial industry represents 19.9% followed by hospital and healthcare, Retail, Hospitality, Higher Education, Insurance, Medical Practice, Accounting. The other 90 industries represent 46.3% of the total breached companies with a low frequency under 3% (Fig. 1).

American Express (5.9%) and Discover Financial Services (1.8%) were the most frequently compromised companies. These companies are required to track all the publicly exposed personal information including debit/credit cards credentials in order to notify their customers and block the hacked cards.

Sectors. Most of the companies that were breached were either privately owned (37.0%) or public companies (34.8%). The rest of the companies were nonprofit (11.5%), educational institutions (6.9%), government agencies (6.0 percent), and 3.8% for other businesses (Fig. 2).

Sizes. Sizeable companies with +10.000 employees are by large the most breached companies representing 30.3% of total breaches. Over half of cybersecurity breaches affect large companies with +1,000 employees. The frequency of data breach in small to medium sized companies range from 6 to around 12% which can be identical to large companies (Fig. 3).

Fig. 1 Frequency of breached companies for the top 8 industries

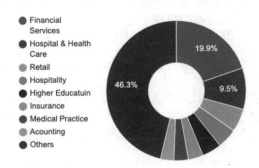

Fig. 2 Frequency of data breaches by company sector in 98 industries

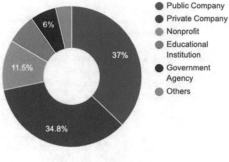

Fig. 3 Frequency of data breaches by company sizes in 98 industries

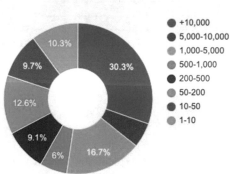

Financial loss. The healthcare sector suffers the most from data breaches. The average cost grew by 29.3 percent between 2020 and 2021, from $7.13 million to $9.23 million.

Data breaches that were discovered and resolved within 200 days cost an average of $3.61 million. However, breaches that took more than 200 days to uncover and contain cost an average of $4.87 million, a $1.26 million difference.

Breaches with at least 50 million records cost 100 times as much as the ordinary data breach. Moreover, Data breaches with 50 million to 65 million records cost an average of $401 million in 2021, up from $392 million in 2020 (Fig. 4).

Only 21.5% of cybersecurity incidents are reported within 30 days. However, prior literature review found that data breach incidents reported within 30 days of the incident may save over 1 million dollars.

Breach detection. There's an average of 108 data breaches per semester. However, almost 19% of businesses that reported data breaches were unable to determine the specific date(s) of the incident.

The amount of data breaches has been gradually increasing, with 1.08% more data breaches occurring each semester than the previous one.

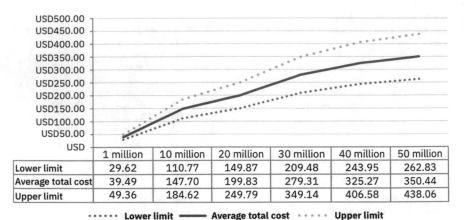

	1 million	10 million	20 million	30 million	40 million	50 million
Lower limit	29.62	110.77	149.87	209.48	243.95	262.83
Average total cost	39.49	147.70	199.83	279.31	325.27	350.44
Upper limit	49.36	184.62	249.79	349.14	406.58	438.06

······ Lower limit ——— Average total cost ····· Upper limit

Fig. 4 Cost of cybersecurity data breach measured in USD million by breached records (http://www.ibm.com)

5.2 Attack Analysis

Due to missing explanation in some reports, Unauthorised Access is the most common attack with 387 frequencies. Software Vulnerability is the second most common attack representing 13.1% of total attack vectors as many softwares or third party softwares are unprotected and open to remote access. The 2017 Ukraine ransomware attacks is a great example of third-party vendors where the attacker hacked a vendor that serviced a large number of companies to use the auto-update feature to upload a malicious code and gain remote access to all clients associated with the vendor. Stolen Computer or Data is the third most common attack vector representing 11.4% with 163 frequencies, followed by Data Found Publicly (159), Wrong Data Sent (105), Exposed Data (103), Compromised Machine (78), Phishing Email (63), Insider Theft (43), Stolen Credentials (42), Compromised Email (35), Lost Computer or Data (31), Ransomware (20), Social Engineering (16) (Table 1).

5.3 Personal Information Analysis

Throughout the top 25 industries, the two most common types of personal information stolen were social security numbers (31.1%) and payment cards (30.1%) followed by Medical Record (10.8%), Password (8.5%), Bank Account (7.5%), Health Insurance (6.4%), Driver's Licence (5.8%) (Table 2).

Table 1 Frequency of data breaches per attack vector

Attack vector	Frequency	Rate (%)
Unauthorised access	387	27
Software vulnerability	188	13.1
Stolen computer or data	163	11.4
Data found publicly	159	11.1
Wrong data sent	105	7.3
Exposed data	103	7.4
Compromised machine	78	5.4
Phishing email	63	4.4
Insider theft	43	3
Stolen credentials	42	2.9
Compromised email	35	2.4
Lost computer or data	31	2.2
Ransomware	20	1.4
Social engineering	16	1.1

Table 2 Frequency of stolen personal information stolen across the top 25 industries

Personal information Type	Frequency	Rate (%)
Social security number	609	31.1
Payment card	590	30.1
Medical record	212	10.8
Password	167	8.5
Bank account	143	7.3
Health insurance	125	6.4
Driver's licence	114	5.8

6 Discussion of Findings

Financial services industry received the highest number of data breaches following the healthcare sector because hackers target organisations that have what they want, which is usually money. Therefore, industries with data that can be sold for money are the most targeted industries. The fact that financial services organisations are often breached doesn't mean that they aren't as careful about security as their peers. It means that they are being targeted more than other businesses because of their rich financial and data assets, and a small percentage of attacks are successful because of their cybersecurity issues, which leads to an overall higher number of attacks. conducting biannual comprehensive security awareness training that goes beyond the basics to teach employees how to recognise sophisticated threat strategies. Advanced phishing techniques, a broad spectrum of social engineering strategies, indicators of insider threat activity (along with anonymous reporting systems), and physical security should all be addressed throughout training. The training programme should include custom modules addressing the unique characteristics of each group inside the organisation and how they may be targeted.

Every industry has vulnerabilities, and they are all being attacked. Financial services, on the other hand, face the brunt of attacks due to the financial and data assets they oversee. By focusing security efforts on the right areas and properly training employees, the financial services industry can reduce cybersecurity threats.

Most companies are either public or private companies. Therefore, most affected companies by data breaches are public and private companies. However, public companies tend to be an easier target for attackers and attract cybercriminal organisations due to the poor funding, outdated technology and insufficient staff training as it's essential to implement integrated and comprehensive protection that enables a public sector organisation to identify and effectively respond to multiple threat vectors.

The data demonstrates that no business is too large or too small to be a victim of a cybersecurity attack or data breach. In fact, small organisations can affect large organisations because they're always working together as suppliers, business partners or providing services. Attackers can use privilege escalation to escalate the breach from small companies to large companies. Due to the fact that larger firms invest considerably in network security equipment, hackers are constantly looking for any entry points into large organisations networks. Therefore, several cybercriminals target smaller organisations that collaborate with larger corporations in order to escalate the privilege from small to large companies network using information exchange protocols between the two organisations like auto software update or advanced social engineering and fishing using corporate email credentials.

Data breaches can take a wide range of shapes and sizes. Each of these may result in financial loss. All industries are moving to automation with technology. Therefore, new opportunities are created for hackers and fraudsters to benefit from users' sensitive financial information. Due to data breaches the financial industry lost an average of $5 million in 2021. Even though the financial industry is the most valuable and targeted industry. In 2021, the financial industry cybersecurity cost is slightly lower than in 2020.

Since 2019, There's a high demand for cloud solutions, especially in the banking industry and cloud security has become more frequent and resulted in an increase in compromised credentials.

A growing number of softwares are developed and compiled using commercially available or open-source code. Therefore, hackers can freely analyse the open-source code and find 0 day vulnerabilities which make software vulnerability the most frequent attack vector.

Software vulnerability and related attacks were the most common attack vectors, aside from unlawful access. Ransomware and phishing emails are two relatively new attack vectors that have become increasingly frequent since 2016.

7 Limitations

This analysis is limited to the publicly available dataset provided by the office of the attorney general under the notification law, Other publicly available dataset in the state of California are available and they can be used in order to have more accurate and precise analysis which can be used to identify other data breach factors and patterns, However, this research focused only on analysing companies types, attack vectors and stolen personal information. The result of the attack vector analysis can be further discussed in order to create an effective cybersecurity threat detection process based on the most common attack patterns. Moreover, an effective incident response plan can be extracted from the analysis to help enterprises prepare for, detect, and respond to data breaches incidents. Previous years records can also be used to train a machine learning model in order to predict future data breaches frequencies by companies' profile (Industry, Sector, Size, Financial loss), attack types and stolen personal information.

8 Conclusion

Eight industries found to be the most commonly affected by cyber-attacks represent more than 50% of breaches while 25 industries represent 80% of all California data breaches when looking at the company's analysis. Over half of all data breaches were caused by large firms (those with 1,000 or more employees). The healthcare sector suffers the most from data breaches and software vulnerability is the most common attack vector due to third-party software, the financial industry is the most targeted industry while both large and small, public and private organisations are equally targeted by attackers and credit/debit card information and social security numbers represent the most stolen personal information. Breaches with at least 50 million records cost 100 times as much as the ordinary data breach.

References

1. Data breaches reach record levels worldwide (2019) Netw Secur 2019(3), 1–2
2. Dean A, Agyeman MO (2018) A study of the advances in IoT security. In: Proceedings of the 2nd international symposium on computer science and intelligent control
3. Why you should care about the Target data breach (2016) https://www.sciencedirect.com/science/article/pii/S0007681316000033
4. The Cryptographic Implications of the LinkedIn Data Breach (2017) https://arxiv.org/abs/1703.06586
5. FBI 2020 Internet Crime Report, Including COVID-19 Scam Statistics
6. Statistics of IoT connected devices installed base worldwide from 2015 to 2025. https://www.statista.com/statistics/471264/iot-number-of-connected-devices-worldwid-e/

7. September 2020 Healthcare Data Breach Report: 9.7 Million Records Compromised https://www.hipaajournal.com/september-2020-healthcare-data-breach-report-9-7-mil lion-records-compromised/
8. California Civil Code
9. "Data Breach List" State of California Department of Justice. https://oag.ca.gov/privacy/databreach/list
10. Trautman LJ (2016) Corporate directors and officers cybersecurity standard of care: the yahoo data breach. SSRN Electron J
11. Xie X, Lee C, Eling M (2019) Cyber insurance supply and performance: an analysis of the U.S. Cyber insurance market. SSRN Electron J

Profiling Cyber Attackers by Classification Techniques; A Case Study on Russian Hackers

Eghe Ehiorobo, Sina Pournouri, Setareh Jalali Ghazaani,
and Joel Mathew Toms

Abstract Cyber-attacks have become increasingly common in recent years all over the world. In particular, the ongoing unfortunate events between Russia and Ukraine have been aided using technology. The Russians have used various types of attacks for various reasons, such as reconnaissance to gather information about Ukraine's critical systems. It is critical to be able to profile these types of attacks for identification and other strategies to prevent such attacks in the future. Data mining techniques can be used to profile hackers by nation states, law enforcement agencies, and organizations alike. Profiling hackers can assist in better preparing for and responding to cyber-attacks. The goal of this study is to use data mining techniques to profile Russian hackers. Another important aspect of this study is the dataset that will be used, OSINT. The open-source dataset was chosen because it is free, suitable for research, and widely available. Different classification algorithms will be used to train the proposed predictors, and the most reliable and accurate model will be chosen. This study emphasizes the importance of better understanding well known Russian hacker groups, which could lead to better preparation and handling of cyber incidents involving such threat actors. Nation-state governments, researchers, and law enforcement agencies can use the developed model in conjunction with other techniques to quickly identify attacks from Russia APT groups, recommend controls, and develop a detailed strategy for countermeasures against such attacks, which will improve data security and privacy.

Keywords Profiling · Data mining · Classification · Russian hackers · Cyber-attacks · Open-source intelligence

E. Ehiorobo · S. Pournouri (✉) · S. J. Ghazaani · J. M. Toms
Sheffield Hallam University, Sheffield, UK
e-mail: S.Pournouri@shu.ac.uk

© The Author(s), under exclusive license to Springer Nature Switzerland AG 2023
H. Jahankhani (ed.), *Cybersecurity in the Age of Smart Societies*,
Advanced Sciences and Technologies for Security Applications,
https://doi.org/10.1007/978-3-031-20160-8_11

171

1   Introduction

Cyber security is concerned with the safety of technology on computer systems. It is a computer branch regulating security measures to prevent risk and threats from causing any kind of harm to the system. There are third-party intruders or malicious hackers who steal private, national, or organizational information for their own benefit/gain. Knowing who the intruders are and their motivations for carrying out such activities will aid individuals, organizations, and nation states in mitigating the risk to their information. Profiling and understanding intruders/cybercriminals can significantly reduce the risks involved in the security and safety of both nation states, individuals, and organizations. Understanding the characteristics and behaviors of such criminals' aids in better resolving/analyzing the cybercrime committed. As a result, similar cybercrime does not occur in the future [14].

The necessity of being able to detect and stop Russian hackers has been highlighted by recent incidents between Russia and Ukraine, with cyber-attacks at the core. The goal for a successful Occupation was to increase data collection immediately before the Russian invasion of Ukraine by Russian hackers targeting Ukrainians. Although data collection may not appear to cause much harm at first, it has long-term implications that are often overlooked. The Ministry of Internal Affairs was breached on the eve of the February invasion, and a month earlier to the breach, a national database of motor insurance plans was accessed, which acted as a diversion for the defacement of Ukrainian websites. These breaches, combined with active data collection prior to the war, provided Russia with detailed information on the bulk of Ukraine's population. According to cybersecurity specialists and military intelligence analysts, the information gathered was used to identify and locate Ukrainians who were likely to oppose an occupation and may be targeted for imprisonment or worse [1].

Since October 2021, Russian hackers have attempted to hack into Ukrainian government, military, judiciary, and police agencies, as well as organizations critical to emergency response and safeguarding the security of Ukrainian territory, as well as non-profits, to steal sensitive data. Armageddon, a Russian FSB intelligence unit, has been gathering data in Crimea for years. According to Ukraine, the first goal was to infect over 1.500 Ukrainian government computers to damage the information systems of critical infrastructure and government organizations. Personal Identifiable Information (PII) remains a goal for Russian hackers as they strive to access more government networks, according to the deputy secretary of the Ukrainian Security and Defense Council [1]

In yet another strike on April 1st, 2022, Russia's cyber army damaged Ukraine's national phone center, which handles complaints about domestic abuse, corruption, war vets' benefits, and Ukrainians negatively touched by the invasion. Thousands of people utilize this service, which generates COVID certificates and collects caller information such as email addresses, phone numbers, and addresses. This breach, like others, has a psychological impact rather than an intelligence one, with the goal of eroding Ukrainians' belief in their institutions. The call center was taken offline for three days, preventing employees from working, phones and chatbots from working, and the entire system failing. The hackers claimed to have stolen the personal information of 7 million people, but the center director denied that the database of customers' personal information had been breached but confirmed that the contact list of more than 300 of their employees that the hackers had posted online was legitimate [1].

On March 21st, Ukraine's SBU Intelligence agency announced that it had captured a "bot farm" in Dnipropetrovsk that was operated remotely from Russia and that was solely targeting 5,000 Ukrainian police, military, and SBU members through text, threatening them to surrender or damage their units. While inquiries into how the numbers were obtained continued, the CEO of cybersecurity firm ReSecurity believes it was likely not difficult because large Ukrainian wireless companies' customer databases have been for sale on the dark web for some time [1]. By doing so, Russian special services attempted to conduct a special intelligence gathering operation to destabilize the psychological and morale of Ukrainian security forces. The bot farm, a medium through which 5,000 messages were sent, was easily discovered, contained, and eradicated by cyber experts from Ukraine's Security Service (SSU) [19].

Sandworm, a threat actor linked to Russia's state security agencies, was recently accused by ESET of using the updated version of "Industroyer" to create power outages in Ukraine's high-voltage electrical substations. On April 8th, 2022, the malware was set to execute. The "Industroyer" malware was allegedly used by the Sandworm APT organization to disrupt power in Kiev, Ukraine, in 2016 [9].

Security experts from ESET and CERT-UA, who successfully handled the attack on an undisclosed vital infrastructure network, have announced that they are continuing their investigation. There is currently no information available about how the attackers acquired initial access to the victim's IT network before moving on to the industrial control system network. The attackers were likely aiming to give Industroyer2 control over certain ICS systems at this point in the attack campaign so that electricity might be shut off. According to Ukraine's State Service of Special Communication and Information Protection (SSSCIP), if the attack was successful, a huge chunk of the country's territory would have been blacked out, "leaving a large number of residents without energy" [9].

Another case study of Russian hackers using third-party supplier SolarWinds to obtain access to numerous key US Federal agencies reveals how dangerous it may be to gain access to networks without being detected early. After some Trump administration officials confirmed that additional government agencies, including the State Department, the Department of Homeland Security, and elements of the Pentagon, had been hacked, the scale of a hacking organized by one of Russia's top intelligence organizations became obvious. Investigators were attempting to determine the extent to which the extremely sophisticated attack had affected the military, intelligence community, and nuclear laboratories. The attack was only recently discovered by US officials, after a private cybersecurity firm, FireEye, warned American intelligence that the hackers had gotten past multiple layers of safeguards [16].

Profiling is the collection, storage, and use of information about an individual or group of individuals for a variety of purposes, such as monitoring their online behavior. Cyber Profiling is one of the investigator's efforts to learn more about the alleged offenders through data pattern analysis that incorporates aspects of technology, investigation, psychology, and sociology [18].

OSINT has a significant impact on this study, not only because the dataset that will be used is derived from OSINT, but also because it is necessary when conducting due diligence. It is also significant because this research must be cost and time effective, and OSINT is appropriate in this case. It is also important to note that obtaining a large amount of data from private sources or organizations will be difficult; while this may be advantageous, it is unrealistic. Finally, while OSINT data can be incomplete or inaccurate, it is a reliable source.

There is currently available literature on profiling and data mining techniques, but none on profiling Russian hackers using a data mining methodology, which is the challenge addressed by this study. Profiling Russian hackers could also play a key role in the cyberwar between Ukraine and NATO member states and Russia by assisting in the detection and mitigation of Russian threat actors' cyber intrusions. Finally, it may be useful to security researchers for future research on identification and neutralization of cyber threats.

1.1 AIMS and Objectives

The purpose of this paper is to implement an optimum data mining technique for efficiently profiling and predicting Russian cyber-attacks.

In order to achieve the goal of this study, following steps will be taken:

1. Gather and compile a dataset of reported cyber incidents from 2016 to 2022 using OSINT.
2. Training profiling models based on classification techniques.
3. Investigation of the trained models in order to identify the most efficient profiling method.

2 Literature Review

The purpose of this section is to discuss the current literature on data mining, data mining techniques such as classification and clustering, how they work and how they can be applied, and how data mining techniques can be applied to profiling. Finally, this chapter will review current literature in this subject matter.

2.1 Data Mining

Data mining is the process of gaining new knowledge from a data warehouse. A data warehouse is a centralized database in which data is stored in a single large database. Data mining is a technique used by a user or organization to extract useful information from raw datasets. Certain software is used to find specific patterns in massive datasets (data warehouses), which can help organizations learn about their customers, improve business strategies, and predict behaviors. By categorizing this knowledge into different rules and patterns, a user or organization can analyze gathered data and predict decision processes (Mughal, 2018).

2.2 Data Mining Techniques

The process of determining patterns, data, and trends to gain valuable information from massive data sets for decision making and judgement is known as data mining techniques. There are several data mining techniques being developed and used in various data mining projects, such as clustering and classification, with each technique having its own set of rules and methods that determine the type of problem it solves [11].

2.3 Classification in Data Mining

The process of categorizing a set of data into different classes or groups to produce accurate prediction and analysis from massive datasets is known as classification technique. By identifying different attributes to determine a specific class, classification can be used to develop an idea of the category of customer, item, or object in a dataset. For example, by highlighting different attributes such as height, unit, or structure, one can categorize buildings into different categories based on the type of construction or occupancy. By comparing the established attributes in the database, a new building could be applied to a specific category or class. These principles can also be applied to customers, who can be classified by gender, age, and social group. Furthermore, classification can be used as a feeder to other techniques' results, such as decision tree to determine a classification or clustering to determine clusters by utilizing similar attributes in different classifications [11].

2.4 Current Literature on Profiling and Data Mining Techniques

Reference [4] conducted a study that examined patterns of cyberattacks and hacker personas in South Korean cybercrime incidents using data from 83 cybercrime findings in Korean court cases and news releases published between January 2010 and January 2019. The researcher thoroughly examined each reported case in the sample using the FBI's criminal profiling framework, which includes inductive and deductive methods, and then methodically coded variables such as date of incident, demographic features of the hacker, intentions, presence of an accessory to the crime, attack tools, and whether the attack occurred only in South Korea or globally. Youth hackers were those aged 15–24, whereas adult hackers were those aged 25 and up. The current investigation yields some significant findings that reveal hacker profiles and hacking patterns. According to geographical measures, Korean hackers were more frequently responsible for intra-national level hackings than international level hackings. Furthermore, monetary gain was the most powerful motivator for both youth (55%) and adult hackers to hack (87%). In fact, more than 85% of adult hackers were motivated primarily by monetary gain. Meanwhile, entertainment, hacktivism, vengeance, and exposure were motivating young hackers. Because cybercriminals usually have accomplices, if this profiling technique leads to the arrest of cybercriminals, it may also lead to the arrest of the accomplices, preventing a further crisis such as a cyberwar or cyber terrorism. This has an indirect impact on policy implementation, leading to new policies that could mitigate or combat cyber-attacks in the future. In summary, the findings of this study show that adult hackers' motivations are more monetary than youth hackers' motivations [4]. This paper, like any other study, has strengths and weaknesses, which include: The study has policy implications for South Korean officials fighting cybercrime. The research was both cost and time effective. However, because this data only covers 83 cases in South Korea during a specific period, the results' generalizability cannot be determined and because this study only looked at reported incidents, it cannot provide a more accurate result. The study is pertinent to this paper because it focuses on hacker profiling.

Furthermore, [7] focused solely on detecting insider threats in network data streams by employing a dynamic model based on Markov Models. The author created a framework for identifying malicious insider threats. The model is based on an analysis of each user's behavior. Each employee's action sequence was modelled using the Markov chain. The CERT Division at Carnegie Mellon University has assembled a collection of synthetic insider threat datasets that include both synthetic background data and data from synthetic malicious actors. The experiments were conducted on the CMU-CERT dataset, which contains information about malicious activities such as scenarios enacted as well as the identifiers of the synthetic users involved. The authors selected some distinguishing features based on login, log-off, device usage, online activities such as website visits, downloading, and uploading, and activity timestamps. They used a dual scoring method to evaluate each model against a pre-defined request for a specific threat scenario. Insider threat detection experiments on several datasets reveal every instance of an insider threat. Finally, potential insiders who were extremely certain of the threat were identified.

Reference [7] formulated the corresponding sequence as a request and interrogated the user models by calculating the likelihood of the new sequence, which is defined as a pair of positive and associated negative sequences to detect a threat. The desire to focus on a unique feature by inverting the feature value while leaving the others unchanged motivates the use of the positive/negative sequence. They developed a model for each user based on the data stream. Because each row of data has an exact matching cluster, the transition probabilities deal with the temporal relationships between clusters. They investigated the problem of determining users from the CERT database who may have engaged in internal malicious behavior as a means of experimentation. Each data subset in the Malicious scenarios contains only a few insider threat incidents or scenarios. Each one is generated as synthetic data with the same form and scope as the regular background data generated for all users in the dataset. This paper, like any other, has strengths and weaknesses, including but not limited to: Potential insider threats were discovered with confidence. as well as online and real-time data storage. The scope of this study, on the other hand, is restricted to user activities. Another intriguing area of knowledge is the psychometric feature, which can be effectively integrated into the proposed framework to integrate the user's personality traits. Because it focuses on cyber profiling utilizing a data mining technique, the paper is pertinent to this study.

This study [10] used topic modelling to categorize carding forum member users from three carding forum websites based on comment histories. The unique username was used to download and aggregate user posts and comments. Topic modelling is used to group and classify documents into non-exclusive groups. Topic modelling is a text mining procedure that can analyze the raw key word frequencies in the data being processed. Stop words were removed from the data set before converting it to this matrix. These include common function words (but, and, for, and so on) that, while common across all groups, are unlikely to distinguish differences among users. Words are stemmed as well, which reduces them to their root. This procedure removes suffix characters from the end of words with multiple morphologies, resulting in words like "offending" or "offended" becoming "offend." Punctuation was also removed from the user comments training corpus. Finally, any hypertext markup language (HTML) in user comments was removed. LDA was the specific method employed. LDA uses soft clustering, which means that the cluster categories are not mutually exclusive. That is, a user can belong to more than one category. Each user is assigned to one of the k categories in such a way that the probabilities sum to 1.0 for each user. LDA is known as latent Dirichlet allocation because it estimates latent constructs (the topics) while assuming that the category probabilities follow a Dirichlet distribution, which is a distribution over distributions (documents distributed over topics, with topics distributed over words). Custom Python web crawlers were written to extract data from each of the three forum websites. Each crawler was designed to begin at the main forum portal, which listed all boards/sections. The crawler extracted all board links, saving them to a list and then accessing them in order. Each board had an index that listed all the pages of posts that users had made. The model results revealed 21 user categories. After programmatically analyzing each of the 200,105 comments, the 21 labels described 30,469 users across three carding websites. Like any other paper, this study has its advantages which include: The result is not only a better understanding of the types of dialogue found on such forums, but also a data set with a rich set of 21 non-mutually exclusive binary variables describing the large sample. Because most of the research done on carding forums was qualitative, the goal of this research was met because the results were quantitative. However, While the qualitative description of the topics allows for its own analysis, the approach was mostly exploratory and investigative, and further quantitative analysis of the topic data is possible. The difference in size of the three different websites will have skewed findings more toward the larger BitsHacking, which had 23,440 users. There is also the possibility that some users will visit multiple websites in the sample and register under different usernames, resulting in duplication. Finally, the study's approach to the data was descriptive rather than hypothesis-driven or inferential. These are some of the limitations of this study [10]. The study is relevant to this paper because it implements a data mining approach to its cyber profiling process.

Reference [13] investigates the effectiveness of some classification techniques in profiling and predicting cyber attackers and potential future cyber-attacks. OSINT and historical data on cyber-attacks are being used in the investigation. Because of its low cost and ease of use, OSINT data was chosen for this study. The information includes reported cyber-attacks in which the hackers were identified or claimed responsibility for the attack. To build the classifiers, a training set and a test set for validation were created. The training set includes 1432 attacks from 2013 to the end of 2015 that were linked to previously established hackers, while the test set includes 484 attacks from 2016 to the end of the first quarter of 2017. The dataset was restructured and classified into five different columns before being preprocessed by a Google tool called open refine. Open refine includes several data cleansing and transformation features. The preprocessing method for open refine was divided into two self-explanatory steps:

I.   Removing records that are repetitive or redundant.
II. Integration and categorization.

Decision trees, K nearest neighbor, Naive bayes, Support vector machine, and artificial neural network were investigated for their effectiveness in predicting cyber breaches. These classification algorithms were used by dividing the training data into ten equal random samples. One subsample is used as the test sample for validation, while the others are part of the training set. This process is repeated ten times, and the average result provides the estimation. This procedure is commonly known as tenfold cross validation. In terms of accuracy, Support vector machine was the best with a rate of 61.34%, followed by K nearest neighbor with a rate of 61.27%. C4 5, Random Forest, and recursive partitioning are three decision tree algorithms that were used. C4 5 and Recursive partitioning came in third and fourth, with 61.25% and 60.48%, respectively. With 59.71% and 59.51%, respectively, Artificial Neural Network and Random Forest are the next highest on the accuracy table. The algorithm with the lowest accuracy, 58.17%, is Naive Bayes. Given that the Support vector machine has proven to be the most reliable model, this study evaluated the results by applying the model to the test dataset, which is 484 records of cyber breaches from 2016 to the end of the first quarter of 2017 against one of the limitations of this study, which is the identity of the hackers, because they may change identities or simply stop performing cyber-attacks. The following parameters were used to assess the model's success: true positive (recall), false positive (FP), precision, and ROC area. The ROC area is a two-dimensional graph with X and Y axes representing the false positive and true positive rates, respectively. This model's overall accuracy is 25.92%, which is not reliable enough for cyber experts as a single predictive method, but by comparing various benchmarks, the following points can be highlighted:

1. The highest TP rate, 0.627, belongs to the Anonymous hacker group, demonstrating that the model, despite its insufficient accuracy level, has made reliable predictions about this cyber attacker group, which can be concluded with a contribution of significant level precision, 0.824. The Chinese hackers rank second in terms of TP rate, with a level of 0.538, demonstrating the predictive model's accuracy in detecting these cyber attackers. DERP hacker group has the highest FP rate.
2. The anonymous group has the highest ROC with 0.897. Chinese hackers rank second in terms of ROC area, with 0.84, demonstrating that the model is accurate in predicting them.
3. This step shows that despite unreliable overall accuracy, precision, and ROC values of 0.26, 0.324, and 0.684, Anonymous and Chinese hackers are the most well predicted among other cyber attackers.

This study has its benefits and drawbacks. Data mining and predictive analytic techniques are used in everyday life, including cyber security. Predictive analytic techniques can assist law enforcement agencies in identifying and apprehending criminals before they cause further damage. On the other hand, As our proposed model based on SVM classifier using OSINT indicated, the accuracy for current and past prediction of cyber attackers is higher than prediction of future by nearly 35%, which can be attributed to changing identities and motivations of cyber criminals over time; however, this method can be used in conjunction with other techniques to make more reliable and accurate prediction [13].

The study by [17] used latent profile analysis to identify three teacher profiles with consistent readiness (high or low) or inconsistent readiness, considering three key variables: the teachers' gender and Online Teaching and Learning (OTL) experience, the context of OTL shift and Innovation potential in education, and cultural orientation. The study began by discussing several theoretical frameworks, including conceptualizing teachers' readiness for OTL in higher education, online teaching presence, Determinants of readiness for OTL in higher education, gender differences, culture and innovation, and the current study. Reference [17] conducted a survey in primary, secondary, tertiary, adult, and vocational education between March and May 2020 to assess teachers' readiness for online teaching around the world during the COVID-19 pandemic. An anonymous survey inviting teachers to participate was created and distributed through various channels such as high schools, universities, and social media. Teachers' TPACK self-efficacy, perceptions of the online presence they create, and perceptions of institutional support were used to measure the core construct of teachers' readiness for OTL in higher education. These indicators were derived from the survey questionnaire and served as the foundation for profile identification. To better understand the nature of these profiles, [17] assessed additional teacher level predictors, such as variables characterizing teachers' background, academic discipline, and context of the shift to OTL. In addition, two country-level predictors representing cultural orientation and individual innovation potential in education were extracted. This study used a methodical approach with three steps. In the first step, the authors used confirmatory factor analyses to assess the measurement models describing teachers' TPACK self-efficacy, perceived online presence, and perceived institutional support. This step was critical for assessing the psychometric quality of these scales, establishing several factors representing the respective constructs, and extracting factor scores that would later serve as indicators in subsequent latent profile analyses. In a subsequent step, [17] used factor scores of TPACK self-efficacy (gTPACK), the three dimensions of perceived online presence (POPCLA, POPFED, POPCOG), perceived institutional support (gPIS), and the two grand-mean centered items PISCO1 and PISCO2 as profile indicators to identify the (latent) profiles of higher education teachers' readiness for OTL. In the final stage, they transformed the optimal LPA model with teacher-level explanatory variables into a multilevel LPA model in two stages: To begin, they permitted the intercepts of the latent categorical variables indicating profile membership for each teacher to have intercepts and variances at the country level. Finally, they included country-level variables to explain this variation and, ultimately, to investigate the relationship between average profile membership probabilities, innovation potential, and cultural orientation [17]. Overall, the findings suggest that higher education teachers are not a homogeneous group in terms of their reported readiness for OTL. However, different subgroups of teachers may necessitate different approaches to support. Identifying such profiles is critical for bringing to light the diversity of teachers and, ultimately, facilitating tailored support for OTL implementation. To summarize, teachers' readiness for OTL is dependent on institutional, cultural, and innovation contexts in addition to their own self-efficacy and teaching presence [17].

This paper has benefits and drawbacks. Personal and contextual readiness did not always go hand in hand in the profiles identified in this study. Furthermore, this research is an important step in determining how to best support teachers as they transition to online learning. However, rather than a random and stratified sample, a convenience sample of university teachers was used. In addition, instead of the three core components of TPACK self-efficacy, perceived institutional support, and perceived online presence. More factors could have been considered in this study to represent teachers' readiness for OTL. Finally, this study only assessed teachers' readiness as well as their individual and contextual characteristics once. This research is relevant to this study because it focuses on profiling.

The table below compares the current literature on this study.

Table 1 compares the five papers discussed in this section, including the paper's title, author, and publication date. It also shows where the dataset used in each paper came from and discusses the various strengths and limitations of each paper.

To summarize, this section discusses data mining, data mining techniques such as clustering and classification, as well as defining and discussing how it works. This section also discusses current literatures on the subject, with a table comparing the six papers discussed in this section. The next section will go over the tools, data collection and analysis, which includes data pre-processing and processing.

3   Methodology

The purpose of this section is to go over the tools, and techniques, data collection methods, data categorization, and data analytics for this study. The first goal will be addressed based on the aims and objectives, which is to gather and compile a dataset of reported cyber incidents using OSINT.

3.1   Tools

3.1.1   Open Refine

Open refine has several data cleansing and transformation functions. This web application tool includes features such as support for various dataset formats [12].

There are other tools that could be used for data cleansing and transformation, such as Cloudingo and Mitto, but none of them have a free version, making this tool the most cost effective.

3.1.2   Weka

This is the tool being used in this study to verify the accuracy of models, as well as in the evaluation stage of this research. Weka is free and open source, and it can be run on any computer running any operating system [12].

Table 1 Comparison of current literatures in this topic

Name, author and date	Dataset source	Strengths	Limitations
Youth hackers and adult hackers in South Korea: an application of cybercriminal profiling [4]	Published court cases and incident reports	I. The study has policy implications for South Korean officials fighting cybercrime II. Because they used an already developed criminal profiling framework, the research was both cost and time effective	I. Because this data only covers 83 cases in South Korean during a specific period, the results' generalizability cannot be determined II. Because this study only looked at reported incidents, it cannot provide a more accurate outcome
Combating insider threats by user profiling from activity logging data [7]	CMU-CERT dataset	I. With high confidence, potential insider threats were discovered II. The method could handle online and real-time data storage	I. This approach focuses solely on log data, and as such, it may not provide an accurate picture of the threat scope because other contents, such as email, may play a significant role
Profiling Cybercriminals: Topic model clustering of carding forum member comment histories [10]	Three online carding forums	I. The topic model analysis produced a large data set of 21 types of cybercrime users II. The findings highlighted many topics that were consistent with previous research, as well as some aspects of cybercriminal roles that were novel in comparison to previous research	I. The differences in size between the three websites will have skewed results more toward the larger forum, BitsHacking II. There is also the issue of cross-site dependency. Some users in the sample may visit multiple websites and register under different usernames III. The selection criteria for including forums in the sample could be considered narrow, capturing only a small portion of the Internet's cybercrime communities IV. The study's approach to the data was descriptive rather than hypothesis-driven or inferential

(continued)

Table 1 (continued)

Name, author and date	Dataset source	Strengths	Limitations
Predicting the cyber attackers: a comparison of different classification techniques [13]	OSINT	I. The techniques used can aid cyber experts in dealing with future threats II. The techniques used may assist law enforcement agencies in identifying and apprehending criminals before they cause further damage III. Using OSINT data is inexpensive and simple	Using OSINT data can be noisy and inadequate, resulting in lower accuracy More cyber-attack data could have been obtained, and more attempts could have been made to add more variables to the data collected, because cyber-attacks are influenced by more factors than those mentioned in this study
Profiling teachers' readiness for online teaching and learning in higher education: who's ready? [17]	Anonymized online survey	I. The profiles identified in this study demonstrated consistently high and low levels of readiness, highlighting that personal and contextual readiness do not always go hand in hand II. This study is an important step toward determining how to best support teachers as they transition to online learning	I. The convenience sample of university teachers was used rather than a randomly drawn and stratified sample II. Instead of the three core components: TPACK self-efficacy, perceived institutional support, and perceived online presence. This study could have considered more to represent teachers' readiness for OTL III. This study only measured teachers' readiness, as well as their individual and contextual characteristics, on one occasion IV. Regardless of the participants' first language, all items in this study's survey were administered in English

Orange is the best alternative to Weka. Weka is being used for this study because it is a more powerful tool than orange for running Classification and Clustering algorithms.

3.2 Data Collection

The primary data source is Open Source Intelligence (OSINT). Data was gathered from Hackmageddon, a blog that tracks cyber-attack incidents from various sources in the form of Timelines from 2011 to the present. As a result of the owner, Paolo Passeri, granting access to a Google Drive containing reported cyber-attacks from 2011 to present, Hackmageddon was chosen as the initial data source. A large excel sheet containing cyber-attacks incidents that occurred between 2016 and April 2022 was created as the most recent cyber breaches for training and predictive models, totaling 10,860 records. There are several other blogs, websites, and Twitter accounts that report on cyber breaches and incidents, such as hackernews.com, hackread.com, Darkread.com, and so on, but getting access to Hackmageddon was easier and faster than using the other resources mentioned.

Furthermore, because access to data gathered by official agencies and the government is impossible, the primary data source used in this research is OSINT. There are various companies that collect information about cyber breaches, but they are only for commercial purposes, and it would not be cost effective for this study. Using OSINT allows researchers to gain access to information without any ethical concerns and at a lower cost; however, gathering OSINT data comes at a cost of incompleteness, noisy data, and inaccuracy.

3.3 Data Categorization

The initial data set must be examined in greater depth at this stage. The data structure must first be described. According to the data collected, each cyber-attack incident has ten distinct characteristics:

1. Date: The date of the reported cyber-attack.
2. Author: The hacker or threat group who carried out the attack or accepted responsibility.
3. Target: The government agency, organization, or individual who was targeted.
4. Description: This section discusses how the attack occurred.
5. Attack: This refers to a type of attack such as account hijacking, DDoS, malware, and so on.
6. Target class: This is the target's class or category, such as Finance and insurance, Education, and so on.

7. Attack class: This is the category or class in which the attack falls, such as CC (Cybercrime), CE (Cyber Espionage), CW (Cyber War), and so on.
8. Country: The target country, such as RU (Russia), CH (China), US (United States), and so on.
9. Link: This is the link to the reported cyber-attack.
10. Tags: These are common names or acronyms for the reported cyber-attack.

Because the goal of this project is to train models based on a classification algorithm and not a Time series analysis, the Date column will be removed. Furthermore, the Description, link, and tags columns will be removed to eliminate irrelevant and redundant data that could interfere with profiling. The attack, target class, attack class, and country columns, on the other hand, will be retained.

After restructuring the dataset and organizing the attributes into five distinct columns, the preparation and pre-processing of the values will begin. There are various data cleansing tools available, such as Orange, Excel, and others; however, as previously stated, Open refine will be used as a data cleansing tool during the preprocessing stage. The data will be uploaded to Open refine, and the preprocessing method will be divided into the following steps:

1. This phase was carried out using the cluster feature, which divided the country column into multiple countries for attacks that targeted multiple countries and the acronym of the actual country for attacks that targeted only one country, such as CH = China, US = United States, and so on.
2. The attack class was divided into four categories: CC (Cyber Crime), CE (Cyber Espionage), CW (Cyber War), and HA (Hacktivism). As there were only a few classes to group into four of the major classes identified above, this was done manually and easily using the text facet functionality.
3. Using the text facet's cluster functionality, the dataset was automatically grouped into several major sectors such as Healthcare, Education, and Government, and then manually grouped into the appropriate class.
4. Because it was difficult to automatically classify attacks, the Attack column had to be completed manually. This process entailed categorizing the attacks into various attack types, such as DDoS, phishing, and hijacking.
5. This phase was difficult and time consuming because the dataset had to be categorized in three different spreadsheets. The primary dataset, which includes all other threat actors' asides the unknown. The second data set is the validation dataset, which only includes attacks involving unknown threat actors. The final data, which is the training dataset, involved going through each attack and looking for any links to Russia or known Russian hackers, then deleting the rest of the attacks that did not have any Russian links. The result was 683 distinct authors or threat actors.

After the Preprocessing stage, the final results of the Datasets are. The primary dataset contains 801 different threat actors, 13 different types of attack, 17 different target categories, 5 different attack classes, and 109 different countries, including the Multiple countries category.

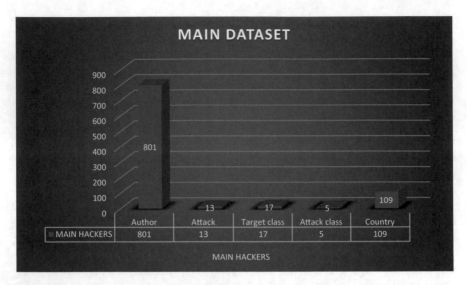

Fig. 1 Visual representation of the main dataset

Malware had the most instances in the attack category (977), followed by Targeted attack (788), Reconnaissance (224), Hijacking (194), and DDoS (119). Blockchain had the fewest instances, with one.

Similarly, the most common motive in the attack class was cybercrime, with 1583 instances, followed by cyberespionage (872 instances), and hacktivism (196 instances).

This dataset is visually represented by the bar chart below (Fig. 1).

The validation dataset contains only one threat actor category called Unknown, which includes 5675 unknown threat actors, 13 different types of attacks, 17 different target categories, 4 different attack classes, and 109 different countries, including the Multiple countries category.

In the attack category, Malware had the most instances in the attack category, with 2708, followed by Hijacking with 1477, Reconnaissance with 367, Targeted attack with 352, and DDoS with 223. The Blockchain and Persistence attacks had the fewest instances, with 11 and 37, respectively.

Similarly, cybercrime was the most common motive in the attack class, with 5337 instances, followed by cyber espionage (292 instances) and hacktivism (45 instances).

This dataset is visually represented by the bar chart below (Fig. 2).

The training dataset, which only contains attacks by Russian hackers, contains 63 different Russian threat actors, 10 different types of attacks, 16 different target categories, 4 different attack classes, and 56 different countries, including the Multiple countries category.

For the attack category, Malware had the most instances (381), followed by Targeted attack (201), Reconnaissance (51), Hijacking (16), and Persistence attack

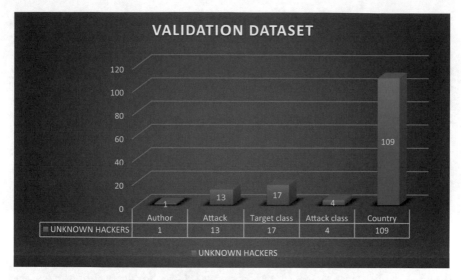

Fig. 2 Visual representation of the validation dataset

(13). Injection attacks and misconfiguration had the fewest instances, with only one and two, respectively.

Similarly, the most common motive in the attack class was cybercrime (458 cases), followed by cyberespionage (189 cases), and cyberwar (30 cases).

This dataset is visually represented by the bar chart below (Fig. 3).

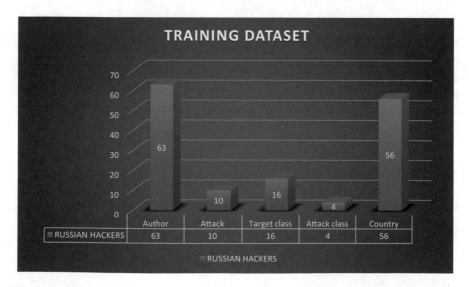

Fig. 3 Visual representation of the training dataset

It should be noted that there is some disparity between the three datasets, which is to be expected because some categories were removed during the preprocessing stage, and what is applicable for one dataset may not be applicable for others. For example, the attack class in the main dataset has five categories: Cybercrime, Cyber espionage, Cyber war, Hacktivism, and 1 > which means a combination of two or more attack classes, whereas the validation and training dataset has only four attack classes: Cybercrime, Cyber espionage, Cyber war, and Hacktivism because these are relevant for profiling.

In conclusion, the most common times of attacks in the datasets are Malware, Hijacking, Reconnaissance, and Targeted attack, while the most common motives of cyber-attacks are cybercrime, cyber espionage, cyber war, and hacktivism.

3.4   Data Analysis Method

Based on the literature review, classification was chosen for the profiling of Russian hackers. For profiling, five main classification techniques will be used: Decision Tree, K nearest neighbour, Nave Bayes, Support Vector Machine, and Multilayer Perceptron. Classification was chosen over clustering because this training must be supervised rather than unsupervised. This means that the dataset must be labelled, and training data (Russian hackers) must be provided for validation. When empirically derived training data is provided, supervised machine learning (ML) can be used to build a predictive and accurate model in datasets that cannot be appropriately modelled with a reasonable set of variables [3].

3.4.1   Decision Tree

As the name implies, decision trees are a tree-like structure in which an internal node represents attributes, branches represent outcomes, and leaves represent class labels. One of the most popular prediction and machine learning algorithms is the decision tree [2].

3.4.2   K Nearest Neighbor

In principle, K Nearest Neighbor (K-NN) works with the closest training samples, those data points that are similar to each other and belong to the same class, also known as instance based learning [20].

3.4.3 Naïve Bayes

Naive Bayes is a popular machine learning algorithm that is a type of Bayesian network. This approach assumes that each feature's value is independent and takes the relationship between the features into account. It includes two types of probabilities: conditional probabilities and class probabilities [2].

3.4.4 Support Vector Machine

Support vector machine is a supervised learning algorithm that is commonly used to solve classification problems. It can be used to extract information, diagnose medical problems, assess credit risk, and categorize text [20].

3.4.5 Multilayer Perceptron

A multilayer perceptron (MLP) is a type of fully connected, feed-forward artificial neural network (ANN) that is made up of neurons arranged in layers. When it comes to making decisions, ANN functions similarly to the human brain. It is based on the principles of evolution and learning [6] [20].

4 Findings

In this section, models will be trained using the training set and classification techniques. Training control is a critical component that must be highlighted during the training process. According to [12], train control is a type of model accuracy validation. Because it is more accurate and efficient, K fold cross validation will be used for this purpose. This is due to the fact that during the training process for k fold cross validation, the dataset will be divided into k equal size sets, 1 set will act as a training set, and k-1 set will be considered as training set, and this set will be repeated k times [12].

Because of the repetitive cycle, this approach yields a more solid and accurate result. For the purposes of this study, k is set to 10 and the training process will repeat 10 times. In the first fold of tenfold cross validation, the first subset is considered the validation set, while the remaining nine subsets are considered the training set. Furthermore, for the second fold, the second subset is the validation set, while the other subsets are the training set, and so on. For example, cross validation is the average of ten accuracies achieved on the validation sets [5].

4.1   Decision Tree Analysis

Following categorization and preprocessing of the initial data, the dataset can now be analyzed using decision tree methods, as discussed in detail in the previous section. The decision tree algorithm is classified into three types: C4.5, Random Forest, and Recursive partitioning [8]. For the purposes of this study, the resulting dataset will be analyzed by these algorithms, and the best algorithm will be chosen. C4.5, Random Forest, and Recursive partitioning will be used to implement the three decision tree algorithms.

C4.5 is a decision tree algorithm developed by Ros Quinlan in 1993 that is an extension of the ID3 algorithm. It is based on information entropy [13].

The C4.5 algorithm will now be used to train the author subset in the training dataset for our model. Cross-validation is set to tenfold for the training control. In Weka, the C4.5 algorithm is now referred to as the J48 under the trees category.

According to Fig. 4, 32.6501% of the 683 instances of Russian hackers were correctly classified (223) while 67.3499% were incorrectly classified (460), which is a rather poor statistic.

Random Forest is an important decision tree algorithm that was developed in 2007 by Breiman and Cutler. To build decision trees, this algorithm employs a process

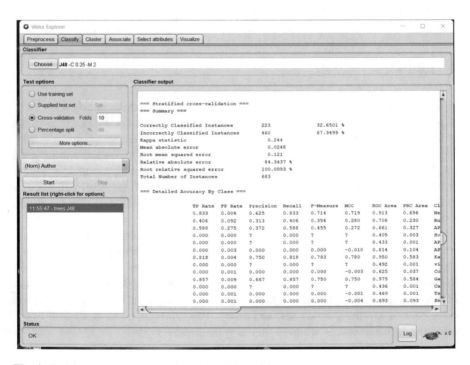

Fig. 4   Training process of authors based on c4.5 model

known as bagging. The process of combining learning trees to improve classification accuracy is known as bagging [13].

The Random Forest algorithm will now be used to train the author subset in the training dataset for our model. For the training control, cross-validation is set to tenfold. In Weka, the Random Forest algorithm is known as Random Forest in the trees category.

Figure 5 shows that 30.4539% of the 683 instances of Russian hackers were correctly classified (208) while 69.5461% were incorrectly classified (475), which is worse than the C4.5 algorithm.

Another decision tree algorithm that is based on greedy algorithms to classify based on independent variables is recursive partitioning. It is best suited to categorical variables, as opposed to continuous variables, where it performs poorly [13].

The author subset in our training dataset will now be trained using the Recursive partitioning algorithm. For the training control, cross-validation is set to 10 folds. The recursive partitioning algorithm is now known as "PART" in the rules category of Weka.

According to Fig. 6, 31.6252% of the 683 instances of Russian hackers were correctly classified (216), while 68.3748% were incorrectly classified (467), which is relatively low when compared to the other algorithms.

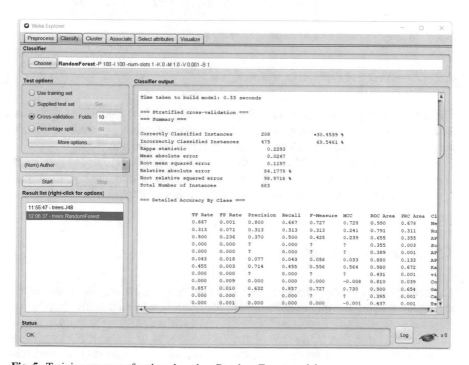

Fig. 5 Training process of authors based on Random Forest model

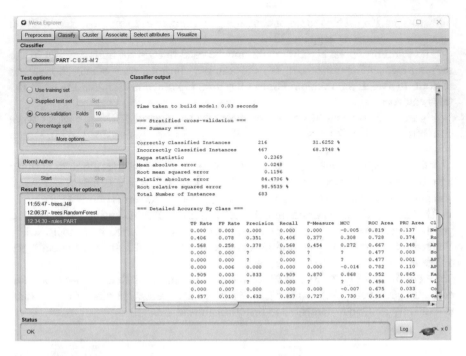

Fig. 6 Training process of authors based on recursive partitioning model

4.2 K—Nearest Neighbour Analysis

The K nearest neighbour algorithm will now be used to train the author subset in our training dataset. Cross-validation is set to 10 folds for the training control. In Weka's lazy category, the k nearest neighbour algorithm is now known as "IBk."

According to Fig. 7, 34.1142% of the 683 instances of Russian hackers were correctly classified (233), while 65.8858% were incorrectly classified (450), which is slightly better than all the algorithms used in Decision tree analysis.

4.3 Naïve Bayes Analysis

The author subset in our training dataset will now be trained using the Naive Bayes algorithm. For the training control, cross-validation is set to 10 folds. The Naive Bayes algorithm is now known as "NaiveBayes" in Weka's bayes category.

Figure 8 shows that 30.8931% of the 683 instances of Russian hackers were correctly classified (211), while 69.1069% were incorrectly classified (472), which is worse than the other algorithms but slightly better than the Random Forest algorithm.

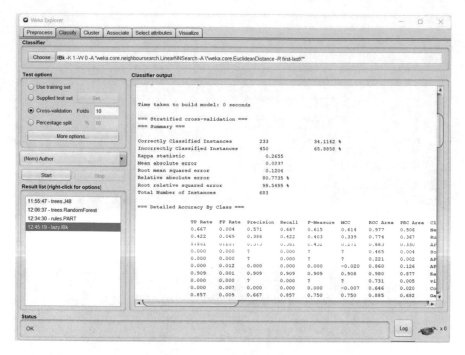

Fig. 7 Training process of authors based on K—Nearest Neighbour model

4.4 Support Vector Machine Analysis

The Support Vector Machine algorithm will be used to train the author subset in our training dataset. Cross-validation for the training control is set to 10 folds. In the Weka functions category, the Support Vector Machine algorithm is now known as "SMO."

According to Fig. 9, 33.3821% of the 683 instances of Russian hackers were correctly classified (228), while 66.6179% were incorrectly classified (455), which is slightly better than the other algorithms except the K nearest neighbour algorithm.

4.5 Multilayer Perceptron

In our training dataset, we will use the Multilayer Perceptron (MLP) algorithm to train the author subset. For the training control, cross-validation is set to ten folds. The Multilayer Perceptron (MLP) algorithm is now referred to as "MultilayerPerceptron" in the Weka functions category.

Figure 10 shows that 31.6252% of the 683 Russian hacker instances were correctly classified (216), while 68.3748% were incorrectly classified (467), which is slightly better than the Random Forest and Naive Bayes algorithms but on par with the Recursive partitioning algorithm.

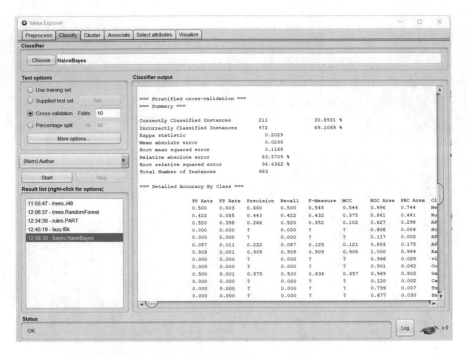

Fig. 8 Training process of authors based on Naïve Bayes model

Based on the findings, it is determined that k—Nearest neighbor has the highest accuracy in predicting Russian hackers, while the other techniques are not far behind. This model correctly predicted 233 instances of Russian hacker groups with an accuracy level of 34.1142%, but incorrectly predicted 450 instances of Russian hacker groups with an inaccuracy level of 65.8858%. While this is not an accurate enough statistic for researchers to rely on, instances of Russian hacker groups that were correctly predicted will be more likely to predict with an unrecognized or unknown dataset.

5 Discussion of Findings

This section will discuss and evaluate the nominated model for cyber attacker prediction and go over the top 5 Russian hacker groups that were correctly predicted in terms of TP Rate, Precision, Recall, F-measure, and ROC Area in greater detail, and as previously stated, k—Nearest Neighbor was nominated as the best performing model (Table 2).

The top five best performing Russian hacker groups that were correctly predicted are listed below, and the statistics and metrics will be discussed in greater detail [12].

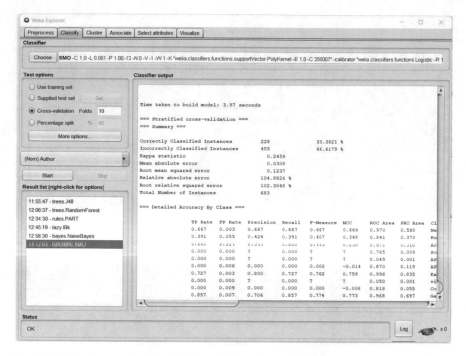

Fig. 9 Training process of authors based on Support Vector Machine model

1. True Positive Rate (Recall): The true positive rate and recall are largely synonymous, and they refer to the number of instances correctly predicted in one class divided by the total number of instances. The formula is as follows: TP rate = TP/ (TP + FN).

2. Precision: Precision, also known as positive predicted value, refers to the number of instances in one class that were correctly classified divided by the total number of classified objects in that class. Precision is calculated as TP/TP + FP.

3. F-Measure: This is a classification technique evaluation metric that is the mean of precision and recall. The following is the formula: 2 * Recall * Precision/Recall + Precision [15].

4. ROC Area: The ROC Area graphic depicts a two-dimensional graph with the X axis labelled false Positive and the Y axis labelled TP rate. The region just below the ROC curve is critical for determining a classifier's success.

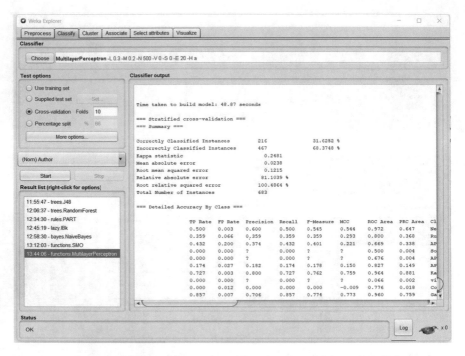

Fig. 10 Training process of authors based on multilayer perceptron model

Table 2 A table depicting the 5 best performing instances

TP Rate	Precision	Recall	F-measure	ROC Area	Class
0.667	0.571	0.667	0.615	0.977	New World Hackers
0.422	0.386	0.422	0.403	0.774	Other Russian Hackers
0.561	0.379	0.561	0.452	0.665	APT28
0.909	0.909	0.909	0.909	0.980	Kapustkiy
0.857	0.667	0.857	0.750	0.885	Gamaredon

Given that the model's overall accuracy is 34.1142% and that it isn't reliable enough for researchers to use as a single profiling method, the following points can be highlighted by analyzing this result by comparing different benchmarks.

1. Figure 11 depicts a bar chart comparing the TP rate, Precision, F-Measure, and ROC Area for the author classes available in the Russian hackers Dataset. The Kapustkiy Hacker group has the highest TP rate, Precision, and F-Measure, all of which are similar with an equal value of 0.909 but with a slightly different

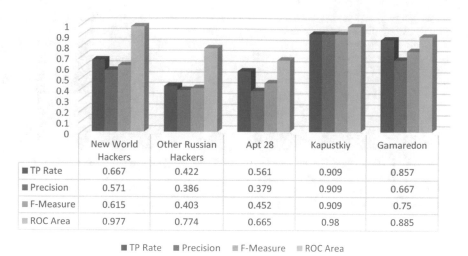

	New World Hackers	Other Russian Hackers	Apt 28	Kapustkiy	Gamaredon
■ TP Rate	0.667	0.422	0.561	0.909	0.857
■ Precision	0.571	0.386	0.379	0.909	0.667
■ F-Measure	0.615	0.403	0.452	0.909	0.75
■ ROC Area	0.977	0.774	0.665	0.98	0.885

■ TP Rate   ■ Precision   ■ F-Measure   ■ ROC Area

Fig. 11 TP rate, precision, F-measure, and ROC area comparison

ROC Area value of 0.980. Other Russian hacker groups, on the other hand, have the lowest TP rate and F-measure, which are 0.422 and 0.403, respectively. In comparison to the other hacker groups, Apt 28 has the lowest Precision rate of 0.379. This means that if an unknown dataset implements the model, it is more likely to profile the Kapustkiy hacker group than any other Russian hacker group. The Kapustkiy hacker group has the highest ROC with 0.980, followed closely by the New World Hacker group with 0.977. Similarly, the Gamaredon hacker group is second to the New World hacker group, with TP rates and Precision of 0.857 and 0.667, respectively. This indicates that this model better profiles the Kapustkiy hacker group, followed closely by the Gamaredon and New World hacker groups, than any other hacker group.

2. The final step in this analysis is to compare Recall, F-measure, and ROC Area. As illustrated in Fig. 12, the Kapustkiy hacker group has the highest value in terms of overall accuracy or recall, with Recall, F-measure, and ROC values of 0.909, 0.909, and 0.980, respectively. Gamaredon ranks second in both the Recall and F-measure metrics, with values of 0.857 and 0.750, respectively, compared to New Word hackers, who have values of 0.667 and 0.615. Although New World Hackers have a higher ROC of 0.977 than Gamaredon, which has a ROC of 0.885, the best profile hacker groups according to this model are Kapustkiy, Gamaredon, and New World Hackers in that order.

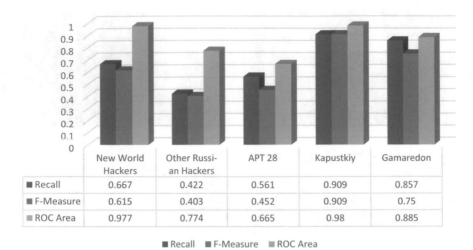

	New World Hackers	Other Russi-an Hackers	APT 28	Kapustkiy	Gamaredon
■ Recall	0.667	0.422	0.561	0.909	0.857
■ F-Measure	0.615	0.403	0.452	0.909	0.75
■ ROC Area	0.977	0.774	0.665	0.98	0.885

■ Recall ■ F-Measure ■ ROC Area

Fig. 12 Recall, F-measure, and ROC area comparison

6   Conclusion

6.1   Contribution to Knowledge

In recent years, data mining has become increasingly important in many aspects of life, particularly in the IT world and cyber security. In terms of cyber security, hackers can be profiled by analysing past and present reported cyber breaches, which can help both researchers and organisations detect and prevent potential future breaches. By profiling Russian hackers, classification as a data mining technique was used in this study to address not only a cyber security challenge but also the ongoing cyber war between Russia and Ukraine. This study's contribution to knowledge in relation to the research question, which is to what extent can a Data mining technique be used to profile and predict potential cyber-attacks by Russian hackers, is as follows:

1. One of the study's novel aspects is the use of Open-Source Intelligence for the training model. The dataset initially contained over 10000 records of cyber breaches, which were then pre-processed into 683 records of cyber-attacks involving Russian hackers that occurred between 2016 and the end of April 2022, as gathered from OSINT. To clean and pre-process the dataset, Open refine was used. The pre-processing was done both automatically and manually. This dataset, which was provided by this study, can be used in future studies. The dataset began with ten different attributes: Date, Author, Target, Description, Attack, target class, attack class, country, link, and tags, which were then pre-processed into five major categories: Author, Attack, target class, attack class, and country.

2. Another contribution to knowledge is that the k-Nearest Neighbour algorithm had the highest accuracy in predicting Russian hackers. While 34.1142% accuracy (233 instances) may not be enough for researchers and organisations to rely on, the Kapustkiy hacker group has the highest rate across all benchmarks (TP rate, Precision, Recall, F-measure and ROC Area). This technique does not have sufficient accuracy because the data is noisy and incomplete, which may result in low accuracy.

To summarise, this study employs five major classification techniques to profile Russian hackers using only the author class category. While this cannot be used as a stand-alone technique, other class categories such as Attack class, target class, country, and attack could be analysed to provide more details for researchers, law enforcement agencies, and organisations when implementing defence strategies or countermeasures for cyber breaches.

6.2   Limitaions

This study, like many others, has limitations in terms of sectional study. The limitations are more focused on the data element, which is concerned with completeness and various aspects of the data. The information was obtained through Open-Source Intelligence, and as previously stated, any data obtained through OSINT contains a significant amount of noise and irrelevant data. The operation was carried out manually in the section of data pre-processing and data cleansing, which included removing irrelevant cyber-attack records, and the only way to validate cyber-attacks was to investigate each record by reading the description and information links. As a result, this operation required time and could not have been more time efficient or faster.

Another constraint of data access was the sensitivity of the cyber security subject; thus, the accuracy and completeness of data can be difficult as most companies and government organizations try to conceal their cyber security incidents to maintain their reputation. A complete and accurate dataset can lead to more precise and reliable results when it comes to forecasting analytics.

6.3   Future Work

Another data mining technique for profiling Russian hackers could be used in future work on this research topic. A classification technique was identified as a data mining technique that could be used for this research after reviewing current literature on the subject. Other data mining techniques, such as clustering or time series analysis, could be used in future research to profile Russian hackers. Clustering algorithms like k—means clustering could be used in future studies because they scale to large datasets like the one used in this study, are relatively simple to implement, and adapt easily to new examples. While time series analysis can help you easily identify patterns and is excellent for prediction, using it can be difficult owing to a couple of factors. When it comes to developing a cyber-security strategy, security incidents, researchers and cyber security experts can benefit from clustering and other data mining techniques.

The dataset is another area in this study that could be improved. As previously stated, the dataset was obtained from OSINT, which may have an impact on the study's results due to the noisy nature of the dataset. Obtaining cyber breaches from organisations themselves, if possible, could greatly improve the accuracy of the findings. Additionally, more attributes, such as the IP address and location of the attack, could be added. With more attributes added to the dataset, a more accurate model could be created.

References

1. A chilling Russian cyber aim in Ukraine: Digital dossiers. (n.d.). NBC news. https://www.nbcnews.com/tech/security/chilling-russian-cyber-aim-ukraine-digital-dossiers-rcna26415. Accessed 8 April 2022
2. Alqahtani H, Sarker IH, Kalim A, Minhaz Hossain SM, Ikhlaq S, Hossain S (2020) Cyber intrusion detection using machine learning classification techniques. In: Chaubey N, Parikh S, Amin K (eds), Computing science, communication and security. Springer, pp 121–131. https://doi.org/10.1007/978-981-15-6648-6_10
3. Andrew DK, Daniel RS (n.d.) Supervised machine learning for population genetics: a new paradigm I Elsevier enhanced reader. https://doi.org/10.1016/j.tig.2017.12.005
4. Back S, LaPrade J, Shehadeh L, Kim M (2019) Youth hackers and adult hackers in South Korea: an application of cybercriminal profiling. In: 2019 IEEE European symposium on security and privacy workshops (EuroS PW), pp 410–413. https://doi.org/10.1109/EuroSPW.2019.00052
5. Berrar D (2018). Cross-Validation. https://doi.org/10.1016/B978-0-12-809633-8.20349-X
6. Car Z, Baressi Šegota S, Anđelić N, Lorencin I, Mrzljak V (2020) Modeling the spread of COVID-19 infection using a multilayer perceptron. Comput Math Methods Med 1–10. https://doi.org/10.1155/2020/5714714
7. Dahmane M, Foucher S (2018) Combating insider threats by user profiling from activity logging data. In: 2018 1st international conference on data intelligence and security (ICDIS), pp 194–199. https://doi.org/10.1109/ICDIS.2018.00039
8. Freund Y (1999) The alternating decision tree learning algorithm. In: Machine learning: proceedings of the sixteenth international conference, pp 124–133.
9. Howler L (2022, April 14). New Industroyer malware targets Ukraine's energy provider. Virus Removal Guides. https://howtoremove.guide/industroyer2/
10. Kigerl A (2018) Profiling cybercriminals: topic model clustering of carding forum member comment Histories. Soc Sci Comput Rev 36(5):591–609. https://doi.org/10.1177/0894439317730296
11. Osman AS (2019) Data mining techniques: review. Int J Data Sci Res 2(1):1–5
12. Pourmouri S (2019) Improving cyber situational awareness via data mining and predictive analytic techniques [PhD, Sheffield Hallam University]. https://doi.org/10.7190/shu-thesis-00202
13. Pournouri S, Zargari S, Akhgar B (2018) Predicting the cyber attackers; a comparison of different classification techniques. In: Jahankhani H (ed), Cyber criminology. Springer International Publishing, pp 169–181. https://doi.org/10.1007/978-3-319-97181-0_8

14. Psychological Profiling in Cybersecurity (2022). GeeksforGeeks. https://www.geeksforgeeks. org/psychological-profiling-in-cybersecurity/
15. Ren J-H, Liu F (2019) Predicting software defects using self-organizing data mining. IEEE Access 7:122796–122810. https://doi.org/10.1109/ACCESS.2019.2927489
16. Sanger D (2020, December 16) Hacked, Again. The New York Times. https://www.nytimes. com/2020/12/16/podcasts/the-daily/russian-hack-solar-winds.html
17. Scherer R, Howard SK, Tondeur J, Siddiq F (2021). Profiling teachers' readiness for online teaching and learning in higher education: who's ready? | Elsevier enhanced reader. https://doi. org/10.1016/j.chb.2020.106675
18. S, S. (2021) Cyber profiling: K-means clustering. Medium. https://saranyasrissss1523.medium. com/cyber-profiling-k-means-clustering-c866e17e31ce
19. SSU exposes another bot farm in Kharkiv (video) (n.d.) SSU. https://ssu.gov.ua/en/novyny/ sbu-vykryla-novu-vorozhu-botofermu-u-kharkovi-video. 8 April 2022
20. Thangaraj M, Sivakami M (2018) Text classification techniques: a literature review. Interdiscip J Inf Knowl Manag 13:117–135. https://doi-org.hallam.idm.oclc.org/https://doi.org/10.28945/ 4066

Developing a Novel Digital Forensics Readiness Framework for Wireless Medical Networks Using Specialised Logging

Cephas Mpungu, Carlisle George, and Glenford Mapp

Abstract Wireless Medical Networks (WMNs) have always been a vital component for the treatment and management of chronic diseases. However, the data generated by these networks keeps growing and has become a potential target for criminals seeking to capitalise on its sensitivity and value. Wireless networks also happen to be more vulnerable to attacks compared to wired networks. In the event of such attacks, it becomes really difficult to conduct a digital Forensics investigation. This paper investigates and suggests a proactive approach of digital forensics readiness within wireless medical networks by suggesting specialised monitoring and logging mechanisms. The research first identifies threats to wireless medical networks. It then undertakes a trajectory of a systematic review of previously proposed digital forensics frameworks and identifies challenges. Finally, it proposes a conceptual framework for Digital Forensics Readiness (DFR) for wireless medical networks. The paper, therefore, makes a novel contribution to the field of digital forensics. It suggests a more streamlined, robust, and decentralised framework that is partially underpinned by blockchain technology at the evidence management layer. The framework contributes to the enforcement of evidential data integrity whilst also securing wireless medical networks.

Keywords Wireless medical networks · Digital evidence · Digital forensics · Digital forensics readiness · Digital investigations · Incident response

C. Mpungu · C. George (✉) · G. Mapp
School of Computer Science Faculty of Science and Technology, Middlesex University, The Burroughs, London NW4 4BT, UK
e-mail: c.george@mdx.ac.uk

C. Mpungu
e-mail: c.mpungu@mdx.ac.uk

G. Mapp
e-mail: c.mapp@mdx.ac.uk

© The Author(s), under exclusive license to Springer Nature Switzerland AG 2023
H. Jahankhani (ed.), *Cybersecurity in the Age of Smart Societies*,
Advanced Sciences and Technologies for Security Applications,
https://doi.org/10.1007/978-3-031-20160-8_12

203

1 Introduction

The world we live in today is transforming rapidly into a digitized network connecting millions of people, businesses, and several vital sectors including healthcare. Healthcare networks have come a long way from the use of papyrus for medical data records [21] to state-of-the-art digitised equipment [4]. This has been bolstered by complex networks harbouring IoT (Internet Of Things) capabilities [51]. As a result, Wireless Medical Networks like Wireless Body Area Networks (WBAN), RFID tagging, GPRS, UTMS, Wireless Area Networks (WANs), Wireless Local Area Networks (WLANs), Wireless Sensor Networks (WSN) and Bluetooth have emerged and continue to evolve [13]. Miniaturized diagnostic equipment and instruments linked to smartphones have instituted a new term in the healthcare paradigm known as 'mHealth' (mobile health).

As the urgent need to manage chronic diseases continues to rise, mHealth technologies and networks have become a sustainable factor in ensuring proper monitored management [58]. The essentiality of IoT-enabled apps has played a vital role in the management of chronic diseases such as diabetes, heart disease and cancer [1]. Examples of these include implantable and wearable medical devices like insulin pumps, glucose monitors, defibrillators, neuro-monitoring systems, and smartwatches [33]. These evolving additions to wireless medical networks (WMNs) have subsequently created a lot of data as well as complexities in handling it. Data privacy enforcers and regulators have also put pressure on the various stakeholders to ensure that healthcare data is processed and managed securely [9, 26].

The Covid-19 pandemic resulted in the healthcare sector experiencing many difficulties. Healthcare facilities that had previously stuck to traditional methods of delivering services had to succumb to poor service delivery due to overwhelming demand for services and cyberattacks. The pandemic also uncovered several healthcare-related system security vulnerabilities [6]. Countries like the USA accused China and Russia of trying to hack into their healthcare research centres [12]. Cybercriminals also capitalised on ransomware, phishing and hacking attacks, amongst others, targeted at healthcare systems [15]. In 2019, Greenfield Hospital (USA) paid $50,000 to hackers as ransom so that 1400 hospital files would be decrypted during a ransomware attack [49]. Kyaw et al. [32] conclude that such incidents are on the rise in the healthcare sector including, even, those that claim to have the very best security technology and resources at their disposal. Protenus [43] publishes up-to-date information on data breaches and in April 2021 reported that over 40 million patient records (worldwide) had been breached in the year 2021. Other security threats to medical systems include Distributed Denial-Of-Service (DDoS) attacks, SQL injections, zero-day attacks, supply chain attacks, human errors, and Man in The Middle attacks (MITM). The current approach to mitigating these kinds of threats has been more inclined toward security enforcement, disaster recovery and business continuity with less emphasis on Digital Forensics Readiness (DFR) [32].

Aside from external attacks, internal attacks have also been witnessed [16]. This is because most healthcare systems have been designed based on centralised logging mechanisms. Attacks on medical systems often leave behind traces of evidence concealed within the compromised systems [32, 45, 48].

This paper argues that such attacks on healthcare systems can be mitigated by introducing more robust security mechanisms. The attacks can also be more easily investigated by streamlining the extraction of evidence to identify perpetrators. This paper proposes that these two objectives can be achieved by implementing a DFR framework in WMNs.

2 Background and Literature Review

2.1 Evolution of Healthcare Networks

Technology has evolved at a super-fast speed over the past years, with this also encompassing digital healthcare networks. According to Bhavnani et al. [4], some healthcare Wireless Medical Networks (WMNs) are already thriving with state-of-the-art digitised medical equipment and technologies that have since adopted some IoT functionalities. Some examples of these IoT devices include implantable and wearable medical devices such as insulin pumps, glucose monitors, defibrillators, neuro-monitoring systems, and smartwatches [33].

With the emergence of miniaturised diagnostic instruments linked to smartphones, mHealth has developed giving rise to the need for more wireless networks linked to healthcare. The World Health Organisation (WHO) defines mHealth as a component of eHealth (electronic health) that is supported by mobile devices such as patient monitoring devices, mobile phones, PDAs, and other wireless devices [58]. The intersection of mHealth and the 'real world' however has raised new insights and questions into the generation and analysis of data logs collected from these systems [4].

2.2 The Sensitivity of Healthcare Data

Data pertaining to WMNs is of intrinsic value to various stakeholders. This data may include, but not limited to, personal health records (e.g., names, phone numbers, age, medical history, drug prescriptions and home address), network traffic logs, employee records, CCTV footage, authentication records and logistical data. Personal health records are also technically referred to as Electronic Medical Records (EMR). EMR is defined as a collection of medical information pertaining to an individual that is stored digitally on a computer [36]. Examples of EMR are patient history, tests, allergies, treatment plans, medical insurance, and biometric data.

Caution must be taken to ensure that the integrity, confidentiality, and availability (CIA triad) of this data is maintained. Article 9 of the General Data Protection Regulation (GDPR), which is incorporated into the UK Data Protection Act 2018, lists data pertaining to health among "special category data" that should be given more protection. The unauthorised disclosure of such data will lead to data protection violations and the possible discrimination against data subjects [26], in addition to criminal sanctions under S170 of the Data Protection Act 2018. Servers and networks handling logs, personal-health data, and other healthcare-related information should therefore be secured with utmost importance. Measures should also be put in place by data controllers to ensure that logging systems on WMNs are well structured to capture and preserve data that may be of evidential value [13, 32, 46].

2.3 Security Challenges of Wireless Medical Networks (WMNs)

The National Institute of Standards and Technology (NIST) defines vulnerability as any weakness within an organisation's internal controls, system, or its system security procedures [39]. Some researchers within the Digital Forensics (DF) field have discussed vulnerabilities discovered in some wireless medical IoT devices. For example, security experts have demonstrated that some wireless commercially available insulin pumps are prone to hacking attacks [44]. Software radio-based attacks targeted at cardioverter-defibrillators have also been shown to be possible [22]. Flaws in wireless security encryption methods like WPA, WEP and WPA2 have also been discussed by researchers as possible backdoors to cyber-attacks [13, 31].

Beyond hacking threats, other challenges like ransomware attacks [49] and data breaches [16] are increasing. Cusack and Kyaw [13] define such attacks under 'misuse of wireless medical devices' as unauthorised behaviour within the system. Davis [15], discussed some of the biggest healthcare data breaches of 2019. She maintains that phishing attacks and third-party vendors were behind most of these attacks which left over 25 million patient records compromised. Such attacks infringe on the confidentiality, integrity, and availability of healthcare data [34]. Many cyber attacks are instigated internally or externally and exploit identified vulnerabilities within these systems [32].

2.4 Digital Forensics Readiness (DFR)

One of the toughest challenges faced by countries within the European Union (EU) today is ensuring effective compliance with the GDPR Article 33 regarding data breach notifications. It requires organisations to report an incident to the relevant supervisory authority (e.g. the Information Commissioner's Office (ICO) in the UK) "without undue delay and, where feasible, not later than 72 hours after having become aware of it".

This incorporates all aspects of the incident that pertain to its occurrence, what it is, and the damage done [41]. Where the data breach could result in risks to the rights and freedoms of data subjects, the GDPR Article 34 requires communication to the data subjects as well, without undue delay. Realistically speaking, proper compliance with Articles 33 and 34 could be a daunting task to achieve (especially for large networks) as it might entail extensive forensics analysis (needing evidence). Challenges like these, among others, have prompted various researchers to propose the need for DFR within various organisations [13, 23, 29, 47].

DFR is derived from the term Digital Forensics (DF). Vidal and Choo [56], define DFR as an organisation's pro-active approach toward the quick collection of digital evidential data at minimal cost or interruption to its day-to-day business. Rowlingson's definition of DFR is analogous to the one given by Vidal and Choo [48]. Vidal and Choo [56] further clarify that an organisation should be in a position to clearly define this digital evidence. This helps in setting up appropriate teams, programs and infrastructure that facilitate timely availability of the evidence[25]. DFR is therefore incidence-anticipation and not incidence-response driven [48].

DFR's objectives seek to make the best out of an organisation's environment by collecting digital evidence of credibility whilst minimising the cost of digital forensics during incidence response [14, 53].

2.5 Centralised DFR Within Wireless Medical Networks

Cusack and Kyaw [13] proposed a centralised architecture of a DFR system for WMNs with security enforcement abilities as well as a capability to investigate post-events. The researchers delved deep into the architectures of wireless medical technologies and discussed their potential security risks. Their work proposed the addition of drones and a forensic server to an existent wireless network of a hospital information system. It resonates with Rowlingson's suggestion of deploying DFR on top of a system's existent logging mechanism [48].

Kyaw et al. [32] later proposed a modified DFR framework for wireless medical devices (WMedSys) based on the work in Cusack and Kyaw [13], with the aim of streamlining digital forensic investigations. The framework's main objective is to reduce the time and cost of performing a DF investigation. This time the researchers emphasize its conceptual design and its evaluation. Evaluation is done using a thematical expert analysis. The conceptual design consists of a Pi-drone that uses kali Linux Operating System (OS) to scan and capture wireless signals. The captured data is then sent to a Wireless Forensic Server (WFS). Other components of this framework are an Intrusion Detection System (IDS), integrity Hashing Server, Centralised Syslog server, Wireless Access Point (WAP), Remote Authentication Dial-In Service (RADIUS) server and a web server.

Other researchers like Rahman et al. [46] also discussed forensic readiness in WMNs focusing on Wireless Body Area Networks (WBAN). The researchers used the concept of Practical Impact Analysis (PIA) to scrutinise potential WBAN threats and vulnerabilities. Based on their findings, they proposed solutions for implementing a centralised DFR system [46].

2.6 Decentralised DFR Within Wireless Medical Networks

The work of the researchers reviewed in Sect. 2.5 above discussed DFR solutions that utilise a centralised logging mechanism. These researchers proposed frameworks based on the assumption that information system administrators and all those with access to the evidence-log servers can be trusted. This cannot be assumed as human beings are subject to compromise and may also get disgruntled and make irrational decisions leading to alteration of server logs. Centralised logging models built on a client–server architecture are also susceptible to becoming a single point of failure [2]. To address this, researchers like Tian et al. [55] proposed a solution to centralise log management, in the form of a secure digital evidence framework based on blockchain technology for the storage of evidential data. Their proposal, however, was limited by its emphasis on the security of hash files and not the actual evidential logs (data). The proposal in this paper builds on the work of Tian et al. [55] and seeks to improve the use of blockchain in that context.

2.7 Blockchain Technology

Blockchain is defined as a public ledger of transactions distributed and stored on several nodes within a blockchain network [20, 55]. It uses a peer-peer architecture with each constituent node having a copy of the blockchain. It comprises blocks of transactions linked together by cryptographic and distributed consensus algorithms [59]. Each block keeps a record of specific transactions and contains a hash value pointing to the previous block [20]. Blockchain networks can be classified as public, private or consortium (mixed) blockchain [55]. Blockchain is believed to be one of the safest means of storing data and transactions in an immutable form due to its cryptographic and hashing capabilities [3, 20, 42, 55]. Blockchain technology, therefore, ensures auditability, decentralization, immutability, security, and transparency [59]. It is for this matter that researchers like Tian et al. [55] and Pourmajidi et al. [42] have proposed logging frameworks based on blockchain technology.

However, blockchain faces a challenge of scalability. As the blocks of data continue to grow, it creates a need for larger storage facilitation and it might slow down the blockchain network [55, 59]. This is known as blockchain bloat [55]. The work done by Tian et al. [55] suggests a lightweight blockchain to counter this challenge within a digital evidence system.

Blockchain is still a new technology and yet it has attracted a lot of attention from various researchers, industries, and some governments. Estonia (Europe) for example, was the first nation to implement blockchain technology within its production systems. The country opted for a technology known as KSI blockchain, similarly used by USA and NATO [28]. The KSI blockchain in Estonia supports the property, healthcare, succession, and business registries. It also facilitates its digital court system [28]. Similarly, the European Union (EU) is undertaking a blockchain research program known as MyHealthMyData (MHMD). The program aims to ensure interoperability between individuals, healthcare providers, and biomedical industries [19].

3   A New Framework for DFR in Wireless Medical Networks (WMNs)

3.1   Introduction

Most organisations invest a lot of money in securing their networks, but this may not necessarily stop security incidents from occurring. Therefore, the framework proposed by this research assumes that an incident might occur regardless of an implementation that ensures a risk assessment of low probability. The design seeks to enforce the security of WMNs whilst also ensuring forensic readiness, hence aligning itself with the characteristics of a good DFR system identified by previous researchers like Tan [53] and Rowlingson [48].

Figure 1 illustrates the structure of the proposed DFR in the context of a generic WMN. It consists of three major layers: (i) Data Collection Layer (DCL), (ii) Network Layer (NL), and (iii) Evidence Management Layer (EML). The design of the framework takes into consideration five major guidelines identified by Rowlingson [48]:

1.   Identification of possible sources of evidential data within a generic healthcare wireless network structure. This is implemented at the Data collection Layer.
2.   Outlining of the technical and legal requirements for the proper and streamlined collection of digital evidence. These requirements are implemented and maintained at all three layers.
3.   Identification and set up of resources for the proper collection of legally admissible evidence to meet the legal and technical requirements. This is mostly enforced at the EML and partly at the DCL.
4.   Ensuring that all monitoring systems can detect major incidents. This is implemented at the DCL using the Incident Detection System (IDS) module.
5.   Establishing the requirements that warrant a digital forensic investigation.

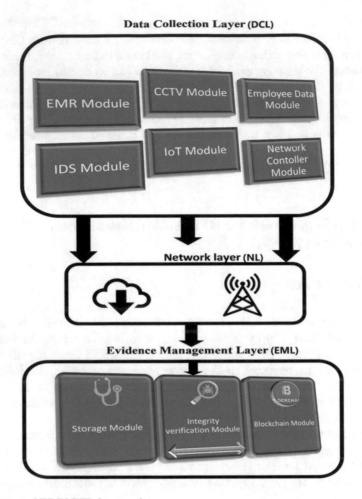

Fig. 1 Proposed WMN DFR framework

3.2 Data Collection Layer (DCL)

The DCL consists of six major modules. These modules were identified as the main sources of evidential data within wireless medical networks by the researcher. These include: Electronic Medical Records (EMR) module, Employee Data Module, IoT Module, CCTV Module, Network Controller Module, and the Intrusion Detection System (IDS) Module. The DCL consists of modules that are implemented and configured to meet the requirements of availing potential evidential data. Collection of logs from these modules can be implemented in about three ways:

(1) The utilisation of existing logging agents within the modules.
(2) Adjustment/reconfiguration of existing logging agents within the modules.
(3) Configuring new logging agents within the modules.

Some potential sources of evidential data considered during the design of this framework include firewalls, switches, wireless access points, proxy servers, DHCP servers, DNS servers, routers, VPN terminators, system logs, IoT devices and CCTV cameras, amongst others.

The **Employee Data Module** collects information relating to employee's day to day activities like patient-diagnosis data, employee-patient interactions, key card logins, etc. and stores them within a MYSQL database. This database(db) is linked to other relational databases like Employee Records db, Medical IoT devices db and Electronic Medical Records db using a Relational Database Management System (RDBMS). The choice to use MYSQL for RDBMS by the researcher is based on its cross-platform support, open-source nature, and compatibility with the Linux operating system.

The **Electronic Medical Records Module** is a database of personal records pertaining to patients e.g. patient history, tests, allergies, treatment plans, medical insurance, and biometric data. This kind of data could be a rich source of evidential data related to medical negligence and fake insurance claims.

IoT Module is used as a central collection point for all IoMT (Internet of Medical Things) data that could be a source of evidential data. Examples of IoT devices include insulin pumps, glucose monitors, cardioverter-defibrillators, and neuro-monitoring systems.

The **Network controller Module** in the framework is configured using Cisco DNA Center software. A Network controller is a hardware or software that implements intermediation between an organisation's needs and its network infrastructure [10]. The choice to use Cisco DNA Center is based on its interoperability, easy setup and resilience attributes. A software implementation also eliminates the need to purchase specialised Network controller hardware. The module streamlines the collection of potential evidential data from network hardware like routers, switches, and wireless access points. The Network Controller software is also paramount for the enforcement of security and tweaking network configurations.

The **CCTV Module** is a collection point for all video footage uploaded from the various surveillance cameras within the hospital premises. This kind of data comes in handy when placing suspects at a crime scene.

The **Intrusion Detection System (IDS) Module** of the framework is configured to monitor malicious or policy violation activity and log data to a SIEM (Security Information and Event Management System). This data can be generated from antivirus logs, proxy servers, database audits and application servers. The IDS can be further configured, and rules set to detect bad IPs, URLs, and elimination of false positives [52]. The rules and configurations are then tightened by setting alerts to the incident response team e.g. through SMS and email whenever incidents are detected. Figures 2 and 3 show the flow of data from the DCL to the EML.

The data collected from the DCL is encrypted and forwarded to the Integrity Verification Module (IVM) of the Evidence Management Layer (EML) as shown in Fig. 2. The IVM server is configured primarily to hash the logs and additionally as a Network Time Protocol (NTP) server (for synchronisation of clocks on all servers). Time synchronisation is a key factor for the proper enforcement of data integrity.

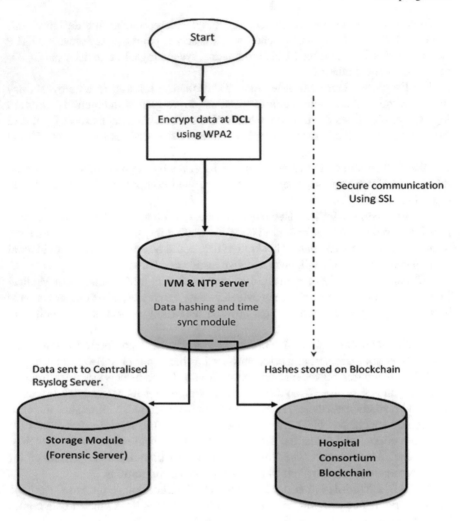

Fig. 2 Data flowchart from the DCL to the EML

Each data source is hashed and then forwarded and logged/stored onto the Storage Module (Forensic Server) within the EML. The data source's corresponding hash is then stored securely on the consortium blockchain. A consortium blockchain is a combination of a private and public blockchain [55]. The researcher's choice to use a consortium blockchain implementation is supported by the fact that a mixed blockchain facilitates easy and streamlined access of data by all third parties relevant to an investigation (e.g., insurers, lawyers, and forensic investigators). It, therefore, enforces interoperability [18] whilst maintaining data integrity.

As an added layer of security, the IVM and the Forensic Server can be configured as virtual machines residing on hypervisor hardware or software. Hypervisors provide a

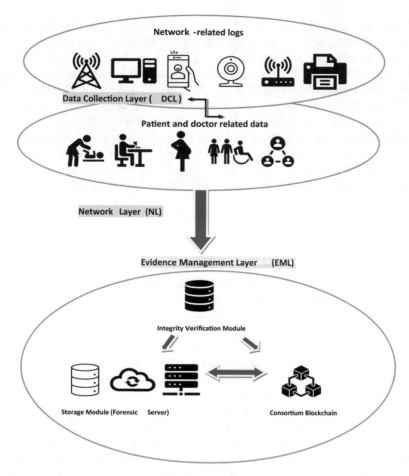

Fig. 3 Data flow from the DCL to the EML

private network ecosystem where two virtual machines can communicate with each other without the knowledge of the physical network they reside on [30]. The two servers can then be assigned two sets of static IPs, one set for private communications with each other and the other set for communication with other relevant devices on the physical network. With this kind of setup, the network traffic generated between the two servers will remain concealed from the healthcare's physical network.

3.3 Network Layer (NL)

The Network layer consists of communication channels needed for ensuring connection setup between different networks and devices for the secure transfer of data packets. The communication channels are configured using a cryptographic protocol known as Secure Sockets Layer (SSL) or TSL (Transport Layer Security) for the safe transportation of data across the network. SSL helps enforce security and ensure data integrity for TCP/IP communications. The researcher also proposes the use of enhanced security protocols WPA2/3 (WI-FI Protected Access 2/3) Enterprise within the WLAN. This is configured using a RADIUS authentication server. The task of network user access and authentication is handled by the RADIUS server [50].

The IoT medical devices within a hospital's premises will also be configured to use dedicated wireless routers and access points whose SSIDs (Service Set Identifiers) are not broadcast. This adds a layer of security [57] by reducing the attack surface for cybercriminals as they wouldn't know the SSIDs of the wireless routers to attack.

3.4   Evidence Management Layer (EML)

The evidence storage and management module is paramount to ensuring the integrity, availability and confidentiality of potential evidential data. It comprises of an Integrity Verification Module (IVM), a Linux Rsyslog server and consortium blockchain nodes. As an added layer of security, these three components should be configured on a separate subnet of the wireless network.

The IVM is run on a Red hat Linux Operating System platform and is configured to serve as a hashing and NTP server. Hashing is necessary for enforcing data integrity, checking data integrity, and speeding up retrieval of evidential data. This research proposes a SHA-256 cryptographic hashing algorithm. SHA-256 is considered the strongest hashing algorithm and is highly recommended by NIST. The hashed data is forwarded to the centralised logging facility which acts as a forensic server. The hashes are sent to the decentralised storage facility on the consortium blockchain.

The Storage Module (Forensic server) is a centralised logging facility also running Red hat Linux OS. It is configured using Rsyslog. The researcher's choice of Red Hat Linux OS is based on its robust security features like systemd log management, Security Enhanced Linux (SElinux), advanced Access Control Lists (ACLs) management and 'Firewalld'. A good logging mechanism should also be supported by data compression and backup capabilities. For this framework, the researcher proposes a periodic archiving and remote backup of the forensic server image on a secured cloud storage. This can be automated by writing scripts on the log server using a facility known as crontab.

Combining local and remote logging helps secure the integrity of the evidential data [48]. Remote logging on the proposed DFR is configured using the Rsyslog facility by editing its configuration file (/etc./rsyslog.conf). Rsyslog also contains a log rotation facility known as logrotate that can be configured using the /etc./logrotate.conf configuration file to align log data retention policies on the server with the laws governing data retention of healthcare-related data. The ICO in the UK maintains that data should be gathered and logged for defined purposes and nothing more [48].

The consortium blockchain is a mixture of a private and public blockchain distributed on several nodes using a peer-to-peer network. This kind of configuration ensures a better consensus mechanism in comparison to a private blockchain. It also offers a more decentralisation setup compared to a private setup [40]. To solve the issue of blockchain bloat, the blockchain is only used to store the data hashes and to provide a secure interface for access and handling of evidential data. The stakeholders that may need access to the evidential data go beyond the scope of just healthcare employees. They include insurance companies, private GPs, forensic investigators, courts of law, government compliance officers and lawyers. A mixed blockchain setup allows for this kind of varied access whilst also enforcing the data integrity of evidential data.

Different stakeholders (authorised to access the evidential data) like the hospital, insurance companies and government institutes can share the blockchain development costs to suite the DFR requirement. Companies like Hashed Health are already offering distributed ledger solutions within the Healthcare sector tailored to the needs of their clients [24]. The researcher proposes the use of smart contracts for the streamlined and transparent access of evidential data by the various stakeholders. Smart contracts are software/program codes written to execute actions when fed inputs or when a specific event is triggered [38, 54]. Smart contracts can also be programmed to enforce private/public key-oriented registrations, access control rules and interfaces for various stakeholders pertaining to healthcare data.

Smart contracts rely on the "if–then" logic of programming and help ensure immutability, security, privacy, reduced costs and automation [38]. The blockchain interface of the proposed framework will constitute a smart contact configuration that checks the public/private key of a user and grants them access to evidential data based on Access Control Rules (ACLs) set by the Rsyslog server System administrator. For example a digital forensics Investigator would be assigned access to all logs on the server whilst a medical insurance company investigator would be limited to accessing only logs pertaining to their client. A smart contract configuration is also set up to trigger alerts to the incident response team when nefarious activity is detected within the Evidence Management Layer. The proposed consortium blockchain is meant to be accessible by all parties relevant to an investigation (e.g. law enforcement investigators, lawyers, digital forensics investigators, and medical insurers, amongst others). Verification, management, and retrieval of evidence are therefore executed at the blockchain interface module.

4 Digital Forensics Investigation Process Using the Proposed Framework

The conceptual design of the proposed framework is partly inspired and guided by previous (validated) research and best practices within the field of DFR (e.g. [37, 55]. It aims to simplify and streamline the process of retrieval and submission of admissible evidential data [5] by a forensic investigator as much as possible. This is further reflected in the researcher's choice to use highly customisable software of high repute like Red Hat Enterpise Linux (RHEL), Rsyslog, Cisco DNA Center and MYSQL. The open-source nature of RHEL and MYSQL also optimises the attributes of scalability, integrated-management and interoperability needed for this framework. Figure 4 illustrates the digital forensic investigation process-flow within the proposed DFR framework.

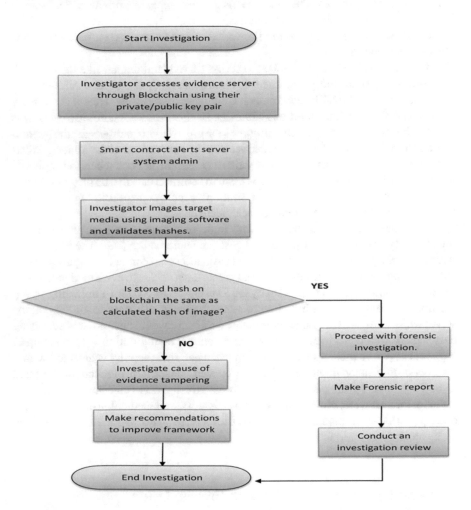

Fig. 4 Digital Investigation protocol on the proposed DFR framework

5 Evaluation of Proposed Framework

5.1 Introduction

This section focuses on the evaluation of the proposed DFR framework using two approaches, a standard-based approach, and a scenario-based approach.

5.2 Standards-Based Evaluation

The standards-based evaluation entails investigating whether the framework conforms to relevant guidelines for proper DFR planning and implementation provided by the International Organisation for Standardisation (ISO/IEC 27,043:2015). The guidelines utilise models that are idealised to represent common incidents encountered during investigations. The ISO/IEC 27,043:2015 defines the three categorical groups of readiness processes as planning, implementation, and assessment process groups [27]. Evaluation of the framework is done based on the planning process group. The implementation and assessment process groups are beyond the scope of this research. The planning process group comprises six major steps:

1. Definition of scenarios
2. Identification of potential evidential data sources.
3. Pre-incident planning, gathering, storage and handling of potential evidential data.
4. Pre-incident analysis of potential evidential data planning.
5. Incident detection planning.
6. System architecture definition.

5.2.1 Definition of Scenarios

Scenario definition is important to help conceptualise simulations of incidents that may trigger a digital forensic investigation or audit of the wireless medical network. A risk assessment is conducted by the incident response team to identify potential vulnerabilities and threats within the WMN. The scenarios can then be defined based on the risk assessment analysis and also based on the current threats facing healthcare networks. The researcher used scenarios like past ransomware attacks on the NHS [35], internal data breaches, DDOS attacks as well as hacking cases during the COVID-19 season. Identification of scenarios enlightens the technical and system administration team on the requirements of achieving digital forensics readiness.

Evaluation results: Appropriate scenarios were identified to guide the design of the proposed framework based on current security threats to wireless medical networks cited in media and articles and literature review of relevant work. This met the scenario definition requirement.

5.2.2   Identification of Potential Evidential Data Sources

The step of identification of potential evidential data sources is informed by the scenario-definition process. For example, a scenario of a data breach on personal health data by authorised personnel informs this step that electronic medical records (EMR) are vulnerable to attacks. EMR is therefore identified as a potential data source. Emphasis can then be put on logging authorized personnel activities within the systems as well as logging EMR metadata information for analysis during an investigation. Access control lists on this kind of sensitive data can also be enforced and monitored keenly.

Evaluation results: The proposed DFR identified five major modules as the main sources of evidential data within wireless medical networks by the researcher. These include: Electronic Medical Records (EMR) module, Employee Data Module, IoT Module, CCTV Module, Network Controller Module, and the Intrusion Detection System (IDS) Module. The identified modules met the requirement of potential data sources.

5.2.3   Pre-Incident Planning, Gathering, Storage and Handling
of Potential Evidential Data

The step of pre-incident planning, gathering, storage and handling of potential evidential data was defined in the proposed DFR. This was defined within the EML. It comprises of an Integrity Verification Module (IVM), a Linux Rsyslog server, and consortium blockchain nodes. The IVM enforces integrity, the Linux Rsyslog server centrally manages evidential log data and the consortium blockchain enforces the confidentiality, integrity, and availability of evidential data.

Evaluation results: The requirement was met by the above-proposed modules for the framework.

5.2.4   Pre-Incident Analysis of Potential Evidential Data Planning

Pre-incident analysis of potential evidential data planning is achieved by the proposed model at the SIEM (Security Information and Event Management System) of the IDS module. This event analyzer implemented at the logging-agent level is configured to alert the incident response team about suspicious activity within the WMN.

Evaluation results: The SIEM of the IDS in the proposed DFR met this requirement.

5.2.5   Incident Detection Planning

Incident detection planning is handled at the IDS module by configurations made by the systems administrators, network administrators and network engineers. These are also part of the incident response team and DFR team and liaise with the forensic investigators during an investigation.

Evaluation results: The IDS in the proposed DFR met this requirement.

5.2.6 System Architecture Definition

System architecture definition has been partially achieved in the framework proposal. The operating system and logging format of the evidence log server have been identified. The operating system of the IVM is defined, the type of blockchain and the database management system have also been identified and justified. However, the network architecture has not been fully defined because the proposed framework is based on a generic wireless medical network.

Evaluation results: The system architecture definition requirement was partially met.

5.3 Scenario-Based Evaluation

A scenario can be defined as a prediction made about a sequence of events [11]. Scenario-based evaluation of the proposed framework was done by first identifying common security threats to wireless medical networks. These include (but not limited to) Distributed Denial-Of-Service (DDoS) attacks, human errors, Man in The Middle attacks (MITM), ransomware attacks, malware, data breaches, and hacking. The researcher then identified the kind of data that needs to be logged to sustain digital forensics investigations and audits of WMNs.

Scenario 1: Unauthorised access

A disgruntled employee Mike, having financial challenges, is suspected of accessing and copying personal medical records of HIV patients. The employee plans to anonymously contact the patients and ask for money or release their medical records online.

Investigation Process:

The investigator will make use of CCTV footage data to help place the suspect at the scene of crime. Logs generated by the IDS module like computer MAC address, IP address and User login credentials will also provide evidence proving unauthorised access. The metadata of the accessed HIV medical records will also be a rich source of evidence for proof of unauthorised access and timeline analysis. Table 1 summarises the data source modules of the framework and logged data that are utilised for this investigation.

Table 1 Investigation process for unauthorised access

Logging agent on DFR framework	Data logged	Benefit to investigator
CCTV module	Time-stamped footage of suspect accessing hospital computer	• Will link the CCTV timestamp to other evidential metadata • Placing the suspect at the crime scene
IDS module	SMS alert to the incident response team, MAC address and IP of the device, medical data accessed metadata, login user credentials, key card login	• Account login details of employee and key card logins • Metadata of accessed records
The server containing medical records	Registry files, personal data access time, copy and paste activity, external device attachment	• Server metadata proving unauthorised activity

Scenario 2: Alteration of log files

The disgruntled employee connives with an assistant system administrator known as Patrick and asks him to delete all the logs on the server that could implicate him. Mike blackmails Patrick and threats to tell his wife about his affair with the receptionist at the hospital. The systems admin, fearing to lose his marriage, deletes all the CCTV footage and logs that implicate Mike in the crime. The systems admin forgets to delete the backup copies of the server in the cloud.

Investigation Process:

Smart contracts customised within the consortium blockchain alert the incident response team about nefarious activity on the log server. Once an investigation is commissioned, the backup copy of the server is imaged and compared against the current server image for traces of discrepancies. The Integrity Verification Module also contains an index of the history of all log ids and hash ids which can be used for further cross-references to prove alteration activity. Table 2 summarises the data source modules of the framework and logged data that are utilised for this investigation.

Scenario 3: Hacking

A hacker obtains access to personal healthcare information intending to sell it to the highest bidder.

Investigation Process:

Unauthorised access alerts triggered by the IDS and blockchain smart contracts alert the incident response team. The investigator images logs generated by the IDS module to obtain rogue IP and MAC address of external threat agent device. Table 3 summarises the data source modules of the framework and logged data that are utilised for this investigation.

Table 2 Investigation process for log alteration

Logging agent on DFR framework	Data logged	Benefit to investigator
Blockchain module	The hashes of the deleted logs and CCTV footage	• Smart contracts alert the incident response team about the deletion of data corresponding to the hashes on the blockchain
Backup server	A full backup image of the server that was compromised by the assistant systems admin	• Restoration of the server backup image from the cloud is used to identify the logs that were deleted using their corresponding hashes
Integrity Verification Module (IVM)	Index of log ids and hash ids	• The IVM module is used to map the compromised data hashes onto the restored logs from the backup

Scenario 4: Malware attack.

Malware is any software that is designed with the intent to damage a computer, network, server, or client. The term is derived from the terms 'malicious' and 'software'. In this scenario, an employee is suspected of unintentionally installing malware by clicking on a link sent through email.

Investigation Process:

The investigator images IDS module logs to identify malicious file-based signatures after an alert is sent to the incident response team by the IDS module. The investigator will then compare the hashes of the suspicious file signatures to the NIST database used by law enforcement. Table 4 summarises the data source modules of the framework and logged data that are utilised for this investigation.

Table 3 Investigation process for hacking

Logging agent on DFR framework	Data logged	Benefit to investigator
Blockchain module	Unauthorised access login attempts	• Smart contracts alert the Incident response team of attempted access by an unauthorised agent
IDS module	SMS alert to data to the incident response team, MAC address and IP of the external device	• SMS alert to the incident response team • IP and MAC address of the rogue device

Table 4 Investigation process for malware attack

Logging agent of DFR framework	Data logged	Benefit to investigator
IDS module	• IDS module is configured to detect and file-based signatures and send alerts to incident response team • IDS is also configured to detect unauthorised data movement and malicious file signatures	• Extracted files are scanned with antivirus software • The investigator can compare suspicious hashes against the NIST hash (updated) database used by law enforcement and forensics experts

6 Conclusion

The proposed digital forensics readiness framework for wireless medical networks makes a novel contribution to the field of digital forensics. The research builds on the work done by [32] and Tian et al. [55] to propose a tamper-proof digital forensics readiness framework for wireless medical networks. It proposes a logging mechanism with an additional layer of security that utilises a consortium blockchain technology for integrity enforcement.

The standard-based evaluation discussed in this chapter shows that the proposed WMN DFR framework addresses the requirements highlighted within the planning readiness process group defined by the ISO/IEC 27,043:2015 standard. Furthermore, the scenario-based evaluation which focused on security threats faced by the healthcare sector demonstrated the effectiveness of the framework. Therefore the proposed DFR framework provides possible solutions to current security threats (e.g. unauthorized access, log alterations hacking, and malware attacks) based on its logging mechanism.

References

1. Albesher A (2019) IoT in health-care: recent advances in the development of smart cyber-physical ubiquitous environments. Available at: https://www.researchgate.net/publication/331642487_IoT_in_Health-care_Recent_Advances_in_the_Development_of_Smart_Cyber-Physical_Ubiquitous_Environments. Last accessed 19 July 2022
2. Atlam H, Alenezi A, Alassafi A, Wills G (2018) Blockchain with internet of things: benefits, challenges, and future directions. Available at: https://eprints.soton.ac.uk/421529/1/Published_Version.pdf. Last accessed 19 July 2022
3. Belchior R, Correia M, Vasconcelos A (2019) JusticeChain: using blockchain to protect justice logs. In: Lecture notes in computer science (online) pp 318–325. Available at: https://link.springer.com/chapter/10.1007%2F978-3-030-33246-4_21. Last accessed 19 July 2022
4. Bhavnani P, Narula J, Sengupta P (2016) Mobile technology and the digitization of healthcare. Eur Heart J 37(18):1428–1438. https://doi.org/10.1093/eurheartj/ehv770
5. Bsigroup.com (2014) BS 10008 Electronic information management (online) Available at: https://www.bsigroup.com/en-GB/bs-10008-electronic-information-management/. Last accessed 19 July 2022

6. Burgess M (2020) Hackers are targeting hospitals crippled by coronavirus. Available at https://www.wired.co.uk/article/coronavirus-hackers-cybercrime-phishing. Last accessed 19 July 2022
7. CDC (2019) About chronic diseases (online) Available at: https://www.cdc.gov/chronicdisease/about/index.htm. Last accessed 19 July 2022
8. Cabinet Office (2016) Security policy framework (online) GOV.UK. Available at https://www.gov.uk/government/publications/security-policy-framework/hmg-security-policy-framework. Last accessed 19 July 2022
9. Cellan-Jones R (2020) Coronavirus: England's test and trace programme breaks GDPR data law. The BBC. Available at: https://www.bbc.co.uk/news/technology-53466471. Last accessed 19 July 2022
10. Cisco (2020) What is a network controller? (online) Available at: https://www.cisco.com/c/en/us/solutions/enterprise-networks/what-is-a-network-controller.html. Last accessed 19 July 2022
11. Collinsdictionary.com (2019) Definition of scenario (online) Available at: https://www.collinsdictionary.com/dictionary/english/scenario. Last accessed 19 July 2022
12. Corera G (2020) Coronavirus: US accuses China of hacking coronavirus research. The BBC. Available at: https://www.bbc.co.uk/news/world-us-canada-52656656. Accessed 21 July 2020
13. Cusack B, Kyaw A (2012) Forensic readiness for wireless medical systems. Available at https://ro.ecu.edu.au/cgi/viewcontent.cgi?article=1107&context=adf. Last accessed 19 July 2022
14. DWP Forensic Readiness Policy (2018) DWP forensic readiness policy Available at: https://assets.publishing.service.gov.uk/government/uploads/system/uploads/attachment_data/file/886724/dwp-forensic-readiness-policy.pdf. Last accessed 19 July 2022
15. Davis J (2019) The 10 biggest healthcare data breaches of 2019, So far. Available at https://healthitsecurity.com/news/the-10-biggest-healthcare-data-breaches-of-2019-so-far. Last accessed 19 July 2022
16. Ehlinger S (2017) Former employee reportedly steals mental health data on 28,434 Bexar County patients. The San Antonio Express-News. Available at: https://www.expressnews.com/business/local/article/Former-employee-reportedly-steals-mental-health-12405113.php. Last accessed 19 July 2022
17. Endicott-Popovsky B, Frincke D, Taylor C (2007) A theoretical framework for organizational network forensic readiness. Available at: https://www.researchgate.net/publication/42803345_A_Theoretical_Framework_for_Organizational_Network_Forensic_Readiness. Last accessed 19 July 2022
18. England.nhs.uk. (no date) NHS England. Interoperability (online) Available at https://www.england.nhs.uk/digitaltechnology/connecteddigitalsystems/interoperability/. Last accessed 19 July 2022
19. Europa.eu. (2020) CORDIS|European commission (online) Available at: https://cordis.europa.eu/project/id/732907. Last accessed 19 July 2022
20. Furneaux N (2018) Investigating cryptocurrencies: understanding the technology. Wiley, IN. Last accessed 19 July 2022
21. Gillum R (2013) From papyrus to the electronic tablet: a brief history of the clinical medical record with lessons for the digital age. https://doi.org/10.1016/j.amjmed.2013.03.024
22. Halperin D, Heydt-Benjamin T, Ransford B, Clark S, Defend B, Morgan W, Fu K, Kohno T, Maisel W (2008) Pacemakers and implantable cardiac defibrillators: software radio attacks and zero-power defenses. Available at: https://scholarworks.umass.edu/cgi/viewcontent.cgi?referer=https://www.google.com/&httpsredir=1&article=1067&context=cs_faculty_pubs. Last accessed 19 July 2022
23. Harbawi M, Varol A (2017) An improved digital evidence acquisition model for the Internet of Things forensic I: a theoretical framework. In: 2017 5th international symposium on digital forensic and security (ISDFS). https://doi.org/10.1109/ISDFS.2017.7916508
24. Hashed Health. (2020). *About*. (online) Available at: https://hashedhealth.com/about/ (Last accessed: 19 July 2022).

25. ISACA (2014) Importance of forensic readiness. Available at: https://www.isaca.org/resources/isaca-journal/past-issues/2014/importance-of-forensic-readiness. Last accessed 19 July 2022
26. Ico.org.uk. (2019) Special category data (online) Available at: https://ico.org.uk/for-organisations/guide-to-data-protection/guide-to-the-general-data-protection-regulation-gdpr/lawful-basis-for-processing/special-category-data/. Last accessed 19 July 2022
27. Iso.org. (2020) (online) Available at: https://www.iso.org/obp/ui/#iso:std:iso-iec:27043:ed-1:v1:en. Last accessed 19 July 2022
28. Karm A (2019) Estonia–the digital republic secured by blockchain estonia -the digital republic secured by blockchain Estonia-the digital republic secured by blockchain PwC 1 (online) Available at: https://www.pwc.com/gx/en/services/legal/tech/assets/estonia-the-digital-republic-secured-by-blockchain.pdf. Last accessed 19 July 2022
29. Kebande V, Venter H (2014) A cloud forensic readiness model using a botnet as a service. Available at: https://www.researchgate.net/profile/Natalie_Walker4/publication/263617788_Proceedings_of_the_International_Conference_on_Digital_Security_and_Forensics_DigitalSec2014/links/0f31753b5cd085c06a000000/Proceedings-of-the-International-Conference-on-Digital-Security-and-Forensics-DigitalSec2014.pdf#page=25. Last accessed 19 July 2022
30. Komperda T (2012) Virtualization security [online]. Available at: https://resources.infosecinstitute.com/topic/virtualization-security/. Last accessed 19 July 2022
31. Kumkar V, Tiwari A, Tiwari P, Gupta A, Shrawne S (2012) Vulnerabilities of wireless security protocols (WEP and WPA2). Available at: https://www.researchgate.net/publication/266005431_Vulnerabilities_of_Wireless_Security_protocols_WEP_and_WPA2. Last accessed 19 July 2022
32. Kyaw A, Cusack B, Lutui R (2019) Digital forensic readiness in wireless medical systems. In: 2019 29th international telecommunication networks and applications conference (ITNAC). Auckland, New Zealand, pp 1–6. https://doi.org/10.1109/ITNAC46935.2019.9078005
33. Lenk W (2020) Wireless mobile medical devices. Available at: https://sites.tufts.edu/eeseniordesignhandbook/2015/wireless-mobile-medical-devices/. Last accessed 19 July 2022
34. NHS Digital (2018) Protecting patient data—NHS digital (online) Available at: https://digital.nhs.uk/services/national-data-opt-out/understanding-the-national-data-opt-out/protecting-patient-data. Last accessed 19 July 2022
35. NHS services hit by cyber-attack (2017) BBC news (online). Available at: https://www.bbc.co.uk/news/health-39899646. Last accessed 19 July 2022
36. National Cancer Institute. (2011). *NCI Dictionary of Cancer Terms.* (online) Available at: https://www.cancer.gov/publications/dictionaries/cancer-terms/def/electronic-medical-record. (Last accessed: 19 July 2022)
37. National Institute of Standards and Technology (2006) Guide to integrating forensic techniques into incident response: publication 800–86. Available at: https://nvlpubs.nist.gov/nistpubs/Legacy/SP/nistspecialpublication800-86.pdf. Last accessed 19 July 2022
38. Neuburger J, Choy W, Milewski K (2020) Smart contracts: best practices. (online) Available at: https://prfirmpwwwcdn0001.azureedge.net/prfirmstgacctpwwwcdncont0001/uploads/dc2c188a1be58c8c9bb8c8babc91bbac.pdf. Last accessed 19 July 2022
39. Nieles M, Dempsey K, Pillitteri VY (2017) An introduction to information security. Available at: https://nvlpubs.nist.gov/nistpubs/SpecialPublications/NIST.SP.800-12r1.pdf. Last accessed 19 July 2022
40. OpenLedger Insights (2019) What are consortium blockchains, and what purpose do they serve? (online) Available at: https://openledger.info/insights/consortium-blockchains/. Last accessed 19 July 2022
41. Park S, Akatyev N, Jang Y, Hwang J, Kim D, Yu W, Shin H, Han C, Kim J (2018) A comparative study on data protection legislations and government standards to implement digital forensic readiness as mandatory requirement. https://doi.org/10.1016/j.diin.2018.01.012
42. Pourmajidi W, Miranskyy A (2018) Logchain: blockchain-assisted log storage (online) IEEE Xplore. Available at: https://ieeexplore.ieee.org/document/8457918. Last accessed 19 July 2022

43. Protenus (2021) 2021 breach barometer (online) Available at: https://www.protenus.com/res ources/2021-breach-barometer/Last accessed 19 July 2022
44. Radcliffe J (2011) Hacking medical devices for fun and insulin: breaking the human scada system. Available at: https://cs.uno.edu/~dbilar/BH-US-2011/materials/Radcliffe/BH_ US_11_Radcliffe_Hacking_Medical_Devices_WP.pdf. Last accessed 19 July 2022
45. Rahman N, Glisson W, Yang Y, Choo K (2016) Forensic by-design framework for cyber-physical cloud systems. EEE Cloud Computing 1(3):50–59
46. Rahman A, Ahmad R, Ramli S (2014) Forensics readiness for Wireless Body Area Network (WBAN) system (online) IEEE Xplore. Available at: https://ieeexplore.ieee.org/document/677 8944. Last accessed 19 July 2022
47. Raju B, Geethakumari G (2016) An advanced forensic readiness model for the cloud environment. In: 2016 international conference on computing, communication and automation (ICCCA). Noida https://doi.org/10.1109/CCAA.2016.7813819.
48. Rowlingson R (2004) A ten step process for forensic readiness. Available at: https://www. utica.edu/academic/institutes/ecii/publications/articles/A0B13342-B4E0-1F6A-156F501C4 9CF5F51.pdf. Last accessed 19 July 2022
49. Ryckaert V (2019) Hackers held patient data ransom, so Greenfield hospital system paid $50,000, The Indianapolis Star. Available at: https://eu.indystar.com/story/news/crime/2018/ 01/17/hancock-health-paid-50-000-hackers-who-encrypted-patient-files/1040079001/.   Last accessed 19 July 2022
50. SecureW2 (2020) WPA2-enterprise and 802.1x simplified. [online] Available at: https://www. securew2.com/solutions/wpa2-enterprise-and-802-1x-simplified. Last accessed 19 July 2022
51. Somasundaram R, Thirugnanam M (2020) Review of security challenges in healthcare internet of things. https://doi.org/10.1007/s11276-020-02340-0
52. Studio Fiorenzi Security & Forensics (2017) GDPR & Forensics Readiness-English (online) Available at: https://www.slideshare.net/AlessandroFiorenzi/gdpr-forensics-readin ess-english. Last accessed 19 July 2022
53. Tan J (2001) Forensic readiness (online) CiteSeer. Available at: http://citeseerx.ist.psu.edu/vie wdoc/download?doi=10.1.1.480.6094&rep=rep1&type=pdf. Accessed 04 Aug 2020
54. Thompson E (2019) Three ways smart contracts are used in healthcare. Available at: https://uk. finance.yahoo.com/news/three-ways-smart-contracts-used-120013678.html?guccounter=1& guce_referrer=aHR0cHM6Ly93d3cuZ29vZ2xlLmNvbS8&guce_referrer_sig=AQAAAB2Bc 57INL-hL55sY-VBV-schgWWyDxqgEOx40lhZSWQDfis2VIALKJ-d-AyHK6GEGXHag 1SY5Lpr09EQntC-IxsCLTx75ejZz3lsqTMRxxUEBHE-HFHfCcbNsPNsubeQtdYLpU1btex vS7tgTmzPVSC-l-rrTbTDonRC1FNHMSR. Last accessed 19 July 2022
55. Tian Z, Li M, Qiu M, Sun Y, Su S (2019) Block-DEF: a secure digital evidence framework using blockchain. Information Sciences (online). Available at: https://www.sciencedirect.com/ science/article/pii/S002002551930297X?via%3Dihub. Last accessed 19 July 2022
56. Vidal C, Choo K (2015) 'The cloud security ecosystem. Available at: https://www.sciencedi rect.com/book/9780128015957/the-cloud-security-ecosystem. Last accessed 19 July 2022
57. Wallace K (2020) Configuring security—wireless networking essential training video tuto-rial|LinkedIn Learning [online]. Available at: https://www.linkedin.com/learning/wireless-net working-essential-training/configuring-security-2?u=42408908. Accessed 30 March 2021
58. World Health Organisation (2011) mHealth, New horizons for health through mobile tech-nologies. Available at: https://apps.who.int/iris/handle/10665/44607. Last accessed 19 July 2022
59. Zheng Z, Xie S, Dai H, Chen X, Wang H (2017) An overview of blockchain technology: architecture, consensus, and future trends. In: 2017 IEEE international congress on big data (BigData Congress). (online) Available at: https://ieeexplore.ieee.org/document/8029379. Last accessed 19 July 2022

Assembly, Deployment and Extension of Blockchain Based Decentralized Autonomous Organizations: A Framework Point of View

Ava Halvai and Umair B. Chaudhry

Abstract Decentralised autonomous organisations, entities built on the blockchain based with no central leadership, are changing the landscape of organisations and governance. Operations governed by smart contracts where bylaws are embedded into code and the decentralised nature of their governance are offering a unique structure to organisations looking for transparency and community engagement. This paper will give an overview of blockchain technologies and DAOs, how computer programming is incorporated into DAOs through the use of smart contracts and how this can be used for governance, in addition to how this differs to centralised organisations. Various DAO frameworks will be explored, compared and analysed such as MolochDAO, MakerDAO and the LAO. From here, the current challenges of DAOs will be explored and recommendations will be proposed followed for DAO frameworks. Finally, having considered the above, the future of DAOs will be discussed, including recommendations, improvements and developments.

Keywords Decentralised Autonomous Organisation (DAO) · Blockchain · Framework · Smart contract

1 Introduction

A decentralised autonomous organisation (DAO) is an organisation built on the blockchain and governed by proposals which are voted on by members of the DAO [1, 2]. Its structure and rules are encoded by smart contracts which take the form of software code built on the blockchain which self-execute once its conditions are met. Transactional decisions are recorded on the blockchain which is not only available for

A. Halvai (✉) · U. B. Chaudhry
Queen Mary University of London, London 1 4NS, UK
e-mail: a.halvai@se21.qmul.ac.uk

U. B. Chaudhry
Northumbria University, Middlesex Street, London 1 7HT, UK

H. Jahankhani (ed.), *Cybersecurity in the Age of Smart Societies*,
Advanced Sciences and Technologies for Security Applications,
https://doi.org/10.1007/978-3-031-20160-8_13

227

anyone to see but is also timestamped and immutable [3, 4]. This underpinning structure results in an entity that is decentralised, autonomous and transparent, shifting away from typical organisational models that are governed by a central body with executive decisions not openly available to the public [5].

Blockchain is the technology on which DAOs are built on. It is a distributed database which acts as a transactional ledger, recording transactions between two bodies and thus consequently tracking the movement and possession of assets [6]. It is composed of 'blocks' which make up an ongoing list of records which are open-source and unable to be modified by anyone. These blocks are joined through each block containing a cryptographic hash of the previous block, along with a timestamp of the transaction, forming a 'chain' as a culmination of transaction records. Therefore, it is impossible to alter one record or block without affecting consequent blocks, making it immutable [7, 8]. Blockchains are decentralised through being managed by a peer-to-peer (P2P) network, a network model that consists of multiple nodes which each store and communicate information with each other, each acting as a peer and bearing equal power and responsibility without the need for a controlling central body (Fig. 1). Each node bears equal power and responsibility as they all interact with each other equally, unlike the traditional central server network where the server acts as the central body [10]. This model, which allows assets to be sent between nodes in a tracked and transparent manner without the need for a middleman, differs greatly from, for example, a bank, where transactions are private and managed solely by the bank itself [9].

Blockchain principles can be traced back to cryptographer David Chaum's 1982 paper "Computer Systems Established, Maintained, and Trusted by Mutually Suspicious Groups" which presents both a proposal and the accompanying code for a decentralised model ran by peers who collectively do not trust each other, but consequently together establish a trustworthy distributed computer system amongst themselves. Using cryptography techniques, a small vault secures the data which can only be controlled through decisions made by the aforementioned peers, otherwise the vault will destroy its internal data in order to protect it [11].

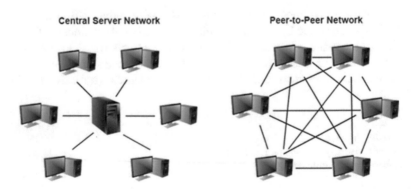

Fig. 1 A central server network compared to a peer-to-peer network

The breakthrough moment for blockchain was the creation of Bitcoin, a decentralised, digital currency and the first cryptocurrency to exist. Invented by an individual under the pseudonym Satoshi Nakamoto who in 2008 wrote and published both the bitcoin white paper and 'Bitcoin Core', its reference implementation, it was a system that included P2P networks, cryptocurrency wallets to hold private and public keys for cryptocurrency transactions and mechanisms to carry out block validation [12]. Nakamoto had referred to 'blocks' and 'chains' as two separate entities in his whitepaper, but the underlying technology that Bitcoin utilised was recognised to have potential in many other fields and use cases beyond currency, leading to the adoption of the term 'blockchain' and the evolution of Blockchain 2.0 around 2014–2015. Blockchain 2.0 established blockchain as a decentralised infrastructure which had the ability to revolutionise a number of industries through decentralising the movement and ownership of assets (both digital and physical), redefining organisational structure and interactions and shifting towards a culture of transparency and decentralisation within these bodies [13].

This evolution led to the emergence of using blockchain to form decentralised, autonomous organisations and consequently led to the creation of the very first DAO in 2016, known as 'The DAO'[14]. The DAO took the form of a set of smart contracts built on the Ethereum blockchain, by several members of the Ethereum community. Founded by Vitalik Buterin, Ethereum is an open source, decentralised blockchain that was launched in 2015. Ethereum expanded upon Bitcoin, adding smart contract functionality which would allow contracts, programs and applications to be built on to it, thus significantly increasing the functionality of blockchain beyond simply an alternative method of payment and data store. The DAO acted as a decentralised venture capital fund, with funds crowdfunded through an initial coin offering (ICO), where tokens are sold by the DAO in exchange for traditional currency or another token such as bitcoin. This was not only a deviation from the traditional forms of raising capital, but at its peak raised a record-breaking $150 million equivalent in ether (ETH), Ethereum's native token [15]. Despite being heralded as a revolutionary project for both crowdfunding and blockchain, it fell victim to what is now known as 'the DAO hack', where a loophole in the DAO's smart contract code was exploited and the DAO was hacked by draining the ETH funds into a smaller DAO with an imitated structure [16]. Not only did this lead to the Ethereum Foundation's controversial decision to hard-fork the entire Ethereum blockchain to restore most of the funds back to the smart contract and the DAO, but also brought up questions about the risks of a DAO's algocracy and the reliance on computer programming, or in Buterin's words, "automation at the centre, humans at the edges [17]."

This led to a near halt in the development of DAOs for a few years, particularly as tokens became the main focus of attention within the Web3 space. This changed in early 2019 with the formation of MolochDAO, kick-starting the re-emergence of DAOs. MolochDAO was created with the similar aim to fund innovation in the space [18], but instead raised funds through offering membership and voting shares in exchange for a submission of a set amount of ETH, provided that current DAO members approved prospective members' membership proposal [18]. However, arguably the most innovative change was MolochDAO's framework that allowed members to 'rage quit', where members could forfeit their membership and voting abilities in exchange for their shares which granted them a proportion of the assets within the DAO's treasury. MolochDAO's framework paved the way for other subsequent DAO's including MetaCartelDAO and Meta Gamma Delta DAO which adopted the same Moloch V1 framework. Since then, this framework has evolved to Moloch V2 and now Moloch V3, with updates to the DAO's framework to improve governance, accessibility and efficiency.

As the framework of a DAO is embedded into smart contracts and written in code, it is particularly intentional and is often decided with thought and care to best benefit the DAO and its shareholders. Various smart contracts for various DAO frameworks have now emerged that serve different missions and DAO structures, from crowdfunding, to collectively investing in non-fungible tokens (NFTs) and physical assets, to simply forming exclusive social communities. Whilst DAOs are still arguably in their infancy and still rather niche, their presence is rapidly growing in today's economy and global cultural landscape. Despite the cryptocurrency market's current decline, the top 4833 DAOs have a total market capitalisation of $6.9 billion, with Uniswap, a decentralised exchange-turned-DAO, maintaining a treasury with a value of $1.5 billion [19]. DAOs have also permeated through social and political current events, including ConstitutionDAO which raised just under $47 million to collectively purchase an original copy of the US Constitution (albeit losing to a higher bid) [20] and UkraineDAO which raised $6.75 million in ETH and was able to send donations to war-torn Ukraine considerably faster than through other crowdsourcing avenues [21].

In this paper, the features of DAOs will be covered in depth, including the use of smart contracts and tokenisation as part of a DAO framework. Various DAO frameworks and their respective versions will be compared, before considering the challenges that current frameworks pose and what recommendations can be made to improve them.

2   Features of a DAO

The formation of a DAO starts with defining its rules and structure via a smart contract, followed by financing the DAO and granting members tokens and voting powers, leading to a DAO which should be autonomous and controlled via a majority vote by DAO members. This is only possible with the following features:

2.1   Decentralisation

DAOs stray away from typical organisational structures which have a central body, board of directors or a CEO deciding the fate of the organisation through a centralised legal entity. Instead, a protocol is written within code on the blockchain which is run

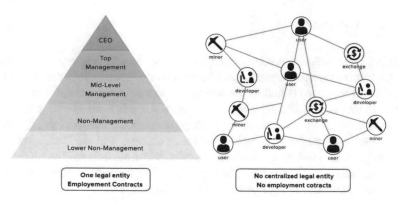

Fig. 2 The different structures between a traditional centralised organisation (left) and a DAO (right)

when a majority of the DAO members vote on it [4, 22]. Rather than a hierarchical structure, each member can put forward a proposal and each member has voting power to vote on it. The network is not left in the hands of a select few, with the hope that this will lead to a more democratic and fair process free of corruption and that members will act in their best interests which will coincide with the best interests of the DAO, as a more robust protocol will lead to higher overall involvement and a consequent increase in the value of the governance tokens that the DAO members own (Fig. 2) A DAO lacks the hierarchal structure of a traditional organisation and instead follows a decentralised model Fig. 2, where each member contributes to the network [27].

2.2 Autonomous

As rules and decisions are encoded in smart contracts, a DAO can operate autonomously largely without the input of humans or external entities. Human involvement is required to a certain extent within DAOs, for example to establish a protocol, create proposals and vote on them, but beyond these a DAO should be able to run by itself as it will only execute operations when the conditions within the smart contract are met. Furthermore, through the use of artificial intelligence (AI), DAOs can be even further automated and autonomous, eliminating the need for human input through automating processes such as managerial operations [23, 24].

2.3   Transparent

DAOs are built using open-source code and are deployed using smart contracts. In addition to an autonomous system, smart contracts also contribute to the transparency of DAOs. As they are essentially software code stored on the blockchain, the transparency of blockchain also pertains to the transparency that exists within DAOs. All assets and transactions including proposals, payments and decision-making made by the DAO are available to view by using a block explorer and the DAO's cryptocurrency wallet address. A block explorer provides real-time information on the assets and transactions owned and carried out by that wallet, a level of transparency that is unheard of in traditional organisations or corporations [4].

2.4   Secure

The inability to alter any single block within the blockchain results in smart contracts that are unable to be censored or altered, ensuring that transactions cannot be carried out or hidden by a single person. This also extends to the treasury, where no single DAO member or shareholder has the authority to access the DAO funds. Issues with the smart contract code including bugs and loopholes can lead to the DAO's security being compromised, but in theory a DAO should be a secure, immutable entity.

2.5   Based on Smart Contracts

In its simplest form, a smart contract is programmable code stored on the blockchain which executes a mechanism involving assets between at least two parties. This mechanism allows a transaction of these assets only when the conditions within the smart contract are met [25]. An example of a smart contract is one that enables the deployment of a multi-signature wallet, where funds from the wallet are distributed as soon as a set number of parties sign off on the transaction proposal, something which is often used for DAO treasuries. However, it is not necessary for smart contracts to be manually initiated at a given time, e.g., conditions for a smart contract can include changes in the market or a particular date.

2.6   Financing and Tokens

Typically, DAOs are financed through the sale of native tokens, of which the profits go towards the DAO's treasury. Native tokens are cryptocurrencies tied to the particular project or DAO, e.g., LEX is the native token for LexDAO. In return, those who buy tokens are given voting rights within the DAO. Consequently, these are also referred to as 'governance tokens' [26]. This system not only gives members equity in the DAO, but also encourages members to maintain and improve the DAO's protocol as this will potentially drive up the price of the token. However, certain DAOs e.g., MolochDAO, do not utilise native tokens, instead taking ETH as a form of payment to join and have voting rights.

3 Frameworks

DAOs are a recent phenomenon and with various founding teams, smart contracts, blockchains and missions, DAO frameworks vary significantly and consequently it can be difficult to pigeonhole DAOs into a unified framework. At the most basic level, DAOs use the following model:

1. **Smart contracts**: Created as the foundation for the DAO ('code-is-law'), including protocols to cover governance, consensus mechanisms, token issuing and on-chain collaboration.
2. **Funding**: ICOs or DAO membership fees to initially crowdfund, followed by long-term raising through dividends from investments made by the DAO amongst other strategies.
3. **Deployment**: Smart contracts are deployed on the blockchain and cannot be altered without shareholders voting on a proposal to do so.

3.1 Minimum Viable DAO

In order to look more in depth and attempt to comprehensively define a basic DAO framework, MolochDAO's first iteration (based off the the Moloch V1 primer) can be used as an example, as it was one of the first DAOs built and was built with the intention of being a "Minimum Viable DAO".

3.2 Smart Contracts

The first iteration of MolochDAO has two simple smart contracts: 'Moloch.sol' which contains code to carry out transactions relating to proposals, shareholder activity and voting rights, and 'GuildBank.sol' which handles the assets of the DAO. Any ETH submitted with a membership proposal is deposited into the GuildBank.sol smart contract, which is only retained if the proposal is successfully voted on by the current shareholders of the DAO [18].

3.3   Shares and Governance

Shares can be bought in exchange for membership, but membership proposals must be put forward by current members on behalf of others. Prospective members can request a certain number of shares as part of their proposal and if successful, have voting rights that are weighted according to the shares they own. These shares, along with the governance powers that accompany them, cannot be transferred to another member.

3.4   Rage Quit

Members can 'rage quit' and forfeit their shares and governance powers in exchange for the equivalent of their shares in ETH from the overall treasury. The [ragequit] function is available for a period of seven days to any member after voting No on a proposal, an innovative addition to the DAO framework which furthermore created an incentive for members to refrain from putting forward controversial proposals which would lead to other members rage quitting.

The impact of this simple framework was vast, being adopted by prominent DAOs such as MetaCartelDAO, Meta Gamma Delta DAO and Raid Guild, which still use the Moloch framework today.

4   Improved and Alternative DAO Frameworks

MolochDAO was upgraded with the Moloch V2 framework, which allowed over 200 ERC20 (a protocol used to issue tokens on the Ethereum platform) tokens to be held by the DAO other than just ETH. The type of on-chain proposals extended to additional proposal types including requests to add a new ERC20 token, fund a new project within the DAO, interact with external smart contracts and membership proposals [28]. Moreover, individuals who were not members could put forward proposals, increasing both accessibility to the DAO and the number of proposals put forward. The proposal workflow changed so that a proposal has to be sponsored then placed in a queue before it reaches the voting period, unlike in V1 where a proposal was moved straight to voting, although still ensuring that a filter is put in place for incoming proposals to manage the influx of proposals efficiently (Figs. 3 and 4) [29].

With the re-emergence of DAOs in the Web3 ecosystem, other frameworks have emerged to suit the particular DAO's mission or community. Protocol DAOs such as MakerDAO, a protocol which allows individuals to lend and borrow cryptocurrency and generates the stablecoin Dai (DAI), have frameworks in place revolving around governance for adjusting key parameters within the protocol. Anyone can make a proposal, similarly to MolochDAO, but those with the native token MKR have voting

Fig. 3 Proposal workflow in Moloch V1

Fig. 4 Proposal workflow in Moloch V2

rights surrounding these additional changes that include altering the stability fees, the Dai savings rate and even triggering an 'Emergency Shutdown' in the event that the DAO infrastructure is under attack [30]. The framework of this DAO largely prioritises governance flexibility to adapt to real-time changes that occur externally to the DAO.

DAOs which were originally built on Moloch V1, such as the LAO, have retained certain features but have partially deviated from the Moloch V2 framework in order to suit their own needs. A venture capital fund DAO which is also registered as a limited liability company (LLC), the LAO is compliant with US law but carries out its proceedings on-chain [31]. Similarly to MolochDAO, voting rights are weighted depending on the shares of native tokens that the member owns, however, unlike MolochDAO the LAO allows voting rights to be delegated to others through the LAO decentralised app (DApp), combatting voter apathy and increasing positive participation of third parties in the DAO.

Moloch V2 allows more individuals, including those who are not shareholders, to submit proposals by following a workflow where their proposal can be sponsored by a current DAO member [29].

4.1 Future Frameworks

In February 2022 Moloch V3 was announced, the latest iteration of the Moloch framework. Yet to be deployed, this framework offers both increased functionality and flexibility as MolochDAO transitions from its purpose of solely funding projects deployed on Ethereum to widening its uses in the ecosystem. One example of a change to further the DAO's flexibility is the ability to carry out various arbitrary external interactions by altering the length of time that a proposal is active, depending on the circumstances, creating flexible governance so that the DAO can interact directly with a range of external smart contracts. The V3 framework has also introduced 'shamans', external contracts that can carry out operations such as minting and burning shares and altering governance parameters without the need for a DAO proposal, significantly increasing the DAO's efficiency and speed. Additionally, the V3 framework allows for a Gnosis Safe (a popular platform for digital assets) to have the V3 framework added onto it to upgrade it to a comprehensive governance system or DAO without the need to move any funds. Through a 'Compound Governor', the delegation of voting power will also finally be carried out in Moloch V3, and both shares and loots (non-voting shares) will be able to be transferred and will be ERC20 tokens, allowing the DAO to interact with a range of ERC20-based DApps for use cases such as token gating [32].

5 Challenges and Recommendations

As DAOs and their frameworks develop to adapt to their changing ecosystems, new challenges will continue to arise. Common issues that DAOs face are outlined below with framework improvement recommendations to overcome these challenges.

5.1 Voter Apathy

A reduction in voter turnout for proposals and a general low interest in on-chain governance has continued to be an ongoing problem as DAOs increase in size but decrease in voter turnout, leading to proposals being passed at a lower and slower rate, a halt in the development of the DAO overall and decisions being influenced by a small number of individual who are large shareholders and thus straying from the principles of a decentralised body [33]. One potential solution is to incentivise voting, whether that is through awarding native tokens or paying governors for 'showing up'. A gamification element could also be utilised within the DAO framework to encourage voter participation. The delegation of voting powers is something that has already been seen in Moloch V3 and the LAO, amongst others, but the challenge of voting apathy is still apparent in many DAOs. One of the most common roadblocks as to why individuals do not participate in on-chain governance is poor voting infrastructure and tools which considerably discourage stakeholders from voting. Improving these tools and the user experience for governors, combined with the solutions above would be a viable solution to increase voter turnout [34].

5.2 Risk of Centralisation

Despite the name, DAOs often run the risk of straying from their decentralised principles. Tools such as multi-signature wallets, requiring a select few to sign off on transactions to distribute funds from the treasury are an integral component of many DAOs' treasury. Relying on a select few to be responsible for this (especially when those select few remain are the founding members of the DAO) in addition to communities suffering from voter apathy can lead to decisions in the DAO being made by very few members who happen to have a large proportion of governance tokens. Possible solutions that have been explored include a system of one vote per DAO member, exploring other decentralised forms of governance or voting rights proportional to a ranking based on reputation or engagement in the DAO [33, 35].

5.3 Not Everyone is an Expert on Everything

As DAOs grow and increase in numbers, it can be counterproductive to give every token holder governance over every proposal as stakeholders will unlikely be experts in every field [35, 36]. Domain knowledge is required to decide on whether to pass certain proposals and not every stakeholder will be equipped to vote on these. Instead, DAOs can redistribute governance in a decentralised manner to give stakeholders voting rights in their particular domain. This grouping of governors can itself be voted on by the DAO and can potentially both increase voter turnout and the quality of decisions decided upon.

Other areas for improvement include lowering the entry barrier in regards to purchasing governance tokens. Whilst the purchase of governance tokens can be used as an effective quality control measure, ensuring that those who wish to have governance powers commit to this cost, it decreases the DAO's accessibility as many cannot afford the price that comes with a DAO membership proposal. Lastly, to further the community-first mission of DAOs and the wider Web3 space, inter-DAO connectivity would be a recommended area to explore and one which could be facilitated through changes to the DAO framework. Examples include token swaps and mergers, where DAOs are able to use each other's infrastructure and benefit from each other's community and overall growth.

6 Conclusion

As the popularity and utility of DAOs increase, their frameworks must evolve to adapt to both changes in the external environment and to internal changes and growth. Many DAOs have emerged since the formation of MolochDAO and have formed their own frameworks to suit their unique goals and community. In this paper, the features associated with DAOs are overviewed before delving into the formation of one of the original DAO frameworks, the Moloch framework, and its evolution from the V1 framework to the recently launched V3. These framework iterations are compared to other DAO frameworks such as those that form MakerDAO and the LAO. Current challenges associated with DAOs are analysed and future solutions related to changes in frameworks are proposed.

References

1. Wang S, Ding W, Li J, Yuan Y, Ouyang L, Wang F-Y (2019) Decentralized autonomous organizations: concept, model, and applications. IEEE Trans Comput Soc Syst 6(5):870–878
2. Liu L, Zhou S, Huang H, Zheng Z (2021) From technology to society: an overview of blockchain-based DAO. IEEE Open J Comput Soc pp 1–1
3. Szalachowski P (2018) (Short Paper) Towards more reliable bitcoin timestamps. In: 2018 crypto valley conference on blockchain technology (CVCBT), pp 101–104
4. Faqir-Rhazoui Y, Arroyo J, Hassan S (2021) A comparative analysis of the platforms for decentralized autonomous organizations in the Ethereum blockchain. J Internet Serv Appl 12:9
5. Law AW, Benjamin N (2021) The rise of decentralized autonomous organizations: opportunities and challenges. Stanf J Blockchain Law Policy
6. Xu M, Chen X, Kou G (2019) A systematic review of blockchain. Financ Innov 5:27
7. Casino F, Dasaklis TK, Patsakis C (2019) A systematic literature review of blockchain-based applications: current status, classification and open issues. Telemat Inform 36:55–81
8. Anascavage R, Davis N (2018) Blockchain technology: a literature review, SSRN
9. Donet J, Pérez-Solà C, Herrera-Joancomartí J (2014) The bitcoin P2P network
10. Nfctouch.com.hk (2020) The beginner's guide to the emerging technologies. [online] Available at: https://nfctouch.com.hk/beginner_guide.html. Accessed 22 June 2022
11. Chaum LD (1982) Computer systems established, maintained, and trusted by mutually suspicious groups. University of California, Berkeley
12. Nakamoto S (2019) Bitcoin: a peer-to-peer electronic cash system
13. Batra G, Olson R, Pathak S, Santhanam N, Soundararajan H (2020) Blockchain 2.0: What's in store for the two ends—semiconductors (suppliers) and industrials (consumers)? Blockchain 2.0: What's in store for the two ends—semiconductors (suppliers) and industrials (consumers)? Blockchain 2.0: What's in store for the two ends—semiconductors (suppliers) and industrials (consumers)?
14. Morrison R, Mazey NCHL, Wingreen SC (2020) The DAO controversy: the case for a new species of corporate governance? Frontiers in Blockchain, vol 3
15. Daian P (2016) Analysis of the DAO exploit. [online] Hacking Distributed. Available at: https://hackingdistributed.com/2016/06/18/analysis-of-the-dao-exploit/. Accessed 22 June 2022
16. Dika A (2017) Ethereum smart contracts: Security vulnerabilities and security tools. M.S. thesis, Dept. Comput. Sci., Norwegian Univ. Sci. Technol., Trondheim, Norway
17. Delmolino K, Arnett M, Kosba A, Miller A, Shi E (2016) Step by step towards creating a safe smart contract: lessons and insights from a cryptocurrency lab. In: Proceedings of the international conference financial cryptography and data security, pp 79–94
18. Soleimani A (2019) moloch/v1_contracts. [online] GitHub. Available at: https://github.com/MolochVentures/moloch/tree/minimal-revenue/v1_contracts. Accessed 22 June 2022
19. Deepdao.io. DeepDAO (2022) Available at: https://deepdao.io/organizations. Accessed 22 June 2022
20. Daian P (2016) Analysis of the DAO exploit. [online] Hacking Distributed. Available at: https://hackingdistributed.com/2016/06/18/analysis-of-the-dao-exploit/. Accessed 22 June 2022
21. Roberts D (2022) What DAOs can Do: $6.75M in Ethereum for Ukraine. Decrypt. Available at: https://decrypt.co/94386/ukraine-dao-millions-in-ethereum-shows-what-dao-can-do. Accessed 22 June 2022
22. Hassan S, De Filippi P (2021) Decentralized autonomous organization. Internet Policy Rev 10
23. Rauff JV (1999) Multi-agent systems: an introduction to distributed artificial intelligence. Addison-Wesley, Reading, MA, USA
24. Wang S, Wang FY (2018) A general cognitive architecture for agent-based modeling in artificial societies. IEEE Trans Compu Soc Syst 5(1):176–185
25. Beklemysheva A (2020) Making effective use of smart contracts. [online] Steelkiwi.com. Available at: https://steelkiwi.com/blog/making-effective-use-of-smart-contracts/. Accessed 22 June 2022

26. Kondova G, Barba R (2019) Governance of decentralized autonomous Organizations. 15
27. Ivan on Tech (2022) DAO versus Traditional Organization. [online] Moralis Academy. Available at: https://academy.moralis.io/blog/dao-vs-traditional-organization. Accessed 22 June 2022
28. GitHub (2020) GitHub - MolochVentures/moloch. Available at: https://github.com/MolochVentures/moloch. Accessed 22 June 2022
29. Turley C (20200) Moloch evolved: V2 primer. [online] Medium. Available at: https://medium.com/raid-guild/moloch-evolved-v2-primer-25c9cdeab455. Accessed 22 June 2022
30. Makerdao.com (2020) The maker protocol white paper|Feb 2020. Available at: https://makerdao.com/en/whitepaper/. Accessed 22 June 2022
31. GitHub (2021) GitHub—openlawteam/lao-docs. [online] Available at: https://github.com/openlawteam/lao-docs. Accessed 22 June 2022
32. GitHub (2020) GitHub—Moloch-Mystics/moloch-v3: MolochDAO smart contracts Available at: https://github.com/Moloch-Mystics/moloch-v3. Accessed 22 June 2022
33. Lassen D (2005) The effect of information on voter turnout: evidence from a natural experiment. Am J Polit Sci 49(1):103–118
34. Learner R (2019) Blockchain voter apathy. [online] Medium. Available at: https://medium.com/wave-financial/blockchain-voter-apathy-69a1570e2af3. Accessed 22 June 2022
35. Rachmany G (2020) The good, the bad and the DAOs only a founder could love in. coindesk. Available at: https://www.coindesk.com/tech/2020/12/18/the-good-the-bad-and-the-daos-only-a-founder-could-love-in-2020/. Accessed 22 June 2022
36. The Defiant. (2021) The Defiant: the problem with voting in DAOs. Available at: https://thedefiant.io/the-problem-with-voting-in-daos/. Accessed 22 June 2022

Artificial Intelligence Techniques in Cybersecurity Management

Mercy Ejura Dapelⓘ**, Mary Asante, Chijioke Dike Uba, and Michael Opoku Agyeman**ⓘ

Abstract The rapid development in internet services led to a significant increase in cyberattacks. The need to secure systems and operations has become apparent as cybersecurity has become a national concern. Cybersecurity involves techniques that protect and control systems, networks, hardware, software, and electronic data from unauthorised access. Developing an effective and innovative defensive mechanism is an urgent requirement as traditional cybersecurity solutions are becoming inadequate in safeguarding information against cyber threats. There is a need for cybersecurity methods that are capable of making real-time decisions and respond to cyberattacks. To support this, researchers are focusing on approaches like Artificial Intelligence (AI) to improve cyber defence. This study provides an overview of existing research on cybersecurity using AI technologies. AI technologies made a remarkable contribution in combating cybercrimes with significant improvement in anomaly intrusion detection.

Keywords Artificial intelligence · Cyberattacks · Cyber threats · Cybersecurity

1 Introduction

The rapid development in information and communication technology (ICT) created positive implication to the global economy. The internet has improved the quality of life by providing a platform that facilitates knowledge sharing, communication and interaction which is important for development and daily life [1]. In view of the benefits, the dark side abound as cybercriminals exploit the vulnerability of individuals who use computer networks and rely on third party and cloud based data storage [2]. Providing security for systems has become difficult. Hackers are

M. E. Dapel · C. D. Uba · M. O. Agyeman (✉)
Centre for Advanced and Smart Systems (CAST), University of Northampton, Northampton, UK
e-mail: Michael.OpokuAgyeman@northampotn.ac.uk

M. Asante
University of Warwick Coventry, Coventry, UK

becoming smarter and more innovative in exploiting individuals and organisations. With cyberattacks and data breaches coming to light daily, cyberattacks have been ranked among the top 5 most likely sources of severe global risk [3]. Cyber fraud have become complex to track as cyber theft can originate from any part of the world. Organisations have become challenged with the complexity of cyberattacks which calls for the adoption of AI or intelligent methods to mitigate them. AI is a rapid growing technology that can analyse millions of datasets to track down cyber threats and prevent data breaches. AI is a thriving field that has been deployed in application areas such as manufacturing [4], healthcare [5], education [6], agriculture [7] and Cybersecurity. According to Abraham et al. [8] AI algorithms can predict previously seen and unseen attacks, they have demonstrated effectiveness in detecting cyber-attacks with low false alarm rate.

Advancement in AI has produced technologies that can learn from past patterns to improve future experiences. Researchers and developed countries have adopted cybersecurity solutions like AI to improve cyber defence [9].

2   Related Surveys on AI in Cybersecurity

Vinayakumar et al. [10] proposed a highly scalable and hybrid deep neural network to monitor network traffic and host level events to raise alert for unforeseen cyber-attacks proactively. They employed distributed and parallel machine learning algorithms with optimization techniques making them capable of handling a high volume of network and computing resources. Their scalability and real-time detection of malicious activities from early warning signals made their framework stand out. Artificial neural network (ANN) approach was employed as the computational model. To increase training speed and avert over fitting, batch normalization and dropout approach was used. Machine learning techniques were compared, deep neural network performed well by detecting and classifying unforeseen and unpredictable cyber-attacks in real-time.

Sokolov et al. [11] analysed cybersecurity threats in cloud applications using deep learning techniques to monitor data. Suricata engine and module based on Google tensor flow framework was used. They proposed a system that used neural classifiers for network traffic, spam comments, spam email and images. The suricata engine is capable of real-time intrusion detection, inline intrusion prevention and network security monitoring.

Maimo et al. [12] explored a self-adaptive system for anomaly detection that identifies cyber-threats in 5G mobile networks, deep learning techniques was used to analyse network traffic by extracting features from network flows. The authors proposed a high-level cyber defence architecture consisting of virtualized infrastructure (VI), virtualized network function (VNF), management and orchestration (MANO), operations and business support systems. Anomaly symptom detection (ASD) and network anomaly detection (NAD) were proposed to achieve effective network anomaly detection. Once an anomaly is produced from traffic generated, it is communicated to the monitoring and diagnosis module. The experimental result showed that the architecture can self-adapt to anomaly detection based on the volume of network flows gathered from users in real-time.

A botnet is one of the significant threats infecting devices today. Abraham et al. [8] compared the performance of five (5) machine learning approaches and identified useful features to classify malicious traffic. Random forest proved to be more robust, it could generalise unseen bots' types.

Intrusion detection technology is a mechanism that monitors and prevents system intrusion. Zhang et al. [13] introduced a multiple-layer representation learning model for accurate detection of network-based attack and proposed a new data encoding scheme based on P-Zigzag to encode network traffic into two dimensional gray-scale images for representation. Comparing the combination of gcForest and CNN allowed detection of imbalanced data with fewer hyper parameters, which increased computational efficiency. The experimental results showed that the combined algorithms outperform single deep learning methods in terms of accurate detection and false alarm rate thereby demonstrating its effectiveness in attack detection. The authors proposed a new intrusion detection method by combining random forest and LSTM to address the above challenges.

In view of the vast amount of data generated daily, and the increased interconnection of the internet infrastructure, Zhong et al. [14] proposed big data based on a hierarchical deep learning system that utilizes behavioral features. Companies can adapt it as a solution for the detection of intrusive attacks. The authors defined the hierarchical structure in five (5) phases. In the first phase, behavioral and content features are extracted using big data techniques. In the second phase, the dataset is separated into clusters, in the third phase, the root clusters of each sub tree is combined until the quality of the merged clusters dropped below the given threshold. In the fourth phase the deep learning model for each cluster was trained, while in the fifth phase, deep learning model was merged to select the most confident model and concluded that it increased the detection rate of intrusive attacks when compared to a single model learning approach. Their strategy is effective in capturing data patterns for intrusive attacks.

Dey [15] proposed the effectiveness of attention mechanism for intrusion detection based on Convolutional neural network (CNN) and LSTM model utilizing a 2018 dataset. The authors observed increased performance based on LSTM.

Dawoud et al.'s [16] concept is based on unsupervised deep learning for revealing network threats and detecting anomalies by evaluating the use of restricted Boltzmann machines. This intrusion detection system is used to expose network threat and protect network assets. Their simulation study showed 99% detection accuracy with significant improvement.

Ishaque et al. [17] explored deep learning research by manipulating large amount of data using the functionality of computational intelligence. An important feature which the authors applied for dimensionality and attribute reduction is feature extraction. They concluded that the proposed system can detect attacks that are not hybridized.

Distributed denial of service (DDOS) attack has been a real threat to cyber infrastructure that can bring down ICT infrastructure. Mat et al. [18] adopted deep learning to analyse traffic, focusing on mitigating cyber-attacks with machine learning. Assembly module for statistics collection and adaptive machine learning module for analysing traffic and enforcing policies are the two main functionalities proposed. Auto encoder and random forest algorithm possessed an accuracy of 98.4% with a decreased amount of training and execution time. The result proved that the model is optimally efficient for real-time intrusion detection.

Detecting cyber-attacks requires analysing cyber-threat to match potential attack profiles by filtering malicious connections to improve the accuracy of threat detection and reduce false-positive rates. Lin et al. [19] focused their study on network intrusion detection using enhanced CNN based on Lenet 5 to classify network threats. The authors developed an improved behaviour-based model for anomaly detection by training a CNN to extract enhanced behaviour features and identify threats. Their experiment showed overall prediction accuracy with 97.53% intrusion detection rate. The proposed model improves the accuracy of intrusion detection for threat classification.

Zeng et al. [20] proposed a deep full range (DFR) framework comprising a network of encrypted traffic classification and intrusion detection. Three deep learning algorithms (CNN, LSTM and stack auto encoder SAE) were employed for traffic classification and intrusion detection. CNN was used to learn features of the raw traffic; LSTM was used to learn features from time-related aspects and SAE was used to extract features from coding characteristics. The full range consists of three algorithms capable of classifying encrypted and malware traffic within one framework without human intervention. The authors proved that the DFR could attain a robust and accurate performance on both encrypted traffic classification and intrusion detection.

Dey et al. [21] proposed Gated recurrent unit (GRU)-LSTM using Google's tensor flow that provided options to visualize network design. Their analysis showed that GRU-LSTM provided high accuracy with low false alarm rate. When compared GRU-LSTM showed a strong potential in terms of accuracy for anomaly detection.

Hsu et al. [22] proposed a deep reinforcement learning-based (DRL) for anomaly network intrusion detection. Their design revealed incoming network traffic by data sniffing and a pre-processing data module that checks the quality of data before it is fed for intrusion detection. This method can be adopted for self-updating and detecting abnormal incoming network traffic on real-time basis in company websites. SVM and Random Forest algorithms were used. They showed the highest anomaly detection accuracy and improved performance of processing speed.

Privacy protection and national security in the cyber world depends on safe cyberspace. Network intrusion is one of the sophisticated actors stemming from cyber-threats. Sezari et al. [23] applied a deep feed forward network by modifying the parameters of the anomaly-based network. Their result demonstrated better performance with less complexity and a low false alarm rate. Therefore, their model is trustworthy and can be used to prevent intruders. It can detect unknown attacks based on its network features.

Naseer et al. [24] investigated the suitability of deep learning approaches for anomaly-based intrusion detection. They developed a model based on ANN, Autoencoder and RNN. The models were trained on NSL KDD training dataset and evaluated on the test dataset provided by NSL KDD. A Graphic Processing Unit (GPU) powered test bed using keras with theano backend was employed. A comparison between DNN and conventional machine learning models was carried out where both DCNN and LSTM models showed exceptional accuracy on the test dataset, this demonstrates the fact that Deep learning is a feasible and promising technology for intrusion detection.

Anomaly detection has received considerable attention in cybersecurity. The clandestine nature of cyber-attacks increased considerably where malware is installed through a supply chain. Malware eavesdrops and disrupts information exchange. Khaw et al. [25] proposed a deep learning based cyber-attack detection system to detect cyber-attack 25 min after the cyber-attack begun to enhance cybersecurity at its embryonic stage.

Several industries have adopted the Industrial Internet of Things (IIoT) in smart homes, smart cities, connected cars and supply chain management which introduced new trends in business development. However, these edge devices have become exploitation points for intruders, it raised security and privacy challenge to the trustworthiness of edge devices by compromised devices that transmit false information to cloud servers. An intrusion detection system (IDS) is widely accepted as a technique to monitor malicious activities [26].

Huma et al. [27] proposed a detection approach deployed to secure incoming and outgoing traffic, they utilized the application of deep random neural network with multilayer perceptron and evaluated the scheme using two datasets DS205 and UNSW-NB15. They proposed a deep learning based cyber-attack detection system that detects cyber-attack 25 min after the attack was initiated to improve cybersecurity at its embryonic stage. It provided performance metrics like accuracy, precision, recall and F1 score which can be compared with several state-of-the-art attack detection algorithms. Classification of 16 different attacks was proposed, and accuracy of 98 and 99% was achieved.

The growth of modern cyber infrastructure made network security more important. It is estimated that a trillion devices will be connected to the Internet by 2022 [28]. Intrusion Detection Systems are tools with objective to detect unauthorized use and abuse a host network [29].

3 AI in Cybersecurity

AI was proposed in 1956 by John McCarthy as a science concerned with making computers behave intelligently like humans [8]. AI application has evolved significantly, it has a plethora of benefits in education, biometric systems, Internet of Things and cybersecurity among others. AI algorithms contribute to solving security issues, it utilises algorithms that make predictions and analysis possible of cyber threats in real-time. Neural networks have been used to detect classifying data as normal and abnormal [30]. Swarm intelligence methods handle feature selection to identify new intrusion. Technologies like expert systems and intelligent agents are applied to secure internet networks and improve intrusion detection performance [31]. With AI, complexity and model training times is reduced. Presenting new algorithms is a challenge and an opportunity for researchers [32]. AI is quickly becoming a tool for automating threat detection and responding effectively than traditional human driven methods which are unable to keep up with volumes of viruses generated daily [30]. AI is relevant in threat detection, intrusion detection, fraud detection and Cybersecurity thereby increasing accuracy and speed of cyber response. The major disciplines in AI are fuzzy logic, natural language processing, deep learning, machine learning, robotics and computer vision.

3.1 AI as a Tool in Combating Cyberattacks Why Cybersecurity is Essential

With the pace and increase in cyber-attacks, human intervention is insufficient for timely and appropriate response. AI technology is becoming very essential to information security, it is capable of analysing millions of data to track down cyber threats. It can deduce patterns and identify abnormalities in a computer network expeditiously. AI technologies use behavioral analysis to identify and detect anomalies that are indicative of an attack [33]. This technology gathers large amount of data to identify suspicious behaviour that might lead to cyber threat. Processing and analysing massive amount of data in seconds using AI algorithms makes prediction of cyber threats possible before they occur, it also predicts future data breaches. With AI breaches can be responded to immediately an attack is detected by responding anonymously without human intervention and also by sending alerts and creating defensive patches [34]. According to a report by Capgemini, the effort and cost of detecting and responding to cyber threats is lowered by 15% in some organisations with AI as more data is analysed. This technology learns from past patterns to become proficient in identifying suspicious activities thereby protecting information [35]. AI capabilities and adaptive behaviour can overcome the deficiencies of traditional cybersecurity tools.

3.2 Why Cybersecurity is Essential

In the current digital age, data is susceptible to unauthorized access [36]. We live in a world where data is connected and stored on devices. This data contains sensitive information such as personal information, financial data, intellectual property and other forms of data which are exposed to unlawful access [37]. The world's fastest growing crimes are cyber-attacks. Cybercriminals have become smarter and their tactics are resilient to conventional cyber defense mechanisms. According to cyber-security ventures report, global cybercrime is expected to grow by 15% yearly over the next 5 years with financial loss reaching about $10.5 trillion USD annually [39]. Data breach report that 43% of breaches are targeted at businesses [40]. Not only are businesses and organisations at risk, individuals are also at risk. It is important that everyone is aware of hazards associated with internet network use.

3.3 AI Algorithms in Cybersecurity

Several algorithms were identified from the primary studies. The dominant algorithms were Random forest (RF), Long Short-Term Memory (LSTM), Decision Tree (DT), Naive Bayesian algorithm, Adaptive Boost (AdaBoost), J48, Support Vector Machines (SVM), K-Nearest Neighbourhood (KNN), Convolutional Neural Network (CNN), Artificial Neural Network (ANN), Fuzzy logic, Particle swarm optimization (PSO), Logistic Regression and Recursive Neural Network (RNN).

3.4 Impact of AI in Cybersecurity Management

AI presents advantages in several areas, cybersecurity is one of them. AI technology is capable of analysing millions of datasets to track cyber threats. The most significant contribution of AI is anomaly intrusion detection. This study reported improvement in accuracy, intrusion detection rate with reduced false alarm. Sezari et al. [23] demonstrated the performance of a system while comparing the false alarm rates of models on KDD 1999 Cup dataset, they applied a highly optimized deep feed forward network by the modification of the model parameters. Their model achieved a highly accurate low false alarm and detection rate which can be used to detect and prevent intruders. Utilising deep learning provided a system behaviour model that selects abnormal behaviour and is reliable with less complexity. Hsu et al. [22] monitored network traffic to detect abnormal activities and ensured security of communication and information using network intrusion simulation datasets (NSL-KDD and UNSW-NB15) on a real campus network. They proposed a deep reinforcement learning-based (DRL) system with self-updating ability to detect abnormal incoming traffic. Dawoud et al. [16] explored the applicability of deep learning to detect anomaly in Internet of Things (IoT) architecture. They proposed an anomaly detection framework by evaluating the use of Restricted Boltzmann machines as generative energy-based model against auto encoders. The study showed approximately 99% detection accuracy. Deep learning algorithms showed positive results and achieved highest detection accuracy with high-performance speed that is effective in detecting false alarm rate (FAR), they can detect previously seen and unseen threats, however deep neural network could perform better when given more data.

Securing a large network in real-time is a challenge that was identified. Several studies focused on intrusion detection to analyse network traffic by extracting features from network flows and traffic fluctuation. Deep learning algorithm can self-adapt to anomaly intrusion detection and predict network attacks, this was demonstrated in a study conducted by Maimo et al. [12]. Abraham et al. [8] compared several machine learning algorithms, Random Forest had a superior model, it performed optimally for anomaly detection using cross-validation, and their overall result revealed that previously seen and unseen anomaly-based intrusion can be detected.

An improvement in the reduction of false-positive alerts that enabled rapid response to cyber-threat was observed while using Artificial Neural Network (ANN) [41]. CNN can detect anomalies in industrial control systems by detecting majority of attacks with low false positive rate [42]. A study conducted by Hashim et al. [43] showed that LSTM has high detection accuracy in securing websites from external breaches. Vinayakumar et al. [10] analysed ransomware attacks and focused on Twitter as a case study, they concluded that deep learning can be used to monitor online posts and provide early warning about ransomware spread.

3.5 Cybersecurity Frameworks and Solutions

This section presents analysis of cybersecurity frameworks and solutions (see Table 1). This table present cybersecurity solutions with the viewpoint from 2018 to 2021.

Table 1 Cybersecurity frameworks and solutions

Author	Framework/Solutions	Cybersecurity solution used	Focus	Future development
Johansson [44]	Countermeasures against coordi-nated cyber-attacks towards power grid systems	Cryptography	Intrusion detection	Investigate specific countermeasures against coordinated cyber-attacks that must be tai- lored towards critical infrastructure beyond power grids
Vinayakumar et al. [10]	Deep learning approach for intelligent intrusion detection system	Deep neural networks	Intrusion detection	Adding a module for monitoring DNS and BGP events in the network by adding more nodes to the cluster. Improve performance by training complex DNNs architecture on advanced hardware through distributed approach
Sokolov et al. [11]	Analysis of cybersecurity threats in cloud applications using deep learning techniques	Deep neural networks	Intrusion detection	Build a comprehensive framework for cybersecurity threat detection in the cloud

(continued)

Table 1 (continued)

Author	Framework/Solutions	Cybersecurity solution used	Focus	Future development
Maimo et al. [12]	A self-adaptive deep learning-based system for anomaly detec- tion in 5G networks	Deep Neural networks	Anomaly detection	Extend experimental work related to detection/classification of deep learning models. Train two different levels using real data to evaluate the accuracy of anomaly detection
Abraham et al. [8]	A comparison of Machine learning approaches to detect Botnet Traffic	Random Forest	Anomaly-based intrusion detection	Extend this approach to identify botnet traffic across a large network. Setup an online, quick response system that can identify, trigger and quarantine botnet traffic
Lin et al. [45]	Using Convolutional Neural Networks to network intrusion detection for cyberthreats	CNN	Intrusion detection	Not stated in paper
Karatas et al. [46]	Deep learning in intrusion detection systems	Deep learning	Network intrusion detection	For future work using newest datasets with alternative deep learning approaches will be helpful
Zhang et al. [13]	A multiple-layer representation learning model for network-based attack detection	DNN (CNN)	Intrusion detection	gcForest and CNN can be applied to various datasets for intrusion detection

(continued)

Table 1 (continued)

Author	Framework/Solutions	Cybersecurity solution used	Focus	Future development
Zhong et al. [14]	Applying big data based deep learning system to intrusion detec- tion	Deep learning	Intrusion detection	Future direction is to focus on advanced decision fusion algorithms combining outputs from different deep learning models. How to reduce the required minimum com- putational resources to achieve similar per- formance will be studied in the future
A. Dey [15]	Deep IDS: A deep learning approach for intrusion detection based on IDS 2018	Deep (CNN LSTM) and learning	Intrusion detection	Implementing this methodology holds great scope for the future
Dawoud et al. [16]	Internet of things Intrusion Detec- tion: A deep learning approach	Unsupervised deep learning	IoT Intrusion detection	Further investigation for deep learning in intrusion detection systems
Lin et al. [47]	ERID: A deep learning-based ap- proach towards efficient real-time intrusion detection for IoT	ERID-unsupervised deep learning approach	IoT Intrusion detection	Not stated in the paper
Ishaque et al. [17]	Feature learning system extraction using deep for intrusion detection	Deep learning	Intrusion detection	Not stated in the paper

(continued)

Table 1 (continued)

Author	Framework/Solutions	Cybersecurity solution used	Focus	Future development
Isa et al. [18]	Native software defined networks (SDN) intrusion detection using machine learning	Auto encoder and random forest algorithm	DDoS attack/Intrusion detection	Implementation and adoption of SDN-based intrusion mitigation architecture for further prevention process
Zeng et al. [20]	Deep full range: A deep learning based network encrypted traffic classification and intrusion detec- tion framework	CNN, LSTM and Stack auto encoder	Network intrusion detection	Not stated in paper
Hsu et al. [22]	A deep reinforcement learning approach for anomaly network intru- sion detection system	SVM and Random Forest	Network intrusion detection	Not stated in paper
Dey et al. [21]	Flow based anomaly detection in software defined networking: A deep learning approach with fea- ture selection method	Gated Recurrent Unit-Long Short Term Memory (GRU-LSTM)	Network Intrusion detection	Implement the proposed model in a real environment with real traffic of network
Sezari et al. [23]	Anomaly-based network intrusion detection model using deep learn- ing in airports	Feed forward neural network	Network intrusion detection	In the future simulation of local network of airports including series of modern and malicious network intrusion attacks like ran- somware to test and validate the model under predefined conditions

4 Conclusion

This paper presented a survey of existing research on the application of AI in Cybersecurity. We reviewed the use of AI technologies (Algorithms) in detecting and preventing attacks in cyberspace. The importance and impact of AI in cybersecurity management was discussed. Cybersecurity frameworks and solutions with viewpoint from 2018 to 2021 were summarised. Over the years, information and communication technology has advanced and cyberattack surface have continued to grow rapidly. Increased frequency of cyber-attacks has reinforced the need for cybersecurity initiatives. Traditional techniques have become inadequate in mitigating complex cyberattacks, therefore solutions that are capable of tackling cyber threats in real-time is required. LSTM proved to be effective in terms of computational complexity while maintaining low training time. Random forest showed high accuracy in anomaly intrusion detection. It is important to highlight that there is no one size fits all solution to cybersecurity challenges. It can be considered as a holistic approach. In the future, LSTM and random forest can be combined as protection solutions for cybersecurity.

References

1. Xu S (2019) Cybersecurity dynamics: a foundation for the science of cybersecurity. Adv Inf Secur 74:1–31. https://doi.org/10.1007/978-3-030-10597-61
2. Yar M, Steinmetz KF (2019) Cybercrime and society. Sage
3. Ping P, Qin W, Xu Y, Miyajima C, Takeda K (2019) Impact of driver behavior on fuel consumption: Classification, evaluation and prediction using machine learning. IEEE Access 7:78515–78532
4. Ghahramani M, Qiao Y, Zhou MC, O'Hagan A, Sweeney J (2020) AI-based modeling and data- driven evaluation for smart manufacturing processes. IEEE/CAA J Automatica Sinica 7(4):1026–1037
5. Yu KH, Beam AL, Kohane IS (2018) Artificial intelligence in healthcare. Nat Biomed Eng 2(10):719–731
6. Chassignol M, Khoroshavin A, Klimova A, Bily-atdinova A (2018) Artificial intelligence trends in education: a narrative overview. Procedia Comput Sci 136:16–24
7. Smith MJ (2018) Getting value from artificial intelligence in agriculture. Anim Prod Sci 60(1):46–54. Healthcare Data Breach Report: 9.7 Million Records Compromised https://www.hipaajournal.com/september-2020-healthcare-data-breach-report-9-7-million-records-compromised/
8. Abraham, B., Mandya, A., Bapat, R., Alali, F., Brown, DE, Veeraraghavan M, A comparison of machine learning approaches to detect botnet traffic. In: Proceedings of the international joint conference. Neural Nfile:///C/Users/Admin/Desktop/ARTIFICIAL Intell. Yr Retrosp. two/New/A Hardware-Trojan Classif. Method Util. Bound. net Struct. https://doi.org/10.1109/IJCNN.2018.8489096.
9. Abdulhammed R, Faezipour M, Abuzneid A, Abumallouh A, Deep and machine learning approaches for anomaly-based intrusion detection of imbalanced network traffic. IEEE Sens Lett 3(1):2019–2022. https://doi.org/10.1109/LSENS.2018.2879990Trautman; LJ (2016) Corporate directors and officers cybersecurity standard of care: the yahoo data breach. SSRN Electron J

10. Vinayakumar R, Alazab M, Soman KP, Poor- nachandran P, Al-Nemrat A, Venkatraman S, Deep learning approach for intelligent intrusion detection system. IEEE Access 7:41525–41550. https://doi.org/10.1109/ACCESS.2019.2895334
11. Sokolov SA, Iliev TB, Stoyanov IS, Analysis of cybersecurity threats in cloud applications using deep learning techniques. In: 42nd 42nd International Convention on Information and Communication Technology, Electronics and Microelectronics. MIPRO-Proceedings, pp 441–446. https://doi.org/10.23919/MIPRO.2019.8756755
12. Fernandez Maimo L, Perales Gomez AL, Garcia Clemente FJ, Gil Perez M, Martinez Perez G, A self- adaptive deep learning-based system for anomaly detection in 5G networks. IEEE Access 6:7700–7712. 10.1109.2018.2803446
13. Zhang X, Chen J, Zhou Y, Han L, Lin J (2019) A multiple-layer representation learning model for network-based attack detection. IEEE Access 7:91992–92008
14. Zhong W, Yu N, Ai C, Applying big data based deep learning system to intrusion detection. Big Data Min Anal 3(3):181–195. https://doi.org/10.26599/BDMA.2020.9020003
15. Dey A, Deep IDS A deep learning approach for Intrusion detection based on IDS. In: 2nd international conference sustainable technology Ind. 4.0, vol 0, pp 19–20. https://doi.org/10.1109/STI50764.2020.9350411
16. Dawoud A, Sianaki OA, Shahristani S, Raun C, Internet of Things intrusion detection: a deep learning approach. 2021 IEEE Symposium Series on Computational Intelligence SSCI, pp 1516–1522. https://doi.org/10.1109/SSCI47803.2020.9308293
17. Ishaque M, Hudec L, Feature extraction using deep learning for intrusion detection system. 2nd International Conference on Computer Application and Information Security ICCAIS. https://doi.org/10.1109/CAIS.2019.8769473
18. Isa MM, Mhamdi L (2020) Native SDN intrusion detection using machine learning. In: 2020 IEEE eighth international conference on communications and networking (ComNet), pp 1–7). IEEE
19. Lin WH, Lin HC, Wang P, Wu BH, Tsai JY, Using convolutional neural networks to network intrusion detection for cyber-threats. Proc. 4th IEEE international conference on applied system inventio ICASI, pp 1107–1110. https://doi.org/10.1109/ICASI.2018.8394474
20. Zeng Y, Gu H, Wei W, Guo Y, Deep-full- range: a deep learning based network encrypted traf fic classification and intrusion detection framework. IEEE Access 7:45182–45190. https://doi.org/10.1109/AC-CESS.2019.2908225
21. Dey SK, Rahman MM, Flow based anomaly detection in software defined networking: A deep learning approach with feature selection method. In: 4th international conference on electrical engineering and information and communication technology iCEEiCT, pp 630–635. https://doi.org/10.1109/CEEICT.2018.8628069
22. Hsu Y-F, Matsuoka M (2020) A deep reinforcement learning approach for anomaly network intrusion detection system. In: 2020 IEEE 9th international conference on cloud networking (CloudNet), pp 1–6. https://doi.org/10.1109/CloudNet51028.2020.9335796
23. Sezari B, Moller DPF, Deutschmann A, Anomaly-based network intrusion detection model using deep learning in airports. In: Proceeding of the 17th IEEE international conference on trust, security and privacy in computing and communications/12th IEEE international conference on big data science and engineering, pp 1725–1729. https://doi.org/10.1109/TrustCom/BigDataSE.2018.00261
24. Naseer S et al, Enhanced network anomaly detection based on deep neural networks. IEEE Access 6:48231–48246. https://doi.org/10.1109/ACCESS.2018.2863036
25. Khaw YM, Abiri Jahromi A, Arani MFM, Sanner S, Kundur D, Kassouf M, A deep learning-based cyberattack detection system for transmission protective relays. IEEE Trans Smart Grid 12(3):2554–2565. https://doi.org/10.1109/TSG.2020.3040361
26. Qureshi S et al, A hybrid DL-based detection mechanism for cyber-threats in secure networks. IEEE Access 9:1–1. https://doi.org/10.1109/access.2021.3081069
27. Huma ZE et al, A hybrid deep random neural network for cyberattack detection in the industrial Internet of Things. IEEE Access 9:55595–55605. https://doi.org/10.1109/ACCESS.2021.3071766

28. Santos L, Rabada˜o C, Gonc¸alves R (2018) Intrusion detection systems in Internet of Things: a literature review. In: 2018 13th Iberian conference on information systems and technologies (CISTI), pp 1–7
29. Wang M, Zheng K, Yang Y, Wang X, An explainable machine learning framework for intrusion detection systems. IEEE Access 8:73127–73141. 10.1109.2020.2988359
30. Zeadally S, Adi E, Baig Z, Khan IA (2020) Harnessing artificial intelligence capabilities to improve cybersecurity. IEEE Access 8:23817–23837
31. Aljawarneh S, Aldwairi M, Yassein MB (2018) Anomaly-based intrusion detection system through fea- ture selection analysis and building hybrid efficient model. J Comput Sci 25:152–160
32. Wiafe I, Koranteng FN, Obeng EN, Assyne N, Wiafe A, Gulliver SR, Artificial intelligence for cybersecurity: a systematic mapping of literature. IEEE Access 8:146598–146612. https://doi.org/10.1109/ACCESS.2020.3013145
33. Mishra P, Varadharajan V, Tupakula U, Pilli ES (2018) A detailed investigation and analysis of using ma- chine learning techniques for intrusion detection. IEEE Commun Surv Tutorials 21(1):686–728
34. Parrend P, Navarro J, Guigou F, Deruyver A, Collet P (2018) Foundations and applications of artificial Intelligence for zero-day and multi-step attack detection. EURASIP J Inf Secur 2018(1):1–21
35. L.lazic (2019) Benefit from Ai in cybersecurity. In: The 11th international conference on business information security (BISEC-2019). Belgrade, Serbia
36. Li L, He W, Xu L, Ash I, Anwar M, Yuan X (2019) Investigating the impact of cybersecurity policy awareness on employees' cybersecurity behavior. Int J Inf Manag 45:13–24
37. Choi JP, Jeon DS, Kim BC (2019) Privacy and personal data collection with information externalities. J Public Econ 173:113–124
38. Kshetri N (2019) Cybercrime and cybersecurity in Africa. J Glob Inf Technol Manag 22(2):77–81
39. Prester E, Wagner J, Schryen G (2020) Fore- casting IT security vulnerabilities–an empirical analysis. Comput Secur 88:101610
40. Talesh SA (2018) Data breach, privacy, and cyber insurance: how insurance companies act as "compliance managers" for businesses. Law Soc Inq 43(2):417–440
41. Lee J, Kim J, Kim I, Han K (2019) Cyber threat detection based on artificial neural networks using event profiles. IEEE Access 7:165607–165626
42. Kravchik M, Shabtai A (2018) Detect- ing cyber-attacks in industrial control systems using convolutional neural networks. In: Proceedings of the 2018 workshop on cyber-physical systems security and privacy, pp 72–83
43. Hashim A, Medani R, Attia TA (2021) Defences against web application attacks and detecting phishing links using machine learning. In: 2020 international conference on computer, control, electrical, and electronics engineering (ICCCEEE), pp 1–6. IEEE
44. Johansson J (2019) Countermeasures against coordinated cyber-attacks towards power grid systems. [Online]. Available: https://www.diva-portal.org/smash/record.jsf?pid=diva2:1353250
45. Lin WH, Lin HC, Wang P, Wu BH, Tsai JY (2018) Using convolutional neural networks to network intrusion detection for cyber threats. In: 2018 IEEE International conference on applied system invention (ICASI), pp 1107–1110. IEEE
46. Karatas G, Demir O, Sahingoz OK (2018) Deep learning in intrusion detection systems. In: 2018 International congress on big data, deep learning and fighting cyber terrorism (IBIGDELFT), pp 113–116. IEEE
47. Lin M, Zhao B, Xin Q (2020) ERID: a deep learning-based approach towards efficient real-time intrusion detection for IoT. In: 2020 IEEE eighth international conference on communications and networking (ComNet), pp 1–7. IEEE

The Role of Blockchain to Reduce the Dissemination of Fake News on Social Media and Messaging Platforms

Marta Aranda-Tyrankiewicz and Hamid Jahankhani

Abstract The world is progressing further into the digital era, and as a result, more and more people are getting their initial information from online news portals and social media platforms. The dependence on sources of information that are becoming murkier raises the likelihood of being deceived and the possibility of misguided agendas. When it comes to verifying articles, traditional news outlets adhere to stringent norms of practice. In contrast, people nowadays are able to publish news items on social media and unverified sites without having to provide evidence that they are true. Because there are now no indicators on the Internet that can be used to gauge the veracity of such news pieces, an innovative strategy that makes use of technology is required in order to evaluate the realness quotient of unconfirmed news items. This paper proposes a dynamic model with a secure voting mechanism that allows news reviewers to submit news input. A probabilistic mathematical model is utilised to estimate the news item's veracity based on the feedback obtained from the news reviewers. In order to guarantee that the information being spread is accurate, the concept, which is based on blockchain technology, will be presented.

Keywords Blockchain · Fake news · Social media · Messaging platforms · Deep Learning (DL) · Natural Language Processing (NLP)

1 Introduction

Recently, misinformation and fake news have been a source of enormous social losses that have been brought about as a result of misinformed decisions. Multiple occurrences in the real world that have resulted in the loss of life and reputation on a global scale were precipitated by the spread of false information, which takes the shape of doctored memes, articles, and unconfirmed postings from users who want to remain anonymous [8]. There has been tremendous resentment among the

M. Aranda-Tyrankiewicz · H. Jahankhani (✉)
Northumbria University, London Campus, Newcastle upon Tyne, UK
e-mail: Hamid.jahankhani@northumbria.ac.uk

© The Author(s), under exclusive license to Springer Nature Switzerland AG 2023
H. Jahankhani (ed.), *Cybersecurity in the Age of Smart Societies*,
Advanced Sciences and Technologies for Security Applications,
https://doi.org/10.1007/978-3-031-20160-8_15

257

general public as a result of the dissemination of false information about crucial socio-political problems, which has even led to riots.

The use of expert knowledge and machine learning algorithms are two typical ways that are used to detect the false articles that have been purposely generated to trick readers. Several different Deep Learning (DL) and Natural Language Processing (NLP)-based indexing methods are used in computational techniques. It is possible to produce hoaxes and fake news using any kind of media, including text, images, audio clips, and video clips. Over the years, experts have developed efficient computer algorithms for one sort of news item. However, if one wants to build a real-time system that is capable of handling all the wide varieties of fake news, the system itself will be complicated. As a result, the objective seems to be the development of effective computational algorithms tailored to a certain category of news articles or sources.

On the other hand, although requiring human involvement, an expert-based system could be simpler and more productive. In addition, the date the news was published and the environment in which it was published are also very important considerations when evaluating the value of news articles. It has been noted rather often that an older news article or a clipping of an older event or movie scene gets recirculated as if it were more recent. These problems in and of themselves inspire experts to research into detecting approaches based on expert knowledge.

The innovative approach taken into consideration in this paper consists of analysing the feedback that is provided by one or more subject matter experts who review news items, one by one in a centralised manner, and then they vote, providing scores to the news items, and attempting to determine the degree to which the news item is truthful. Because the expert may have a different point of view, their votes should be weighted according to their area of expertise when the trustworthiness score for a piece of news is being determined.

The goal is to investigate the possibility of using a probabilistic model to compute the score. One of the possible models analysed is the use of a centralised voting approach that enables specialists to offer the score one after the other, which lengthens the review process and shows promising outcomes, despite the fact that all of the strategies listed show promising results and have been utilised extensively. Additionally, there is a risk to the safety of new pieces as well as the votes associated with them since the mechanism used to store them is not secure. Because of this, new parts are readily modifiable, and the reliability of votes is called into question in such a system. Therefore, as a solution to this well-known problem, this study aims to propose a model with the help of blockchain technology and an analysis of the existing literature.

Aims

The aim of this paper is to propose a framework based on blockchain, presented and implemented as a method for detecting false news that is effective, safe, and dependable. The system should employ blockchain technology in the building of a safe infrastructure in which the votes of industry professionals are stored and distributed. The blockchain is an immutable ledger used to keep track of all of the news articles' entries and safely record the votes associated with those articles. The news items may be kept in a safe environment thanks to the cryptographic encryption of the blocks. The distributed ledger technology (blockchain) guarantees that news articles and votes cannot be altered or changed at any time.

A news article, the veracity of which is to be assessed, is subjected to voting by authorised reviewers, who then provide ratings to the article depending on the degree to which they are certain that it is real. Following this step, the votes are weighted according to the qualifications of the reviewer, such as their expertise, designation, and connections. A weighted system is necessary because the reviewers come from a variety of different backgrounds and work experiences, which means that some votes may be more reliable than others. As a result, a weighted system is essential because it enables the system to consider a greater number of factors in order to evaluate them in a more reliable manner. Better knowledge of the credibility of the experts may be gained via the profiling of the experts.

Another proposed model will be a probabilistic mathematical model that evaluates the weighted votes in order to make predictions about how accurate they are. Because of the model's accuracy and dependence on blockchain, experts can create a final score for credible deception detection by taking into account the reviewers' characteristics and the votes. Due to the fact that the truthfulness label of the variable newsPiece (the dependent variable) in this particular scenario can only take on one of two potential values (genuine or fraudulent), developing a hypothetical model is the best approach for this study. Hypothetically, the immutability of the data that is stored on the blockchain is utilised to ensure that no modifications are made to the news pieces that have been posted, as well as the simultaneous casting of votes by the reviewers in a decentralised environment, taking into account the exogenous factors that trust in the system.

Blockchain technology originally gained widespread attention when programmers working under the alias of Satoshi Nakamoto built it as the public ledger for Bitcoin, the very first cryptocurrency ever created. Since that time, technology has been mostly associated with the development of cryptocurrencies. Despite this, the fact that it is secure, transparent, and immutable demonstrates significant potential for the technology to be used in other settings.

Blockchain technology may be simplified to its most fundamental form, which can be characterised as a reliable database that has been developed in such a way that it cannot be altered or damaged in any way. This is its most fundamental form. It is seen as decentralised owing to the fact that the database is not kept by a solitary entity, such as the government or public or private enterprises, but rather is dispersed throughout a network of computers [14]. This may be either public or private. Before data is added to a block and subsequently added to an already existing chain of blocks, the whole network runs a check and validation on the data. This happens before the data is recorded in a block. The contents of each block are kept secret by the use of hash encryption, which is added to each block. Because the distributed ledger follows a specific order of blocks and replicates the same information on all of the other computers that are a part of the network, each transaction is guaranteed to be both transparent and unable to be reversed. This is made possible by the uniform replication of the ledger on all of the other computers.

Digital images and videos have the potential to spread false information since they may be replicated, modified, and reused to support erroneous context. The initiation of an instant fact check that makes use of information that a decentralised organisation safeguards might be one approach to solving this issue as soon as it is discovered. This is not out of the realm of possibilities at all, especially considering how much more objective the media has become in recent years. When comparing and contrasting whether or not the media was taken out of context, the only thing that is often required is the original facts of when, where, and what the media is really depicting. This is because when determining whether or not the media was taken out of context, it is important to have all of the relevant information. A blockchain network can preserve the contextual information of digital files; the structure of the network and encryption safeguard the information, preventing it from being altered or deleted in any manner [7]. The fact that the blockchain is accessible to everyone makes it possible for the entire public to see the environment in which the snapshot was taken in an open and unobscured manner.

The New York Times has been undertaking tests as part of its News Provenance Project to investigate this subject. On a distributed ledger that uses blockchain technology, The New York Times keeps a public record that cannot be altered of the information associated with the images and videos that it publishes [22]. The metadata, in this case, includes information on the environment in which the material was created. These are the facts that provide a solution to the question of when, where, and in what context the media was produced. In addition to that, these statistics provide a history of the many ways in which other news organisations have exploited the media. The information on the origin of a photo or video is then obtained from the blockchain database whenever a picture or video is uploaded. A brief explanation of this information is provided in a popup box that is shown next to the media for viewers to see. This substantiates whether or not the media has been employed in a suitable setting.

2   Literature Review

The term "fake news" is synonymous with "misinformation," which refers to information that is both false and misleading, and "disinformation," which refers to false information that is designed to mislead [5]. A vast quantity of data is generated on social networking sites, as well as across the many platforms that make up social media. There has been a very big volume of postings, which has resulted in an enormous expansion in the amount of social media data on the web. When a certain event has taken place, a lot of individuals talk about it on the Internet via various forms of social networking. As part of their everyday activity, they are either looking for or collecting news happenings in order to debate them. Therefore, users are confronted with the challenge of dealing with an excessive amount of data during the search and retrieval processes due to the extremely large quantities of news and postings.

People who get their information from unreliable sources are more likely to believe hoaxes, false news, rumours, and conspiracy theories, as well as fake news. After the successful development of various online social media platforms, there has been an explosion in the circulation of fake news for the purpose of various commercial and political gain. For instance, the use of Natural Languages Processing (NLP) analytics has been explored. Many analytical studies were carried out to predict the sources of false information or evaluate the news content to classify news as fake or true according to a variety of criteria. The currently available techniques have been able to identify incorrect claims by using many different computer vision models, such as ANN, Naive Bayes, SVM, and others.

The problem of misinformation has been debated for ages. For instance, text analysis, which employs information and news to function based on narrative, content, and context, has been a topic of discussion among scholars, who have discussed it as an example. This subject may also be considered as an example of text analysis, despite the fact that there are a number of tools and approaches that can be used to detect false news sources, such as determining if a website or individual publishes fake news. By using it, one may determine whether or not a statement is incorrect by using deep learning models. When the model is tuned and trained on a consistent basis, using specific parameter estimates will provide a greater prediction accuracy, which will lead to the best possible outcomes. Security techniques that use deep neural networks can identify instances of fake news and stop rogue authors from spreading false information. This section will analyse the existing literature on the use of blockchain technology in combating misinformation and highlight the need for such a research approach, especially after the COVID-19 pandemic.

2.1 COVID-19, Misinformation, and the Risks

Fake news has evolved to the point that it poses a threat to human civilisation as a whole as well as the democracies of individual nations. For instance, the most recent COVID-19 has been the newest focus of false news throughout the majority of social media channels. According to the findings of [18], the COVID-19 virus, when combined with false information, may be very harmful to individuals [18]. Concerning the virus, a number of films, audio recordings, and texts have been packaged together and distributed across a variety of social and conventional media. As reported in the literature, some of these stories include the fact that the virus was the result of a botched laboratory experiment and the refusal of some people to take the vaccine due to false claims about its efficacy and safety of the vaccine.

Carrion Alvarez and Tijerina-Salina referred to the rise of misinformation during the epidemic as "destructive thoughts." This tendency has persisted or even intensified after the pandemic began. [27], in an article on deepfakes, revealed that during the mass shooting that took place in Christchurch, New Zealand, a video that was being circulated depicted a suspect being shot dead by police. However, it was later found out that this video depicted a different incident that took place in the United States. [27] argues that deepfakes may pose a danger to national security and potentially create conflicts. The suspect in the shooting that took place in Christchurch was not killed. He discloses that these tapes may be used to demonstrate a politician receiving a bribe, confessing a crime, or admitting a hidden plot to carry out a crime. He also indicates that such videos can be used to expose a conspiracy to commit a crime. For instance, a political scandal was stoked in Malaysia as a result of a deepfake in which a guy claimed to have had sexual relations with a member of the local government. Stories like this might induce individuals to reject even legitimate information supplied by authorities because they teach people to believe that everything is a lie. This can lead to a lack of confidence [27].

One of the criteria for establishing whether or not the information is reliable is whether or not it comes from an expert or an authoritative source. As a result, the possibility that this, too, may be used for evil purposes is a significant risk. The risk associated with the spread of false information is that individuals have a very high and rapid chance of adopting anything that is congruent with their innate or prior views. On the other hand, the chance of rectifying such erroneous information is quite low, and doing so would take a long time. According to Ngwainmbi [19], individuals are more likely to believe negative information than good information [19]. As a result, false information is stitched together using these presumptions as its foundation.

2.2   Machine Learning

Using machine learning, Sharma et al. investigated a method for detecting false news from vast amounts of text data. In the article, the authors brought the concepts of machine learning and natural language processing together to create a detection framework [23]. The plan even takes into account the many approaches that have been used by previous research carried out on the exact same subject. The system simply utilised a few trials as examples and then investigated the principles of certain tactics employed in the detection of bogus news. Along with discussing the challenges involved in doing research in this field, the authors also discussed the potential long-term ramifications of such endeavours.

Additionally, there are machine learning (ML) algorithms and methods that make use of Natural Language Processing (NLP) to discover and highlight the linguistic patterns in terms of false or authentic news. Classifier models are responsible for the majority of the machine learning process because of their ability to differentiate between actual and bogus information. Counting the vectors that are utilised for words is one method that Vijay and colleagues employed in their use of Random Forest and NLP to identify false news. In addition to this, they used a method known as the RID matrix in order to identify similarities and copy sources in the papers [25].

The use of machine learning and natural language processing in order to categorise the documents on the basis of the words they contain is their primary strategy is a common technique. A survey on the identification of false news using deep learning and natural language processing was provided by Chokshi and Mathew. Several different deep learning approaches are discussed here as potential strategies for detection based on very large amounts of data [6]. In a similar manner, both word classification and picture classification using a convolutional neural network (CNN) were investigated. A framework for the identification of false news was suggested by Wang et al., and it utilises the combination of three components, namely, a fake news detector, reinforcement learning, and an annotator [26]. This procedure has been used to remove the news stories with weak labels and choose news stories with high-quality samples in order to detect false news. The current proposed work differs from the related works in that there is a lack of trust in the process that is being used, which in this case involves applying blockchain network technology. In this case, the users are required to register in the network, and once they have been identified, they are given the opportunity to share news on social media.

2.3 Cryptocurrencies

In layman's terms, the phrase "Fake Media" refers to the planned or accidental broadcast of incorrect information also on a public platform. It is almost impossible to quantify the amount of propaganda passed through communication channels across the globe. In an article by Fitwi and colleagues, Confidentiality Security Compact (CSC), a cryptocurrency architecture for edge cameras, was the topic of discussion [10]. The system makes it easier for the VSS to conduct an assessment before invading the privacy of the people captured on camera in the films. The Lib-Pri system converts the VSS that is built into a design that functions as a non-federal bitcoin blockchain. This design includes checking for integrity, maintaining obscurity of keys, trading functionality, and punishing video access. The execution of privacy protections based on policies is carried out without putting a load on the system at the edge in the form of real video monitoring equipment.

In their article, Xu et al. offered a cryptocurrency decentralised infrastructure for smart community safety Kubernetes [28]. To ensure the security of data access control in an SPS framework, a protection framework based on microservices was implemented inside a licensed public blockchain. The technology used for intelligence services was decoupled into separate container-based microprograms. These microprograms are built with the help of a consensus mechanism, and they are deployed utilising cloud technology and nodes on the edges. A thorough observational design was used to validate the hypothesis that the hypothetical BlendMAS is able to provide a distributed IoT-based SPS framework that is hierarchical, scalable, and reliable in its data exchange and intrusion detection capabilities. Intelligent systems for national security employ a variety of different embedded technologies and store multichannel audio gathered from the perimeter of monitoring. Cloud services that are implemented on a private cryptocurrency blockchain have their user authentication information translated from its original analogue form into a digital format and stored in a hash index. The functionality of intelligent monitoring and services is separated into a separate container-based micro-service architecture, and it is distributed on decentralised edge or cloudlet nodes.

Wang and colleagues came up with a two-dimensional model, starting with a centralised SaaS service platform. The platform is centred on the requirements that the public cloud must meet, including recognising the characteristics of virtualised environments, computer-controlled implementation and distribution, the consolidation of data structure, exchanging knowledge, and efficient user and strategic efficiency [26]. Computing may be done via the use of the centralised support network for the service. Also, in order to facilitate the sharing of data, these systems make use of peer-to-peer (P2P) networks. This is in part a reaction to the Bitcoin system. The results are consistent with expectations. Since its inception, Bitcoin has served as the network's most fundamental and fundamentally important implementation of an authorised cryptocurrency.

Every node in the Bitcoin network is responsible for reproducing and keeping the location of all bitcoin transactions in the protocol up to date. Despite this, all processors move to the next available state whenever a new order is issued. This new state records all purchases made over the course of the upgrading procedure. This is shown by Kang and colleagues, which shows how accurate the reporting has been in previously implemented systems [12]. MCE can learn the latent space so that the significance of accurate information vectors gets more unbelievable than those of inaccurate information vectors. Because a news vector becomes bigger during the course of training when its component vectors are well matched and when element vectors remain stable throughout the process. When this is done, MCE will be able to capture the performance of each device and the dynamic relationships that exist between the elements. The results of the comprehensive experiments on two benchmark functions show that MCE performs better than any of the timeframes. In addition, the gadget offers a way for doing research on the subspace in order to confirm its capability of fulfilling the requirements. Fake news has the potential to mislead and influence viewers, leading to unintended effects or even taking advantage of big gatherings.

2.4   The Internet and Other Technologies

Many other assertions and points of view, both in favour of and opposed to the use of security cameras, have been brought up recently. Bhoir and colleagues discussed the power of the Internet, which has grown tremendously over the course of the past few decades [4]. This level of growth has given paws to misinformation, and this is a growing problem because it is the misinformation that blurs the actual truth that is geared towards positive change. As a result, the solution to this problem is to develop a well-thought-out and complete proof gadget that monitors the different patterns in channels that pass information and that really can assist us in determining whether or not it is accurate. The solution developed by Bhoir and colleagues utilises hybrid models and seeks to identify bogus news with the use of computer vision. This design cut the amount of time needed for learning by a substantial amount and improved SVM classifiers. When constructed on their own, the Vector Support Operator and the Random Forests (RF) algorithm worked well, but when combined, this hybrid model performed even better [4].

Ai and colleagues described a power exchange decentralised resolving system for smart grid blockchain-based that sought to address the difficulties of knowledge disunity. A difficult trust framework to construct, power unpredictability that wastes resources and value advancement brought about by electrical pre-sale [1]. The findings of the testing demonstrate that this device achieves remarkable results by using a rational set storage system capable of meeting the requirements of applications used in the real world. The term "smart grid" refers to a set of services that may be applied to populations dispersed over an area and storage facilities connected to a transmission network. These services are intended for the Automatic identification of bogus news through social networking platforms was investigated by Qawasmeh et al. The framework suggests an automated way for the identification of misleading information by making use of up-to-date machine learning techniques [20]. The typical technique uses an asymmetric LSTM convolutional model, which may achieve an accuracy efficiency of 85.3 per cent when applied to the FNC-1 data. The traditional methods of disseminating information, such as printed papers, telephones, radios, and televisions, have been rendered largely obsolete by the advent of digital mass media. The rise in popularity of the Internet may be somewhat attributed to the progression of several current technical developments.

The unconstrained accessibility to content production and sharing of information on social media platforms has paved the way for the growth of scams and fake news, both of which are serious dangers to broadcast sites and the social media fraternity. The social media fraternity has benefited from this unrestricted access to content creation and information sharing [4]. On top of that, the unrestricted access to content production and information sharing that is provided by social media platforms has prepared the way for the proliferation of false information. Consequently, the level of confidence that relevant audiences and content consumers have in online sources and media presenters is placed in jeopardy as a result of the erroneous information.

Under typical situations, a person requires references based on factual information in order to ensure the information provided online is reliable and trustworthy. According to the findings of a study carried out by [17], the rapidity with which fabricated material and online content are disseminated should serve as a warning sign since algorithms hook the interest and attention of readers [17]. In the same manner, material that has been modified will result in criticism and degradation. For instance, the news media outlet Buzz Feed reported a trending piece of modified information that was posted on Facebook. This content aimed to undermine the position once held by Barack Obama, who was the previous president of the United States. According to the fabricated information, President Obama had issued an executive directive that would prohibit the practice of loyalty at all educational institutions throughout the country.

On the other hand, advances in technology have turned into a contentious issue. This is because the pace at which accelerated technology develops goes hand in hand with the degree to which altered content moves. However, a recent study conducted by [21] claims that the introduction of blockchain technology has resulted in a more in-depth investigation of scams and fake news on online platforms [21]. This indicates a trend toward the restoration of trustworthiness and credibility within the community of social media platforms. Consequently, a number of well-known technology companies, such as Verizon and IBM, along with a number of significant social media platforms, have acknowledged and embraced blockchain technologies as a preventative measure against the manipulation of content within their respective platforms.

In addition to this, the preventative aspect of the measure has brought about significant reforms in Twitter and Instagram, where blockchain networks are the cornerstones of rendering void the inaccurate proof of elections and scams [29]. Even though people have less time to investigate and filter the authenticity of information and the digital content presented by online media and social media platforms, it is essential to conduct research into the owner of the content that is being disseminated and the background aspects of the content. As stated by [3], the transparency inherent in blockchain technologies paves the way for the verification and authenticity of the content [3]. As a direct consequence of this, its consumption and subsequent dissemination through the Internet are reduced. On the other hand, due to a lack of pertinent knowledge that supports these contents, millions of online users regularly consume fabricated content and news that is made available on television, online platforms, and websites. This is a noble fact.

2.5 Blockchain Integration

Blockchain is made up of digital recordings of data and the transactions that use that data. Blockchain is available to and trusted by all participants running the same protocol. The widespread adoption of this technology can be attributed to digital currencies, as the application of blockchain technology forms the foundation of every cryptocurrency currently available on the market. In addition, blockchain technology is beginning to find applications in a variety of other fields, such as healthcare, the media, energy, and record management, amongst others. Blockchain technology acts as a trusted agent of the transaction itself, eliminating the need for any additional "physical" or active middleman. This results in a decrease in transaction costs as well as an improvement in transaction speed, all while preserving a high degree of security. The verification and authentication of traditional financial transactions, for example, are handled by centralised systems that banks manage, intermediaries, trustees, and escrow agents. On the other hand, Blockchain technology serves as a trusted agent of the transaction itself, without the need for any other "physical" or active intermediary. The decentralised nature of blockchain is the major element that would significantly improve the distribution of news. This is due to the fact that each new share will be "chained" to the one before it in a passive manner, without any active control coming from the firm. This is due to the fact that if blockchain were to be applied in a content sharing system for social networks, for example, every piece of material could be totally traced back to its original source.

Also, a blockchain is a timestamped set of immutable data records that are controlled by a cluster of computers that a single organisation does not own. As discussed, the concepts of cryptography are used to ensure the safety of each of these data units, also known as blocks, and to link them together in a chain. A SHA256 encrypted blockchain makes up the blockchain architecture of a proposed voting model. In addition to the hashes and the timestamps, the proposed model calls for each block to include two essential fields: the newsPiece field and the uploaderName field. Everyone is allowed to add content to the newsPiece component. Everyone who uses the service has unrestricted freedom to submit any newspiece they choose. When applied to users, cryptographic principles may give a level of security that can lead to the system gaining widespread popularity and widespread confidence throughout the world. Blockchain technology relies heavily on public-key cryptography as an essential building piece. High-order cryptographic methods are used inside the blockchain to establish a public–private pair of keys that anybody may use. The only way to effectively preserve security is to keep the private key hidden, but the public key may be distributed so that users can have access. The model for weighted majority voting that has been presented makes advantage of this kind of cryptographically encrypted security. As illustrated in the literature review, there are a number of important survey papers on blockchain privacy and security that illustrates the feasibility of the technology in reducing misinformation.

At the same time that the newsPiece, together with the uploaderName and timestamp, is being uploaded, the cryptographic hash for the block is being produced. This leads to the formation of the blockchain, in which each block is connected to the next by the reference of the hash of the previous block. The created hash is determined by the newsPiece, the uploaderName, any previous hashes, and the date. These hashes are regenerative and cannot be changed in any way. Any modification made to the data being entered will result in new output. As a result, any modification to the data that is read in will result in a comprehensive shift in the order in which the blocks of the chain are processed. The hashes of each block are updated, beginning with the block in which the modification was made and continuing forward.

Proof of Trust is a trust integration system that has found broad use in the settings of the service industry as well as crowdsourcing. This is because it has the potential to discourage dishonest conduct on the part of those who make use of public service networks. The Proof of Trust algorithm is a consensus procedure that selects validators using Shamir's secret sharing strategy in combination with preset criteria. The classic Paxos and Byzantine Fault Tolerance (BFT)-based algorithms have a scalability problem. However, the Proof of Trust protocol tackles this problem while also avoiding the low throughput and resource-intensive flaws of Bitcoin's "Proof-of-Work" (PoW) mining. The consensus algorithm known as Proof of Work was expanded into what is now known as the Proof of Trust consensus algorithm. The "election process" for a trustless leader in a blockchain network is a method called Proof of Work, which is based on the demonstration of computer capability. It is a security protocol for the blockchain that is implemented in peer-to-peer (P2P) systems that do not rely on trust.

On the other hand, using Proof of Work will result in the loss of a significant quantity of energy. As part of the Proof of Trust (POT) protocol, trust between peers is evaluated using a decentralised trust graph that grows and changes over time. This trust graph is stored in the blockchain and maintained by the network. Utilising efficient consensus BFT algorithms such as RCanopus allows for the provision of dedicated fast peer server channels, which in turn assists in the acceleration of the process of extracting requests and transactions. In light of the fact that the news articles being taken into consideration are the only data items accessible in the public domain, generic smart contracts have been used in this particular instance. In addition, the votes, assessments, and ratings associated with these data items are sent across in a strictly confidential manner.

3   Research Method

The Literature review provides a number of blockchain-based models developed over the years to curb misinformation. This section will utilise this secondary data to determine which model is the best fit to curb the spread of misinformation. The paper analysed a total of 20 journals where authors have developed models geared towards combating misinformation. These models include models such as a power exchange decentralised resolving system for smart grid blockchain-based that sought to address the difficulties of knowledge disunity. In most cases, the goal of the authors was to eliminate the human factor, which in most cases fuels the spread of misinformation. For instance, everyone, including the experts, is biased. Therefore, complex systems that eliminate this human factors seemed to be one of the solutions. In this section, the paper will list three models (including the blockchain and keyed watermarking-based method) based on the literature review above. Then, the paper will compare the feasibility of these models with the blockchain and keyed watermarking-based method, which was initially proposed.

The model includes a weighted majority voting system, which will include the input of journalists and experts. Since the model uses a weighted voting system rather than a majority voting system, human factors will have minimal impact on the results. The other model is the DeHide model, which is a score-based model with players such as Reporters, Analysers, and Validators, who all play a role in building the credibility score.

3.1 Voting Model

A weighted majority voting system is a decision that picks options that have a majority, that is, upwards of half the votes. Every vote counts the same, and there is no difference between them. For the purpose of determining what constitutes fake news, one method under consideration is that of voting by the majority. A majority voting methodology for the detection of fake news would involve a panel of journalists, specialists, and auditors who would research and evaluate news items before they are posted to the site and offer boolean judgements on their authenticity. This would be done in advance of the news items being made public. After doing exhaustive research into the news item, each and every reviewer and expert has the opportunity to cast a vote about the reliability of the news.

In an ideal scenario, a voting system based on a pure majority would consist of only two votes; hence, the outcome of the voting process would be predetermined, with the exception of a circumstance in which the votes were split evenly. It is assumed that each vote counts the same. In spite of the fact that the voting method based on an absolute majority has a few advantages, it cannot be used successfully in this circumstance. The fact that all evaluators' or professionals' opinions could not be on the same level is one of the aspects that makes using a voting system based on simple majorities difficult to detect fake news. Some experts are superior to others in terms of their honesty and their level of expertise. Depending on the structure of the connections among the components, biases might eventually become embedded inside the system. Their capacity for judgment and accuracy of assessments may be different. As a result, there is no way for each vote in a voting system based on a majority to have the same value. This research presents a weighted majority voting approach for the detection of fake news using blockchain technology. Specifically defined criteria and components would be used to define the weight of each vote. A rule system that has been established in advance will be used to determine how much weight to give each vote.

The blockchain's weighted majority voting model was implemented using the Proof of Trust (PoT) consensus protocol, which is the means through which the weighted majority voting model was created. When proof-of-work is used, each user on a network receives a digital token as part of the transaction. In this scenario, a subset of users known as experts are presented with a "puzzle" that has to be solved (usually in the form of a hash function or a plain integer factorisation), and the results of their efforts are compared. The response that garners the most votes is considered to be the correct one, and the corresponding block is added to the chain as a direct consequence of this fact. Rather than relying just on a single expert's vote (which is given varying weights according to the calculations), a group of experts must pool their collective knowledge to reach a consensus on the best solution. When it is established that the solution to a specific block can be located in the sections that make up the totally confident range, that block will be added to the chain. The chain will continue until all of the blocks have been solved. The Proof of Transactions (PoT) protocol eliminates the poor throughput and resource-intensive concerns that are associated with Proof of

Work while at the same time resolving the scalability challenges that are connected with conventional Byzantine Fault Tolerance (BFT)-based protocols (PoW).

Using a voting system, a group of people gathers to evaluate the advantages and disadvantages of a thing or situation, assigning labels or points based on the attributes of the object or scenario under consideration. Reviewers, people who have considerable expertise in a specific subject, and journalists will all be included in the class of users that the proposed model employs to develop its voting mechanism. These individuals are also going to participate in the voting process using the majority voting technique. This group of users goes through the process of studying and analysing each newspiece, and then thereafter, they each cast a vote using a semi-deterministic mechanism to express their opinion on whether or not the newspiece is telling the truth.

In this scenario, the weighted majority voting model is utilised by computing a score based on the experts' interpretation of its fakeness and career statistics. In addition, a score that is based on the level of confidence that the experts have regarding their vote in relation to the news piece is also utilised. At the conclusion of each iteration of the review cycle, an expScore is calculated for each of the experts. This score is an integer, and it is constructed using a static basis in addition to a dynamic basis. The static component of the expScore receives a score, the value of which is determined by an analysis of the reviewers' experience as well as their organisation and designation. This score is then allocated to the component. The relative score given to each criterion is used to aid in building the expert profile, which, in turn, serves to attach a greater degree of responsibility and confidence to the vote. The incorporation of the background and experience of the expert assists in improving the amount of responsibility as well as the level of confidence connected with the vote. The integer score is derived by adding up the points received from each of the components discussed before and the points received from the dynamic component. After each review, the dynamic score component of the expScore goes through a new round of calculation.

Blockchain technology, which enables simultaneous voting and speedy feedback, is used to build the whole method. This helps relieve scaling concerns. The model also demonstrates that probabilistic analysis of the authenticity of news items can be done with improved accuracy by taking into account aspects affecting the reliability of reviewers using the model over a distributed platform such as a blockchain. This can be accomplished by taking into account aspects affecting the reliability of reviewers. By analysing both the quantitative and the qualitative data provided in the article, specialists may obtain a high level of accuracy. In terms of the outlook for the future, it is possible that the model described for the identification of fake news using a probabilistic analysis carried out via a blockchain would become more effective if an increased number of servers are deployed inside the network. It is possible to install the model on the back end of a front-end web application so that it may be employed by a greater number of people.

3.2 DeHide

The Deep Learning-based Hybrid Model (DeHide) is a great example of a reliable file-sharing platform that makes use of blockchain technology and gets support from deep learning models. When using DeHide, each and every registered user has the opportunity to become either a Reporter, an Analyser, or a Validator. However, the action that a user makes will define the function that they are assigned to play in the system. When a person publishes an article, that individual is considered a Reporter. Additionally, when they analyse the article and validate the source, they are promoted to the role of Analyst. When they read an article and provide their opinion on it, they also perform the Validator role. Every action done, regardless of role, in relation to an article, news item, or piece of content, can affect the Credibility score given to each user who is associated with the published material and the content itself. In addition to this, all of these data are registered and kept safe because of the blockchain, and deep learning models based on language analysis contribute to the Analyser role. The Analyser role involves continuously checking and evaluating the content that is published and learning from the interactions that users have with the content. Therefore, if an item obtains a large number of bad ratings because it contains bogus news, it will be concealed or deleted from the platform, and every user who has shared the information will earn a poor assessment of their credibility score. This includes the publisher. As a direct consequence of this, in the future, the items that are shared or published by these people will have a lower beginning dependability score and will have a reduced degree of visibility.

3.3 Blockchain and Keyed Watermarking-Based Model

My proposed method is Blockchain and Keyed Watermarking-based Model. The problem of network scalability has to be addressed for the system to achieve its aim of increasing its throughput. It is necessary to have a fast network that allows blocks to be propagated across the whole of the network extremely quickly so that there is no delay in the process of confirming the blocks. The throughput will subsequently grow as a direct result of the process of generally enhancing the scalability of the network. Distributed bloXroute servers with a large capacity are used across this system. The scalability constraint that exists at the network layer is addressed by bloXroute via the use of the blockchain distributed server concept. Their suggested network is a blockchain distributed network, a worldwide network of computers geared for distributing blockchain data as soon as possible. These BDN servers take advantage of innovative network protocols. For example, when a bloXroute server gets a data packet, it instantly broadcasts this data to the rest of the network. This enables bloXroute servers to spread data up to 100 times quicker than traditional servers. The bloXroute server can alleviate the scalability difficulties of blockchain by reducing the bottleneck caused by networking. The whole the system is made up of two distinct kinds of networks:

bloXroute servers—These are servers with low latency and a large capacity, and they are designed to propagate transactions and blocks for many blockchain systems in a short amount of time. They function similarly to cluster servers that are linked to other clusters. These bloXroute servers reduce both the overhead of the network and the latency. Take note that these servers do not manage any other tiny nodes or operate as a central server in any capacity. The addition of bloXroute servers has been done with the intention of accelerating the rate at which blocks are propagated.

Peer networks P2P networks of computer or mobile nodes employ bloXroute servers to disseminate transactions and blocks while simultaneously auditing the behaviour of bloXroute. These networks are referred to as peer-to-peer networks. These peer-to-peer networks make use of a particular consensus method. These networks are organised into clusters, and each cluster has one blockchain server that is responsible for the propagation of transactions and blocks on behalf of peer nodes, which are smaller nodes.

Following encryption, the blocks that have been received from the various peers are sent to the bloXroute servers so that they may be further propagated to other networks. Additionally, peers can either send these encrypted blocks directly to bloXroute servers or send it through some other peer nodes. Both of these options are viable. Because of these two features, the servers cannot be prejudiced toward certain nodes or cheat in any way with regard to those nodes. The bloXroute servers operate in the dark without any knowledge of the contents of the blocks, and the secret key is only released after the blocks have been delivered to their final destination. Because the speed of propagation of these bloXroute servers is quite rapid, blocks in the network are swiftly passed to other networks for the purpose of verification.

4   Critical Discussion

All these models are capable of processing any kind of news material, whether it is in the form of a photo, text, audio format, or video. Also, because they use expert knowledge when evaluating news items and employ a dynamic weight voting mechanism that considers the credibility of reviewers, the models' level of authenticity is rather high. Using the suggested technique, there is a likelihood that news stories will be further classified as fake or genuine. Also, it is possible for the examined technique to perform better than the fundamentally weighted approach. The models suggest making use of a distributed ledger so that the notion may be put into action in a way that is both more secure and quicker.

However, the blockchain and keyed watermarking-based method is superior to the strategies currently being used in many respects and the two shortlisted models. One of the biggest factors that hinder efforts to fight fake news is the high level of human factors. The voter and DeHide methods, although utilising the safety benefits of blockchain, are susceptible to human biases. All the players in the models can manipulate it to suit their agendas.

The platform that the blockchain and keyed watermarking-based method is based on is one that makes use of a blockchain network and is comparable to social media sites such as Facebook, LinkedIn, and Twitter. Over the blockchain network, a user or news agency may create a profile for themselves. However, in order to utilise the blockchain, each user must first authenticate their identity by presenting either a national identification number, a national identity card, or media credentials. Another way that might be utilised is a digital signature that is already associated with a national ID. This is a method that is becoming more popular. These tidbits of knowledge are kept secret from members of the wider public. On the other hand, the platform that utilises blockchain technology may validate information at any moment.

Note that in the proposed model, we do not recognise false news by utilising any automated Machine Learning (ML) algorithms; rather, the model focuses on identifying the source of news and verifies credible news based on users' reports on social media. This is important to keep in mind. Within the system, there are two distinct kinds of transactions that take place. The first is distributing information that has already been uploaded. On the social networking website, every user who has established an account may publish their own digital news articles. Prior to sharing, the user is required to choose the kind of share they want to do, indicating whether the material will be public or private. If the material is news, then the privacy around it will be made public, and any registered user has the ability to report it as either false or actual news. During this predicament, in which it is unclear if the news is phony or authentic, the news is denoted with an orange question mark. After the participants in the blockchain have verified the news, the material will be marked with a tick if it is authentic and a cross if it is fraudulent.

Note that we are using a private permissioned blockchain, and as a result, it can only be confirmed by people who have been granted permission to do so. The transactions include timestamps of data sharing for those individuals who have previously shared this information at an earlier period, including the person who first provided the material. When a user publishes news to the blockchain, the specifics of the news item, including the user ID, hash value, and timestamps, are preserved in the blockchain in the form of a transaction. When a different user shares the same post, the blockchain will again record the transaction with additional information. Because of this, the transaction will be traceable inside the blockchain based on the information that is included within the transaction itself. When a user edits the content of a post and then shares it, the blockchain also saves the transaction indicating that the content was amended; as a result, it is simple to determine who changed the content of the post. This improves credibility and is in a similar manner to the judges in a Supreme Court, albeit they are helped by a system to deal with the huge amount of data.

Every user has the ability to publish digital news items. However, the user is required to specify whether the data being uploaded is news or some personal information. Suppose the data is about a news item. In that case, its transaction will be logged in the blockchain, and our social networking platform will immediately apply a keyed digital watermark to the information in order to prevent the manipulation of digital content. Because there are so many strong tools for altering multimedia, millions of photographs, videos, audio, and news stories that have been edited or otherwise altered may be found on social media platforms. This is done in an effort to alter people's perceptions of certain topics. Therefore, it is necessary to take precautions to ensure the authenticity of such multimedia information.

These days there are a lot of challenges but the biggest challenge that exist is about cybersecurity and the other challenge is to being able to keep up with the technology. SME have the challenge of reducing the spread of fake news through their social websites. Blockchain has been identified as one of the best ways SME's can use to protect themselves from spread of fake news. Introducing things like a blockchain technology requires the use of different languages, different ideas and even different architecture in order to configure. The blockchain makes use of a variety of consensus methods, including Proof of Authority (PoA), Proof of Work (PoW), and Proof of Stake (PoS), among others. Every procedure offers a number of advantages and disadvantages that are unique to it. On the other hand, the scalability of these protocols is limited, and their throughput is poor. Consequently, implementing these consensus procedures into the proposed system is not a viable option at this time. On the other side, algorithms that are based on Byzantine Fault Tolerance (BFT) have limited network scalability but great throughput. Therefore, the criteria for the network's performance in pBFT are quite rigorous. However, in our situation, we improved the scalability of the network by using bloXroute servers. Hence, this consensus process is an appropriate fit for the requirements of our organization. We make use of a private blockchain, also known as a permissioned blockchain, in which only nodes that have been granted permission may participate in reaching a consensus. There is no such thing as an anonymous node that is able to verify transactions or a node that is able to get mining rewards. As a result, there is no mining cost associated with our system.

One of the things that are clear on the issue of fake news and misinformation is that we are dealing with a complex issue that does not have one definite solution. For instance, if an SME was to utilize the blockchain and keyed watermarking model, there is a need to establish trust since the model will be integrated into their web. This is a potential issue with cybersecurity concerns often overshadowing all other problems in the digital world.

Therefore, in order to foster user trust while also filtering information on the internet, platforms need to find a happy medium between preserving the right to freely express views and deleting anything that is offensive or otherwise objectionable. To achieve this delicate equilibrium, the platforms must first identify the specific issues it wants to address and the criteria by which judgments about specific news items. It is not always easy to get a consensus on which issues should be prioritized for the solution. As a result, it is essential that platforms and any other relevant parties achieve a consensus about the nature of the issues that should be the focus of moderation efforts and those that should be ignored. However, to properly address these issues, fact checkers need to detect instances of those problems in a reliable manner. This is true even when the users reach a consensus over which problems the platform needs to tackle.

If an SME implemented this technology they would be able to store all the messages in the block chain. Because of this, it would be much more challenging for imposters to create fake news involving the company. However, this will be challenging for people who use the forum on a regular basis. In addition, because of the limitations of blockchain technology, the storage of images or photographs would be challenging and would result in even higher costs. Each individual message, as well as any replies to those messages, will incur a fee. The mere storage of user communication on a large website would cost a fortune, even if it were only a penny, because of the volume of data involved. If this is going to be a long-term solution, then I can see the blockchain being used to encrypt messages. The user would then be able to choose whether or not he wishes to sign the message that is presented to him. Because of this, it will be possible to verify both the sender and the content of user-to-user communications. Additionally, it has the potential to be implemented as a paid service for users, such as the production and transmission of signed messages.

4.1 Conclusions and Future Work

The model proposed above puts into consideration the complexities of dealing with misinformation documented in prior sources. In the context of this method of investigating facts, the idea of a fact is not an individual's or a third party's perspective of what is true and untrue; rather, the notion of a fact is the information underneath the images and videos. The "method" that information is recorded in the blockchain using is the source of the validity of the "fact," which is generated from that methodology. This "fact" is the source of the legitimacy of the "fact" that is recorded in the blockchain. This model provides the necessary background information for accessing the accuracy of the media and preserves the capacity for an individual to create their own interpretation and cognitive judgments of truth or untrue in real-time. This model also provides the necessary background information for accessing the accuracy of the media. This approach gives the required background information to assess the media's correctness instead of relying on a basic "X is not true" retraction to assess accuracy.

When deciding what to share on social media, many people do not pay sufficient attention to the quality and veracity of the news. According to research, the primary reason for the spread of misinformation is not that individuals intentionally want to transmit false information; rather, it is because individuals do not pay sufficient attention to these factors [15]. Consequently, enhancing the provenance of information supplied via the use of blockchain technology could be a viable method for shifting our connection with the Internet from one based on misinformation to one based on trust.

It is also conceivable to use blockchain technology to begin the authentication process for news items, which may offer news and media companies a higher sense of validity and assist in putting a halt to the dissemination of incorrect information. When a journalism company generates a particular news item, both the meat of the story and the data pertaining to the time it was published will be recorded in a blockchain network that is open-source [13]. This will occur whenever the news item is distributed. The articles' modifications and any other text-based changes are being tracked, which helps to offer a clear record of every correction that has been done. This documentation strategy helps validate the origin of news stories, making it simpler to evaluate whether or not sources have amended a story from the outside.

In certain nations, the process of putting this into practice on a wider scale has already been started. After a number of incidents in which readers were misled into believing that false stories about COVID-19 originated from a reliable source, the Agenzia Nazionale Stampa Associata (ANSA) byline, ANSA, introduced the system of a digital seal through a public Ethereum blockchain network [2]. This was done after a number of incidents in which readers were misled into believing that false stories about COVID-19. The digital seal works as evidence that the item is stored in the blockchain, verifying that a particular piece of news or interpretation came from ANSA itself with complete transparency and trust. It does this by acting as a proof of storage in the blockchain. Using the seal is one way to accomplish this goal. Even while this technique does not investigate the story's veracity, it prevents the manufacture of fabricated narratives that mislead readers into assuming that information comes from reputable sources such as ANSA. Along with their Full Transparency initiative, companies such as Verizon are among the many initiatives focusing on adopting blockchain technology to officially record their content within an open system [9]. This is one of the many reasons why blockchain technology is becoming increasingly popular. The reason for this is due to the promise offered by blockchain technology.

The fact that this blockchain application provides a multitude of benefits should not come as much of a surprise to anybody. It is feasible for it to protect the brand reputation of media companies, increasing the amount of trust between such organisations and the consumers that patronise them. When news organisations quote the work of other news organisations, it is beneficial for both of them because it assures that viewers are provided with sources that are legitimate and unique, and it also leads viewers to such sources [24]. It has the potential to convince readers that the news they are reading is authentic and that the media they are consuming originates from a trustworthy source. The readers may get something from this as well. This technique assures that the news being read by an individual has not been distorted in any manner by a third party in any form while at the same time protecting the autonomy and subjectivity that an individual has in picking the news source from which they acquire their interpretations.

Right now, it is too early to say whether or not blockchain technology will completely revolutionise the industry in combating the information plague [16]. To begin, blockchain introduces a fundamentally new sort of social contract that does not exist between humans but rather between people and computer algorithms. This kind of contract is called a smart contract. There is still a long way to go until the general public is conversant with blockchain technology, and it will take some time for the novelty of such a contract to propagate throughout society. As a direct result of this, a statement such as "blockchain can ensure the integrity of truth in a decentralised manner" still has very little meaning, and the applications of blockchain that are presently being employed in journalism may be useless to a significant percentage of individuals.

Even with the proposed model defined in this research paper, there are still notable challenges. For instance, experts and journalists are still biased, and although the votes are weighted, there are loopholes to trick the model. Recent studies indicate blockchains, which are employed for the reason of cryptocurrencies, have been shown to be hackable, and trust is another issue that has evolved due to this challenge. The degree of trust required to use blockchain technology for journalistic purposes is far lower than the level of trust required to use cryptocurrencies due to the broad use of blockchain technology [11]. This technology is not applied to carry out personal financial transactions; rather, it is utilised by news organisations to retain records and data in a form that is both open and immutable.

It is possible that widespread adoption of blockchain technology may be delayed for a few years due to blockchain's relative unfamiliarity and lack of reliability. However, it has already shown its ability to be used in a wide range of imaginative ways to curb the spread of false information. In addition, it has proposed ways to address approaches that may be used to carry out fact-checking in real-time and in a decentralised manner. The outbreak serves as a perfect illustration of the disastrous impacts of incorrect information, and blockchain technology provides a fresh method of tackling an age-old problem.

References

1. Ai S, Hu D, Zhang T, Jiang Y, Rong C, Cao J (2020) Blockchain based power transaction asynchronous settlement system. In: 2020 IEEE 91st vehicular technology conference (VTC2020-Spring). https://doi.org/10.1109/vtc2020-spring48590.2020.9129593
2. ANSA (2020) ANSA leveraging blockchain technology to help readers check source of news—English. ANSA.it. https://www.ansa.it/english/news/science_tecnology/2020/04/06/ansa-using-blockchain-to-help-readers_af820b4f-0947-439b-843e-52e114f53318.html
3. Asaf K, Rehman RA, Kim B (2020) Blockchain technology in named data networks: a detailed survey. J Netw Comput Appl 171:102840. https://doi.org/10.1016/j.jnca.2020.102840
4. Bhoir S, Vinit S (2020) An efficient fake news detector. In: 2020 international conference on computer communication and informatics (ICCCI). https://doi.org/10.1109/iccci48352.2020.9104177
5. Chen X, Sin SJ (2013) Misinformation? What of it? motivations and individual differences in misinformation sharing on social media. Proc Am Soc Inf Sci Technol 50(1):1–4. https://doi.org/10.1002/meet.14505001102
6. Chokshi A, Mathew R (2021). Deep learning and natural language processing for fake news detection: a survey. SSRN Electron J. https://doi.org/10.2139/ssrn.3769884
7. Christodoulou P, Christodoulou K (2020). Developing more reliable news sources by utilising the blockchain technology to combat fake news. In: 2020 second international conference on blockchain computing and applications (BCCA). https://doi.org/10.1109/bcca50787.2020.9274460

8. DiCicco KW, Agarwal N (2020) Blockchain technology-based solutions to fight misinformation: a survey. In: Lecture notes in social networks, pp 267–281. https://doi.org/10.1007/978-3-030-42699-6_14
9. Do Val RB, Viana TD, Gouveia LB (2021) O USO de blockchain Na identificação de fake news: Ferramentas de apoio tecnológico para o combate a desinformação/The use of blockchain in the identification of fake news: technological support tools to combat disinformation. Brazilian J Bus 3(3):2726–2742. https://doi.org/10.34140/bjbv3n3-050
10. Fitwi A, Chen Y, Zhu S (2019) A lightweight blockchain-based privacy protection for smart surveillance at the edge. In: 2019 IEEE international conference on blockchain (Blockchain). https://doi.org/10.1109/blockchain.2019.00080
11. Foong KY, Ang SY, Chia DZ, Tee WJ, Murugesan RK, Hamzah MD (2019) Ambient: a blockchain social media to build trust and discredit fake news. J Adv Res Dyn Control Syst 11(11-SPECIAL ISSUE):889–894. https://doi.org/10.5373/jardcs/v11sp11/20193112
12. Kang S, Hwang J, Yu H (2020) Multi-modal component embedding for fake news detection. In: 2020 14th international conference on ubiquitous information management and communication (IMCOM). https://doi.org/10.1109/imcom48794.2020.9001800
13. Katal A, Singh J, Kundnani Y (2021) Mitigating the effects of fake news using blockchain and machine learning. In: 2021 2nd international conference for emerging technology (INCET). https://doi.org/10.1109/incet51464.2021.9456205
14. Khurshid A (2020) Applying blockchain technology to address the crisis of trust during the COVID-19 pandemic. JMIR Med Inf 8(9):e20477. https://doi.org/10.2196/20477
15. Loey M, Taha MH, Khalifa NE (2022) Blockchain technology and machine learning for fake news detection. Blockchain Technol 161–173. https://doi.org/10.1007/978-981-16-3412-3_11
16. Tarun N, Sivanahul R, Pushpavali M (2022) Identification of fake link using blockchain technology. Adv IoT Blockchain Technol Appl 1(1):11–17. https://doi.org/10.46610/aibtia.2022.v01i01.003
17. Narwal B (2018) Fake news in digital media. In: 2018 international conference on advances in computing, communication control and networking (ICACCCN). https://doi.org/10.1109/icacccn.2018.8748586
18. Neto M, Gomes TD, Portp O, Rafael R, Foncesca R, Nascimento J (2020) Fakenews in the context of the Covid-19 pandemic. C. Cogitare Enfermagem
19. Ngwainmbi EK (2022) Dismantling cultural borders through social media and digital communications: how networked communities compromise identity. Springer Nature
20. Qawasmeh E, Tawalbeh M, Abdullah M (2019) Automatic identification of fake news using deep learning. In: 2019 sixth international conference on social networks analysis, management and security (SNAMS). https://doi.org/10.1109/snams.2019.8931873
21. Rashid MM, Lee SH, Kwon KR (2021) Blockchain technology for combating deepfake and protect video/image integrity. J Korea Multimed Soc 24(8):1044–1058
22. Reddy P (2021) Could we fight misinformation with blockchain technology? (Published 2020). The New York Times—Breaking News, US News, World News and Videos. https://www.nytimes.com/2020/07/06/insider/could-we-fight-misinformation-with-blockchain-technology.html
23. Sharma S, Sharma DK (2019) Fake news detection: a long way to go. In: 2019 4th international conference on information systems and computer networks (ISCON). IEEE, p 201
24. Ush Shahid I, Anjum M, Hossain Miah Shohan M, Tasnim R, Al-Amin M (2021) Authentic facts: A blockchain based solution for reducing fake news in social media. In: 2021 4th international conference on blockchain technology and applications. https://doi.org/10.1145/3510505
25. Vijay JA, Basha HA, Nehru JA (2020) A dynamic approach for detecting the fake news using random forest classifier and NLP. Comput Methods Data Eng 331–341. https://doi.org/10.1007/978-981-15-7907-3_25
26. Wang G, Shi Z, Nixon M, Han S (2019) ChainSplitter: towards blockchain-based industrial iot architecture for supporting hierarchical storage. In: 2019 IEEE international conference on blockchain (Blockchain). https://doi.org/10.1109/blockchain.2019.00030

27. Westerlund M (2019). The emergence of deepfake technology: a review. Technol Innov Manag Rev 9(11):39–52. https://doi.org/10.22215/timreview/1282
28. Xu R, Nikouei SY, Chen Y, Blasch E, Aved A (2019) BlendMAS: A blockchain-enabled decentralised Microservices architecture for smart public safety. In: 2019 IEEE international conference on blockchain (Blockchain). https://doi.org/10.1109/blockchain.2019.00082
29. Yazdinejad A, Parizi RM, Srivastava G, Dehghantanha A (2020) Making sense of blockchain for AI Deepfakes technology. In: 2020 IEEE globecom workshops (GC Wkshps). https://doi.org/10.1109/gcwkshps50303.2020.9367545

Profiling Suspected Chinese Cyber Attacks by Classification Techniques

Joel Mathew Toms, Setareh Jalali Ghazaani, Sina Pournouri, and Eghe Ehiorobo

Abstract Day by day the global reliance on internet connectivity grows and so does various dangers or consequences of cybersecurity threats. China has top 1 position currently for the number of hacker groups in the world. Security experts require even more data and tools to keep up with these cyber threats from Chinese Cyber-Criminals. They need to acquire top strategies to fight back in the warfare involving cybercriminals and defences. Cybersecurity monitoring is a technique which tries to fix cyber security strategic planning by studying cybersecurity risks. Since this rate of cyber threats has caused an expansion which substantially, machine learning and data-mining techniques have become an essential part in recognizing security risks. According to this research study, classification methods will be implemented to correctly categorise as well as forecast future cyber attack vectors. The research has aimed to cyber profile and improve the current strategies of cyber attack prevention from Chinese attack groups. This will be done by evaluating previous studies on implementing data-mining techniques in the field of cyber-security. This research uses the Classification Data-Mining framework on the dataset which has been gathered from an open-source data blog Hackmageddon. The dataset has been set to focus on the attacks that took place from 2016 to 2022. 5 of the major classification techniques has been used in order to find the best suitable method to predict and classify future cyber-crime instances from China. Out of the 5, the best one was found to be Naïve-Bayes algorithm.

Keywords Cyber-profiling · Chinese cyber criminals · OSINT · Data mining · Classification

J. M. Toms · S. J. Ghazaani (✉) · S. Pournouri · E. Ehiorobo
Department of Computing, Sheffield Hallam University, Cantor Building, 153 Arundel Street, Sheffield S1 2NU, UK
e-mail: S.J.Ghazaani@shu.ac.uk

S. Pournouri
e-mail: s.pournouri@shu.ac.uk

© The Author(s), under exclusive license to Springer Nature Switzerland AG 2023
H. Jahankhani (ed.), *Cybersecurity in the Age of Smart Societies*,
Advanced Sciences and Technologies for Security Applications,
https://doi.org/10.1007/978-3-031-20160-8_16

281

1 Introduction

The modern world is rapidly doing several of its operations online, from trade and commerce to network and educational socializing. People are dependent on the world wide web more than ever before, due to the recent growth in work being remote owing to the COVID-19 epidemic. A lot of people being online leads to an increase in digital dangers. Intrusion attack targeting, victim exploitation and techniques range from a teenage seasoned script-kiddie hacker to a sophisticated criminal acquiring confidential information. There has also been an increase in the state-sponsored cyberwarfare and intervention from governments worldwide. One such country where the government funds to create elite hacker groups is China.

China seems to have the world's biggest Internet community, and Chinese cyber criminals have grown in importance internationally. The research [18] investigated cybercriminals from China on both social and technological aspects, employing a variety of methodologies such as a research analysis, passive surveillance of internet forums and groups, and physical involvement with Chinese hackers. Cybercrime was shown to be extremely trending in China. The 19 internet forums for hackers had a large amount of users registered, the numbers show that it was around 3.8 million registered individuals.

The major reasons for Chinese hackers were discovered to be political and financial considerations. Insights on Chinese hacktivism webpages during the attacks on Japan have revealed some interesting information about the genuine nature of cyberterrorism in China. Also, it showed the Chinese government's tolerance for such acts. According to assessments by Chinese security specialists, the Chinese shadow market is significantly greater than all that witnessed compared to all in the West.

Lawbreakers have a huge indecent history of carrying out cyberattacks in favour of China. Perpetrators converted governmental cybercriminals perform much of China's reconnaissance activities, shielded against punishment by the association to China's Ministry of State Security (MSS). However frightening though that may appear, there is nothing new in this scenario. An allegation published by the United States Department of Justice in 2020 stated that its concurrent illegal operations of 2 Chinese cyber-attackers dated back to 2009 [6]. The security researcher firm FireEye claims APT41 to be a distinct group of MSS attackers, originated as a crime organization in 2012 then moved to simultaneously executing government surveillance in 2014. However, there seems to be assumption that China is already preparing the younger generation to be trained with cyber-security in a more professional manner.

According to the NSA, FBI, and CISA [2], state-sponsored criminals manipulated by the Chinese authorities has attacked a wide variety of digital technology over the last 2 years, spanning from wireless devices, routers to huge telecommunications infrastructure. These Chinese attackers has exploited widely known programming weaknesses in devices and integrating hacked systems for the purpose of their own offensive network since 2020. The assaults were usually composed of five stages. China's attackers likely employed openly existing means to search networks for flaws. Hackers then would establish primary connection via web channels, obtain login information on networks, connect directly to networks, then capture network activity while exfiltrating the victim's information. By targeting such flaws, they have been able to build large infrastructure connections and abuse a variety of public or personal organizations.

The researchers first became aware of the activity in April 2021, when a corporation raised the possibility of an infiltration when they were having a corporate sales meeting with the security company [3]. Investigators reverse analysed the assault and learnt about every movement hostile force made within the network, noting hackers APT 41 had ready control to anything in the system, allowing attackers to select which material to steal. APT 41 known as "Winnti" is amongst the most influential and effective Chinese government backed hacker groups. The history of introducing Chinese espionage operation and motivated by profit threats to the United States of America and many global objectives that are frequently affiliated with China's Financial Expansion Plans.

This research, in particular, aims to expose the most prevalent attack tactics, targets, and objectives of Chinese computer hackers through the use of Classification algorithms to conduct a cybercriminal profile analysis. This result of this research will help us understand deeper as to how the Chinese Cyber Criminals choose their victims and conduct an attack. Although there have been studies done using data-mining techniques to profile cyber-attackers and their targets in the past, there haven't been any study or research on profiling Chinese Cyber Criminals conducted yet. Based on previous research on profiling users, this study will conduct a cyber profiling analysis by employing a classification algorithm on an OSINT dataset containing cyber-attacks that occurred between 2016 and 2022. The advantage and future application of this study may be obtained by delving further into Chinese-related hacking occurrences in order to better forecast future hacking incidents and develop a more effective preventative plan for Intelligence and Cyber Defence centres.

Open-Source Intelligence OSINT is a collection of datasets from publicly widely accessible resources used by a number of objectives. Also, open-source intelligence encompasses all information regarding organisations or entities that one can obtain legally from various public sources. OSINT methods are used by security agencies, security experts, even cybercriminals to sift across enormous amounts of information and identify valuable data.

OSINT contains a substantial influence for this research. It's indeed particularly crucial since this study has to be both time and cost efficient, so OSINT becomes ideal for such situations. It is therefore crucial to highlight because collecting significant amounts of information through anonymous sources or organisations would be challenging. Ultimately, although OSINT datasets could be termed inadequate or erroneous, they seem to be a trustworthy resource. So, this paper aims to answer the question, "Can Classification Data-Mining techniques or approaches be utilised to identify and anticipate prospective cybercrimes by Chinese Cyber Criminals?".

2 Literature Review

This Section will go through the already existing research on data mining techniques like classification, the functioning and also how it may be used to do cyber profiling using data mining techniques. Furthermore, depending also on goals and targets, the section addresses the purpose that is to conduct a research review on previously done literatures to select a data-mining approach to identify Chinese cyber-criminals. This section will also provide information on what Data-Mining is and what are the different techniques under classification.

2.1 Introduction to Data-Mining

Data mining methods could be implemented to information security elements to make better judgments and responses while doing a cyber profile study. Data mining generally can be described as the process of examining, collecting, and uncovering useful data and insights. Rutgers [13] major corporations use data mining tools to assess the present condition also forecast future strategies in the businesses.

The following are the two primary ways of data-mining. The first one will be 'Supervised'. It is the technique where it derives a model to a function using structured labelled training dataset.

The next one is called 'Unsupervised'. It is the method where the process attempts to identify underlying characteristics using unlabeled and unobserved datasets.

There are four important Data-Mining techniques which will be discussed [13]:

– Clustering: The approach attempts to cluster comparable things depending on the whether there are similar criteria.
– Classification: They are primarily used to divide the statistical model into distinct subgroups, with the results regarded as a prediction method.
– Association: The strategy for discovering meaningful correlations between independent factors inside a data collection.
– Regression: The approach attempts to construct a variable that will contribute to a representation of the dataset.

2.2 Data-Mining Related Works

In Paper 1 with the title 'Predicting the Cyber Attackers: A comparison of different Classification techniques' [11] examines the efficacy of several classification algorithms in identifying and forecasting cyber criminals and probable future computer hackers In the inquiry, OSINT along with previous datasets regarding cybercrimes have been used in this paper. Open-source dataset has been utilised for the investigation since it has a lower cost also is easier to work with. These datasets contain documented cybercrimes has some where the attackers are recognized or taken credit. The training dataset along with a test dataset was applied to conduct validation with development to create various classes. Its training dataset contains 1432 assaults related to recently identified cybercriminals between 2013 till 2015. In the test dataset recorded 484 cyberattacks between 2016 till 2017.

The research covers both advantages and disadvantages. Most of today's applications use data mining techniques which are currently implementing cyber defense. They make use of data-mining and prediction modelling approaches. Prediction modelling approaches involvement in police task force organisations discover or solve crimes prior to the consequences Nevertheless, the suggested framework model focused on SVM approach to be the best on using OSINT Dataset. The accuracy rate for profiling cyber criminals was higher compared to the other approaches which were taken. This could also mean that these varying field of cyber tactics can cause changes in the overall accuracy rate. The use of OSINT datasets is excessive and insufficient this results in reduced accuracy rates. The reason being cybercrimes tend to be impacted by various different elements. A good solution would be that additional cybercrimes dataset should be gathered, also new methods and procedures of a hack being conducted could add more value to the variables in the collection of the dataset.

In Paper 2 with the title 'Cyber Profiling using Log Analysis and K-Means Clustering' [19]. The main goal of the paper is to conduct a study or an analysis on the behaviour of internet log activities of people in the educational sector of Indonesia by using Log Analysis and K-Means Clustering to determine the percentage of usage for suspicious activities. The paper starts off by talking about how the amount of internet users in Indonesia are increasing each year and this has led to a certain behavioural change. The author has referred to a survey conducted in 2014 with respect to this. The author states that internet log activity plays a major role in studying behaviour of the internet users. In this paper, the author specifically wants to study or determine and conduct cyber-profiling of behaviours using log of internet activity in a higher educational institution. So, their main motivation is that they state that educational sector makes use of the internet the most and that there hasn't been any research on this topic among Indonesians. Therefore, cyber profiling would be extremely valuable in learning about the behaviour of Internet users in the higher education sectors of Indonesia. The author also wants to determine whether the result is affected by environmental and daily routine activities or factors. Internet activity in schooling should be used to help the teaching experience, however occasionally the facts acquired show that they used online services for purposes other than education. As a result, there is evidence of such users at higher education institutions contributing to cybercrime. The author wishes to learn more about whether the Internet is actually made use by students to be consistent with the scope of task in the teaching curriculum. The authors have used Log analysis using K-Means Clustering techniques. This research also touches upon profiling cyber criminals. When this technique is used for its right purpose, it could also help in predicting cyber-crimes by targeting the right people. They have taken the effort of explaining Data Mining, Clustering, and K-Means technique. This could be helpful for people who are beginners to read the paper to gain an understanding of the Data Mining. But they have put too much focus on making it much easier for the readers that they have forgotten to explain in depth about the process of how they analyse their data. The research has met its intended result of profiling suspicious activity among the educational sector of Indonesia.

The paper could have expanded their data with respect to the fields. Instead of just limiting this research to the educational sector, they could have included other sectors as well. And instead of just limiting the study on Indonesians, we include other countries as well. The source of data for the profiling procedure in this investigation is limited. For the profiling to be flawless, the procedure should include data from the computer's actions and the dataset should be considered for more than just five days. As a result, additional study is necessary to give a better cyber profiling solution with the help of more comprehensive data sources. The future work for this project could be that they could use this for predicting cyber-crimes with the help of analysing the patterns.

In Paper 3 with the title 'Youth hackers and adult hackers in South Korea: An application of cybercriminal profiling' [1]. The primary purpose of this research is to generate cybercriminal profiles in order to better forecast future cyber-threats and develop a more tailored South Korean preventative approach. The authors conducted the research using the FBI's cyber-criminal profiling analysis. From 2010 to 2019, data were gathered from court records and mainstream media papers detailing computer hacking events in South Korea. Using the cybercriminal profile analysis, this study aimed to highlight the most frequent features and motives of the Black Hat computer hackers. The authors were able to portray that the results of the research conducted reveal that older generation of hackers are more motivated towards money than the younger generation of hackers and their motivations. Professional hackers comprised individuals aged 25 and over, whilst youth hackers have been those aged between 15 and 24. The ongoing investigation has produced several noteworthy results, including hacker profiles and hacking practises. Korean based cyber criminals were commonly dependent to these interpersonal and national level hackings than global level hackings which had been put together with respect to the geographical measurements. Financial gain was the most potent motive for both youth hackers which had 55% and seasoned hackers which was 87% to hack. Indeed, more than 85% of these professional older generation cyber criminals were primarily motivated due to monetary goals. On the other hand, young cyber criminals were motivated by amusement, hacktivist activities, revenge, and exposure. The limitations were that the generalizability of the study's results could not be determined since the dataset only covered 83 cases in South Korean dialect during a specific time period. Along with this, the research only included the top documented or reported cyber-attack incidents. This can be solved by increasing the dataset and using a pre-processing tool to clean the data. Hence increase the amount of data collected and the result will be far more reasonable. They could also gather data from honeypots and log information from the deep web to analyse the pattern of their research target.

In Paper 4 with the title 'Using Classification Techniques to Create a Predictive Framework for Cyber Attacks' [16]. The main goal of the paper is to find a suitable classification approach that can be used to help the gather information and better forecast the upcoming cyber crimes. The open-source dataset that has been used for this project is for the year 2017–2019. The paper only focussed on the European region for the classification of the attacks to understand the attack vectors. So, the dataset contained 1989 lines of cyberattacks which had both known and unknowns in them. The paper firstly gave information about the current world issues with regards to cyber attacks that had taken place due to the increase in global need for internet and connectivity. It then goes on to give a brief introduction to the various Classification algorithm approaches with a literature review for each approach. The paper has also provided detailed overview of the dataset that has been used this project also giving statistics of the data being used. The paper has used five classification approaches for this research. Namely SVM, K nearest neighbour, Random Forest, Naïve Bayes and Neural network. The SVM classifier brought a TP rate 0.633, FP rate 0.297 and Recall rate 0.633 for the dataset. The final accuracy put forth by SVM classifier was 63.3% and the inaccuracy rate was 36.6%. The random forest classifier brought a TP rate 0.675, FP rate 0.216 and Recall rate 0.675. The final accuracy put forth by Random Forest classifier was 67.4% and inaccuracy rate was 32.5%. The Naïve Bayes classifier brought a TP rate 0.643, FP rate 0.224 and Recall rate 0.643. The final accuracy put forth by Naïve Bayes classifier was 64.3% and inaccuracy rate was 35.6%. The K Nearest Neighbour classifier brought a TP rate 0.672, FP rate 0.222 and Recall rate 0.672. The final accuracy put forth by KNN classifier was 67.2% and inaccuracy rate was 32.7%. The Neural Network classifier brought a TP rate 0.669, FP Rate 0.212 and Recall rate 0.669. The final accuracy put forth by Neural Network classifier was 66.8% and inaccuracy rate was 33.1%. Based on the data that the author received, the Random Forest classifier was declared the best or most suitable approach for the validation model.

Next the paper takes into consideration about the validation dataset, where it has 1777 records between January to May 2020. In the study, the author mentions that when Random Forest classifier was applied the targeted attack vector had the greatest section of accurately classified cybercrimes. Thus, the validation dataset model is good for predicting Targeted Attacks.

Overall, it was a well detailed paper but there are some errors or hurdles which can be corrected in the future. The author could have first talked about why they decided to select only classification models to use for this project. The open-source data is not completely correct because there were some assumptions made in order to prepare it for the classifiers. More amount of data could have been used to better predict cyber attacks.

In Paper 5 with the title 'Use of Classification Techniques to Predict Targets of Cyber Attacks for Increased Cyber Situational Awareness during the COVID-19 Pandemic' [4]. The main aim of the paper was to provide a better cyber-situational awareness using data-mining techniques more into classification by detection and prediction of future attack vectors. The open-source dataset that has been used for this project is for the year 2017–2019. The paper only focussed on the European region for the classification of the attacks to understand the attack vectors. So, the dataset contained 3974 lines of cyberattacks which had both known and unknowns in them. The paper has given a proper introduction using statistics and information regarding different approaches to providing and spreading cyber security awareness. The aims and objectives have been clearly mentioned. The author has demonstrated the ways on how they have collected the data from open source and has described the pre-processing stage very well using the tool OpenRefine. The paper has also provided detailed overview of the dataset that has been used this project also giving statistics of the data being used. And has also mentioned about using WEKA to conduct the research. The paper has used five classification approaches for this research. Namely Decision Tree, SVM, K nearest neighbour, Naïve Bayes & Neural network. The focus was put on the target class alone. The Decision tree classifier brought an average F-measure 0.293 and average ROC Area 0.670. The accuracy rate put forth by Decision Tree was 33.89%. The Artificial Neural Network classifier brought an average F-measure 0.33 and average ROC Area 0.683. The accuracy rate put forth by Artificial Neural Network was 34.72%. The K nearest neighbour classifier brought an average F-measure 0.313 and average ROC Area 0.682. The accuracy rate put forth by K nearest neighbour was 34.3%. The Naïve Bayes classifier brought an average F-measure 0.209 and average ROC Area 0.684. The accuracy rate put forth by Naïve Bayes was 32.63%. The SVM classifier brought an average F-measure 0.269 and average ROC Area 0.674. The accuracy rate put forth by SVM was 32%.

The paper has achieved its main purpose but there are some limitations that can be worked upon for future works. The major limitation would be the quantity of data to work with, it was low to give accurate feedback that can prove the aim. The source of the data could have been revaluated especially from medical fields and other sectors which could have provided in depth analysis of the project.

Table 1 contains the five research papers on which critical analysis has been conducted to gather their strengths and weaknesses. The table gives information regarding the author, aim, method used and dataset used for the specific research paper.

Table 1 Overview of data-mining related works

Author	Aim	Data	Method
Pournouri et al. [11]	Prediction of cyber attackers	Open source	Classification
Zulfadhilah et al. [19]	Cyber profiling indonesian hackers	Online survey Log data	K-means clustering
Back et al. [1]	Cyber profiling youth hackers in South Korea	Incident reports and court cases	Criminal profiling framework
Wass et al. [16]	Creating predictive framework for cyber attacks	Hackmageddon-open source	Random forest-classification
Crowe et al. [4]	Improve CSA using classification prediction techniques	OSINT Data	Artificial neural network-classification

By reading and critically analysis these five research papers, the study was able to formulate a plan and gain ideas on how to implement their methodologies for the betterment of this research. This research project will be using Classification data-mining technique in the study but with a much extensive dataset to work with.

2.3 Classification

For this project, the classification techniques have been used so this section will explain with immense detail as to which of major classification models are being implemented to this study.

The classification technique is the act of classifying a collection of data into multiple groups or sub-classes in order to create reliable forecast and evaluation using enormous datasets. Classification may be implemented to establish a notion of the categorization of a client, item, or entity in a database via detecting unique characteristics to select a certain class. The techniques under classification which are going to be used for this research project include 5 main knowledge-based models. Artificial Neural Networks, Support Vector Machine, Decision Tree, K Nearest Neighbour, Naïve Bayes [10].

2.3.1 Artificial Neural Networks

It is a form of densely integrated, input driven Artificial Neural Network (ANN) composed of interconnected layers of nodes. ANN labels entities at first, then compares these to genuine classifications and calculates its error margin. The procedure is then repeated until fault achieves its minimal amount. It is a frequently used neural network architecture. Mostly It is built on evolutionary and training concepts. There are 3 stages in a multi-layer perceptron: input, output, also the hidden units. The quantity of nodes within the input stage should remain equal to quantity of nodes in the input data. Its output stage has one neuron this becomes the Perceptron's value [14].

2.3.2 Support Vector Machine

SVM termed supervised classification approach which can be used to reduce classification tasks because of the high degree of accurate results in comparison to the algorithms previously discussed. SVM used for many varieties of tasks such as classification, categorization and optimisation algorithm. The SVM method was created to be focussed on selecting the best hype—plane that divides 2 or greater features by the largest displacement known as 'Margin' among the nearest neighbours. Objects on the borders are referred to as support vectors training sets. SVMs are the most difficult things to categorise, and these employ an essential part in establishing as well as identifying the best hyper—plane. Support Vector Machine algorithm approach has few key points. The approach creates efficient allocation for memory where it can do analysis on datasets under training. SVMs are essential in the case of number of items being fewer than digits or dimensions. But the only downside to SVMs is that it is highly time consuming [17].

2.3.3 Decision Tree

Decision tree algorithm, generally referred called prediction trees, is composed of a series of choices plus its results as implications. Creating a prediction model using nodes plus associated branching could help with predictions. Every node sees a new input vector, so each fork takes a unique choice mechanism. A term "leaf node" applies for nodes which don't have any branches and instead yield classification even the likelihood values. Decision tree algorithms are widely utilised in prediction machine learning algorithms since those are simple to create, visualise, and communicate. Discrete and continuous input parameters are also possible. There are 2 aspects of decision tree algorithms. They are used for both classification and regress. Whenever the performance parameters are categorized, the decision model is referred to the classification model so whenever the result is consistent like digits is referred to the regression model [17].

2.3.4 K Nearest Neighbour

KNN often used for classification problem solution. This method takes K closest training examples for input and outputs the result like a classifier. An object's classification would be determined by the maximum votes of its neighbours. The training dataset inside the KNN approach is a collection of variables in subspace that have been allocated to distinct categories. The users would then specify K with in classification step, as well as the item would be allocated toward the applicable category with regards to highest occurrence of that type. The KNN method is extremely basic and straightforward to apply. It performs admirably on multi-class classification issues and scenarios. It performs admirably with chaotic datasets. If indeed the learning dataset has a greater number of data, KNN efficiency could be stronger and consistent with the efficiency. KNN is referred to as a complacent classifier since it cannot imply using the training dataset and instead just classifies the sample data using the already existing dataset. KNN is costly to operate due of the calculation of K variable, which is dependent on calculating the space and displacement by different items [17].

2.3.5 Naïve Bayes

Amongst the top significant data mining strategies that lead towards classification analysis and regression is Naïve Bayes. Email spam detection and messages classification are two popular implementations of Naïve Bayes.

The approach of this algorithm makes the assumption that every feature in the data source is distinct of one another calculates the likelihood of multiple activities measurement of the variables. Nevertheless, some others recognise it to be the weakness since it is nearly unobtainable data with characteristics which are immaterial in everyday world. Another key benefit of this approach is that it just requires a few classification models to produce classifiers. In certain classification methods, the Bayes was substantially greater in efficiency, precise, as well as quicker than alternative techniques like decision tree algorithm approach.

The whole Naïve Bayes has a much easier and user-friendly implementation method when put in comparison to its competitor algorithms. It is efficient enough to produce precise models based on small training datasets [17].

3 Methodology

This Section will create an emphasis on how the study intends to provide a framework for conducting a digital profile analytics on Chinese cybercriminals. It will stop potential virtual issues with events using Data-Mining and predicting statistical approaches related to prior previous datasets of computer hackers. The purpose of this section can show the data features with format, architecture, and pre-processing phase. The two tools that were employed would be discussed. Pre-processing is an important phase in all Data-Mining endeavours. This Section will also contain the ethical considerations section.

3.1 Data-Mining Tools Used

3.1.1 WEKA

WEKA originally developed by University of Waikato in 1997 and is centred upon using Java coding [15]. The main purpose of the tool is to be utilised throughout this research to check overall correctness of predictions including the research assessment and analysis phase. WEKA contains a wide range of Data-Mining algorithms, including clustering, classification, and association. This in turn enables pre-processing functions like fixing misplaced data also normalisation. WEKA being an open to all and free option, the application could operate on all devices which are using latest OS. Thus, this makes WEKA the most common place to enable users who are into data analytics. The way the results are displayed with the various

Fig. 1 WEKA application

different information and visualisations of results are essential when presenting to people in high positions of a company. This research will be using WEKA tool since it highly efficient for classification than its alternative option called Orange application (Fig. 1).

3.1.2 Open-Refine

OpenRefine also an open-source tool which is made easy to use for data analyst. This application can be used for dataset cleaning and transform datasets to another form [9]. It accepts file types like CSV because it is similar to a database. It works on rows in the dataset which contains various cells below the feature columns. It was created by Google using Java coding framework. This tool will be essential for the pre-processing phase of the research. OpenRefine can be used for correlating different data. It can fix each cell data based on whether it is missing or invalid. The competitors for this application are Rapid Miner and Cloudingo (Fig. 2).

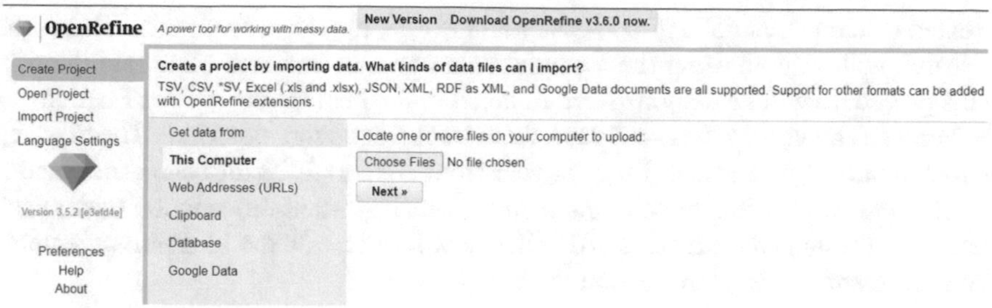

Fig. 2 OpenRefine application

3.2 Gathering of Datasets

As mentioned under the aims and objectives, this research will be leveraging OSINT (open-source intelligence) dataset for the study. It is hard to obtain dataset about cybercrimes updates from government agencies and various companies. Thus, employing OSINT have had a no cost method to obtain the information this research wants. The study requires datasets from not just a month or two, it requires datasets from multiple years. But creating these and producing this by hand by looking at online article stories and other sources will consume a lot of time. Hackmageddon (https://www.hackmageddon.com/) became the primary open-source of datasets for the study. Hackmageddon basically the blog that offers all cybercrime timelines also information regarding each. The Hackmageddon dataset used for this study comprises cybercrimes that happened from 2016 and 2022. This dataset covers 10,836 cells of information throughout this time-frame.

Unfortunately, there are difficulties with OSINT information. To begin, the research relies on different dataset this causes many inaccuracies, misspelled words, or contradictory labelling. Secondly, the file sets contain symbols, texts, or other information which this study does not need and some of them the study requires in a multiple format. Hence the datasets need to be pre-processed to eliminate such errors. Also, because the study relies on Hackmageddon dataset which someone else has identified and registered cybercrimes there is a possibility that the author may only classify on assaults faced by English-speaking globe. Although when assaults are recorded, the knowledge may not always be full. To give an example, a cyber-attack may be registered however the contents remain vague.

3.3 Categorisation of the Dataset

In this section, the paper will provide a deeper understanding on the structure of the dataset that is being used for the purpose of this research. This is the raw structure that has been taken from the Hackmageddon OSINT dataset. For each of the cyber-incident or event that took place, 10 different and unique columns have been created contain essential and non-essential information. Now from this dataset, the research will only be using the essential data columns and will eliminate the rest. This procedure will be demonstrated in detail in the next section of this Section.

Figure 3 shows how the raw dataset from Hackmageddon looks like. The first step is to compile all the datasets for each year from 2016 to 2022 into one excel sheet.

There were 11 different columns with various important information under each cell. To understand the structure in detail, below listed are all the 11 different features from the compiled Hackmageddon dataset.

1. ID: This provides each row record's unique identifier.
2. Date: The day when the cyberattack had occurred.
3. Author: Name of the individual or group responsible for the strike, for example, APT10.
4. Target: Name of the organisation or the victim on whom the attack had happened.
5. Description: The synopsis of the incident.
6. Attack: What category of attack took place, for example, Web Application Attack.
7. Target Class: Type of corporate sector that has been targeted by the attackers.
8. Attack Class: What category for each cyberattack falls under, for example, Hacktivism.
9. Country: Name of the country where the cyberattack took place not always from where the attack started, for example, United Kingdom.
10. Link: Online reference to know more about the cyberattack and its details.
11. Tags: Important terms associated with the cyberattack.

Out of all these columns, this research is only interested in five important sections. They are as follows:

- Author
- Attack
- Target Class
- Attack Class
- Country

This study has implemented and used training models using the five classification algorithms. To do that, it had been decided to eliminate the ID and Date column. To avoid irrelevancy and data redundant issues, the study has eliminated the columns like Description, Target, Links, Tags as well. This provides the research with a much clearer and efficient dataset to work with (Fig. 4).

3.4   Data Pre-Processing Stage

In Sect. 3.4, six different columns were eliminated and only five important columns are remaining to be used as a dataset. This dataset now needs to further pre-processed

ID	Date	Author	Target	Description	Attack	Target Class	Attack Class	Country	Link	Tags
1	1/2/2016	New World Hacktivists (NWH)	donaldtrump.com	The hacking group New World Hacktivists (NWH) takes down the official Election Campaign website of American Presidential candidate Donald Trump (donaldtrump.com). The same attackers claim responsibility for the DDoS attack that crippled the BBC website during the New Year's Eve.	DDoS	Single Individual	H	US	https://www.hackread.com/hacker s-shut-down-donald-trump-electio n-campaign-website/	New World Hacktivists, NWH, Donald Trump, donaldtrump.com
2	1/3/2016	Anonymous	Several Saudi Arabian Government Websites	The Anonymous protest against the execution of 47 people in Saudi Arabia and take down several high-profile Saudi Arabian government websites under the banner of operation #OpSaudi and #OpNimr.	DDoS	Government	H	SA	https://www.hackread.com/anony mous-takes-down-top-saudi-arabi an-govt-websites/	Anonymous, #OpSaudi, #OpNimr
3	1/3/2016	ScOrpln Att@ck3r from Muslim Cyber Army	Goa University unigoa.ac.in	ScOrpln Att@ck3r from Muslim Cyber Army hacks the Goa University (unigoa.ac.in) and dumps 10,380 records with hashed passwords.	Unknown	Education	H	IN	http://goindiareborn.blogspot.co.u k/	ScOrpln Att@ck3r, Muslim Cyber Army, Goa University, unigoa.ac.in

Fig. 3 OSINT dataset structure

1	Author	Attack	Target Class	Attack Class	Country
2	New World Hackers	DDoS	Individual	HA	US
3	Anonymous	DDoS	Government	HA	SA
4	Anonymous	DDoS	Government	HA	TH
5	Chinese Hackers	Hijacking	IT and Communication	CC	CN
6	ISIS	Hijacking	Individual	CW	LB
7	GeNiuS-JorDan	Web App Attack	Government	HA	UG
8	DeleteTheDamnElite	Persistence Attack	Government	CC	Multi
9	Anonymous	DDoS	Government	HA	NG
10	Indian Hackers	Web App Attack	Government	CW	PK
11	Sonny	Persistence Attack	IT and Communication	CC	CY
12	Crackas With Attitude	Hijacking	Individual	HA	US
13	Cyber TeamRox	Web App Attack	Multiple Industries	CC	KH
14	Anonymous	DDoS	Government	HA	TH
15	Crackas With Attitude	Hijacking	Individual	HA	US
16	Russian Hackers	Targeted Attack	Transportation Industry	CW	UA

Fig. 4 Dataset with relevant columns

and cleaned in order for it be easy to use and to deem it as ready for the implementation of various classification algorithms.

The dataset had preparations along with pre-processing procedures done to it so that the results would commence following reconstruction of the dataset and arranging the characteristics into five unique sections. Information purification solutions such as Rapid Miner are available but this project has used OpenRefine to be the dataset cleansing method for the whole pre-processing step. After uploading the compiled dataset with only 5 columns to OpenRefine, the first pre-processing approach can begin. These have been separated into a few phases which are discussed below:

For the first stage of the pre-processing phases, the study has chosen to clean or remove duplicated characters from the dataset. Elimination of the rows which had important values missing or blank had been done.

The main focus will be towards the 'Author' column. Text Facet helps in viewing the various features in each column. Using the 'Cluster' feature in OpenRefine, using this feature quite a few of the attacker group were automatically grouped using similarity mechanism. After using the Cluster feature, there were many more of the information present in the dataset which had repeating values, ex: 'APT10' 'APT 10' 'Red Apollo' all are the same so they all had to be manually categorised into one group. Finally, they were grouped into 778 choices.

The next focus was towards the 'Attack' column. Again, the basic steps of removing duplicates and irrelevant data entries were done. After that, the clustering technique in OpenRefine was applied to reduce the repeating entries due to capital letters or spaces. The grouping of the remaining data rows was done according to MITRE ATT&CK categorisation of attacks [8]. Even the use of Rapid7 categorisation was used as reference to efficiently classify the 'Attack' [12]. Finally, they were all grouped into 12 choices. The following focus was onto the 'Target Class', were the above-mentioned basic cleansing steps were again conducted. They were grouped into 16 choices in the dataset.

The section 'Attack Class' was grouped into 4 categories. Namely Cyber-Crime (CC), Cyber-Espionage (CE), Hacktivism (HA) and Cyber-War (CW). The next focus was onto the 'Country' column. The countries have only been mentioned with the ISO 3166–1 Alpha-2 standard issue 2-digit country codes. But clustering feature was used to group recurring or repeating row entries. Then some countries had been mentioned in the dataset with 'Great Britain', 'United Kingdom', 'GB' and 'UK'. So, these have been categorised into one 'UK' country class. One common pattern noticed in the dataset is that 'US' or United States of America had the greatest number of attacks being conducted against them.

As mentioned in Sect. 3.3, the compiled dataset contained 10,836 records. After conducting the pre-processing phases on the dataset, there was 8272 rows of data. But for the purpose of this research, this dataset has been split into 3 datasets namely Main Dataset which contains all the attack authors not limiting to Chinese Cyber criminals alone but the unknown authors have been removed. The Main dataset now contains 2583 rows of data. These removed unknown authors have been compiled into another dataset called validation dataset. The Validation dataset contains 5689 rows of data. Out of the Main Dataset, the dataset has been scrutinized and only rows which are related to China or Chinese Cyber Criminals tags have been selected. After that, all those selected rows alone have been compiled to form the Training dataset. The training dataset contains 432 rows of distinct attack authors in the dataset.

3.5 Dataset Statistics

In this section, the study will provide detailed statistics and information regarding the three datasets that are going to be used in the project. The three distinct datasets are:

- Main Dataset
- Validation Dataset
- Training Dataset

3.5.1 Statistics of the Main Dataset

In the Main Dataset, a total of 2583 rows of information are present. There are 777 unique Cyber Attackers also known as 'Author', 12 unique Cyber Threats also known as 'Attack', 16 unique types of victims also known as 'Target Class, 4 unique classes of Attack also known as 'Attack Class' and finally 104 unique 'Country' instances in the Main Dataset.

To provide more detailed review of the dataset, the 12 types of attack have been closely looked into. 'Malware' has the highest number with 979 different rows,

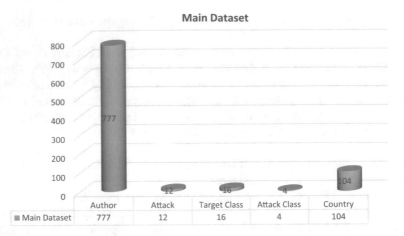

Fig. 5 Main dataset statistics

'Targeted Attack' had the second highest number with 789 different rows, 'Reconnaissance' had the third highest number with 223 different rows, Hijacking attacks with 190 and Web Application Attacks with 119.

If the 'Attack Class' is looked into, 'CC' = Cyber-crime had the highest with 1420 rows of data, 'CE' = Cyber-espionage had the second highest with 863 rows of data, 'HA' = Hacktivism had the third highest with 194 instances and finally 'CW' = Cyber-warfare with 106 instances.

Found below is an overview representation of the statistics of the Main Dataset (Fig. 5).

3.5.2 Statistics of the Validation Dataset

In the Validation Dataset, a total of 5689 rows of information are present. There are 5689 Cyber Attackers also known as 'Author' which contains only Unknown or Unidentified attackers, 12 unique Cyber Threats also known as 'Attack', 16 unique types of victims also known as 'Target Class', 4 unique classes of Attack also known as 'Attack Class' and finally 104 unique 'Country' instances in the Validation Dataset.

To provide more detailed review of the dataset, the 12 types of attack have been closely looked into. 'Malware' has the highest number with 2730 different rows, 'Hijacking' had the second highest number with 1476 different rows, 'Reconnaissance' had the third highest number with 409 different rows, Targeted Attacks with 259 and DDoS Attacks with 221.

If the 'Attack Class' is looked into, 'CC' = Cyber-crime had the highest with 5303 rows of data, 'CE' = Cyber-espionage had the second highest with 285 rows of data, 'CW' = Cyber-warfare had the third highest with 56 instances and finally 'HA' = Hacktivism with 45 instances.

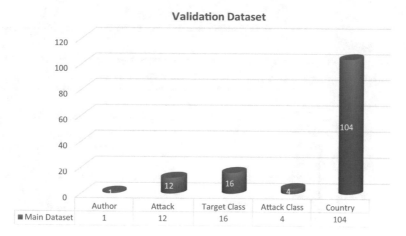

Validation Dataset

	Author	Attack	Target Class	Attack Class	Country
■ Main Dataset	1	12	16	4	104

Fig 6 Validation dataset statistics

Found below is an overview representation of the statistics of the Validation Dataset (Fig. 6).

3.5.3 Statistics of the China-Based Dataset

In the Chinese Dataset also known as the Training Dataset, a total of 432 rows of information are present. There are 101 Cyber Attackers also known as 'Author' which contains only China-based or Chinese attackers, 9 unique Cyber Threats also known as 'Attack', 15 unique types of victims also known as 'Target Class', 4 unique classes of Attack also known as 'Attack Class' and finally 44 unique 'Country' instances in the Training Dataset.

To provide more detailed review of the dataset, the 9 types of attack have been closely looked into. 'Targeted Attack' has the highest number with 240 different rows, 'Reconnaissance' had the second highest number with 80 different rows, 'Malware' had the third highest number with 55 different rows, Hijacking with 31 and DDoS Attacks with 11.

If the 'Attack Class' is looked into, 'CE' = Cyber-espionage had the highest with 294 rows of data, 'CC' = Cyber-crime had the second highest with 122 rows of data, 'CW' = Cyber-warfare had the third highest with 14 instances and finally 'HA' = Hacktivism with 2 instances.

Found below is an overview representation of the statistics of the Training Dataset (Fig. 7).

In fact, it must be observed where certain difference exists between 3 datasets. This is a natural observation given that certain classes were deleted when the pre-processing step had been conducted, and what applies to 1 dataset need not apply toward the 3 datasets. The categories where multiple attack classes were mentioned has been removed in order to present a clear dataset fit for cyber-profiling. The

Training Dataset

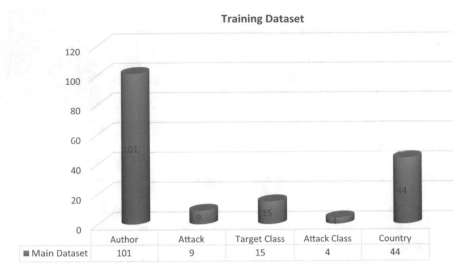

	Author	Attack	Target Class	Attack Class	Country
■ Main Dataset	101	9	15	4	44

Fig. 7 Training dataset statistics

prevalent attack types were noted to be Malware, Reconnaissance and Hijacking. The most prevalent reasons of attack classes were noted to be Cyber-Crime and Cyber-Espionage.

3.6   Ethical Considerations

Since it has been previously indicated, the datasets used for this research is OSINT open-source data. Therefore, none of the ethical issues create any problem areas. Nonetheless, the datasets access has been given by Paolo Passeri and allowed upon approval from the proprietor of the website Hackmageddon.

3.7   Section Conclusion

In the research methodology Section, the paper discussed what strategy was used for this research. It even talked about the 2 tools being used namely, WEKA and OpenRefine. The methods used for dataset collection, dataset categorisation along. The data pre-processing stages were clearly mentioned as well. After completing the pre-processing stage, an overall statistic of each dataset was provided. This Section mentioned about the ethical considerations in particular for this study. The next portion of this paper shall give the wider picture of the actual results of the data analysis by applying classification data-mining techniques.

4 Research Findings

In this Section of the research project, the actual results after applying each classification approaches using the tool WEKA only to the training datasets will be provided. The best classification method can then be selected based on the accuracy rates to further use it to classify the Validation Dataset. Modelling of the datasets have been trained throughout this part utilising the training dataset including classification methods and approaches. The importance of training supervisions of dataset should be emphasised throughout the entire training phase. Training supervision is a sort of approach and model accuracy assessment when studied from earlier research articles. K-fold Cross Validation has been utilised in this research since it contains greater precision and effectiveness for the particular objective.

4.1 K-Fold Cross Validation

Once the classification approach modelling has been done with the help of Training dataset, the accuracy of the result has to be determined with the help of another dataset to gain the correct accuracy. That's not always possible with small datasets like the ones being used for this project. Therefore, Cross Validation technique has been utilised that separates the dataset to predetermined feature set of subgroups also known as Folds. These K-Folds in which K considered as the selected number [7]. The research has utilised the most common number of ten K-Fold cross validations. Although the procedure for this means that there will be a few steps which happens when the project executes each of the classification approaches using WEKA. Dataset gets split to 10 similar sets to be randomised. Rest nine sets gets compiled and has been used to act as a Training Dataset for the implementation of classifier to every set. All the single sets are used as a data to be compared to and verify each result. The outcomes of every one of the 10 examinations are scrutinized and finally produces an accuracy rate.

Hence considered as a result of the outcome that the dataset has split into each K similar sized groups when the training phase was conducted using K-Fold Cross Validation. This process ensures that one group serves as the training dataset, while K minus one group gets treated as a training group therefore the dataset would be replicated K number of iterations.

This method produces a rather reliable and precise outcome due to the repeating rhythm. K has been adjusted at 10 in the case of this research's objectives. This makes the training procedure to replicate 10 different instances of time. The initial subgroup in ten K-Fold Cross Validation is referred to as the validation group, whereas the subsequent 9 parts are referred to as the actual training group. In addition, the subsectors for the 2nd K-Fold seems to be the verification collection, whereas the subsequent subgroups seem to be the training group. Cross Validation represents the 10 levels of accuracy obtained on the validation group procedures for evaluating as an aggregate.

Hence, by doing this the research also would like to note the fact that K-Fold Cross Validation has been an advantage for this project due to fact being the dataset used for this project is comparatively small.

4.2 Application of Various Classifiers

In this section, the application of 5 major types of classification approaches has been discussed and the results as accuracies have been recorded. The 5 major classification approaches that have been used in the order are:

- Naïve Bayes
- Support Vector Machine
- K-Nearest Neighbour
- Decision Tree
- Artificial Neural Network

4.2.1 Naïve Bayes Classifier

The Naïve Bayesian classification approach has been used on the Training dataset for this level of research to forecast the 'Author' class for each cyberattack. In Sect. 2.4.5, the Naïve Bayesian classification approach was covered. The classifier will be implemented using WEKA Analyser, then 10 K-Fold cross validation has been used to test the dataset. The research has selected the 'NaiveBayes' option under 'bayes' category of classifiers.

Figure 8's results indicate the fact, out of the 432 occurrences of Chinese Cyber Criminal groups, 38.1944 percent were properly classified with a count of 165. The wrongly classified had a count of 267 which meant it had 61.8056 percent. The mean absolute error was 0.0143. Out of all the five different classifiers that has been used for this research project, Naive Bayes algorithm had the best accuracy rate with 38.1% for the Training Dataset.

4.2.2 Support Vector Machine Classifier

The Support Vector Machine classification approach has been used on the Training dataset for this level of research to forecast the 'Author' class for each cyberattack. In Sect. 2.4.2, the Support Vector Machine classification approach was covered. The classifier will be implemented using WEKA Analyser, then 10 K-Fold cross validation has been used to test the dataset. The research has selected the 'SMO' option under 'functions' category of classifiers.

Below is the Fig. 9, the results indicate the fact, out of the 432 occurrences of Chinese Cyber Criminal groups, 37.037 percent were properly classified with a count of 160. The wrongly classified had a count of 272 which meant it had 62.963 percent.

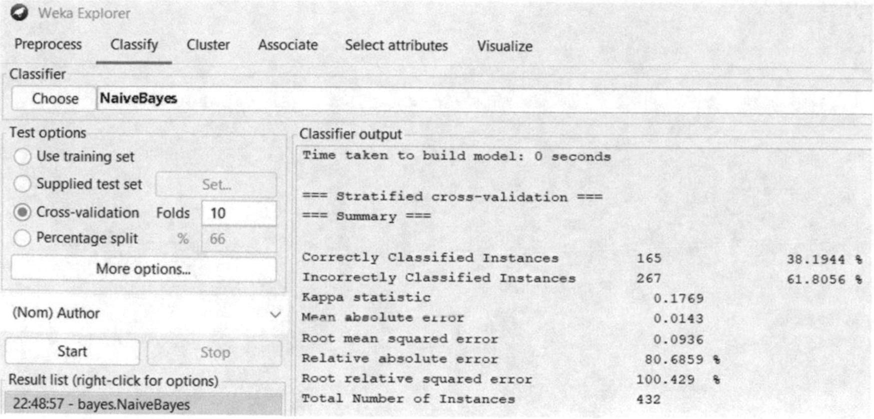

Fig. 8 Naïve Bayes classifier results

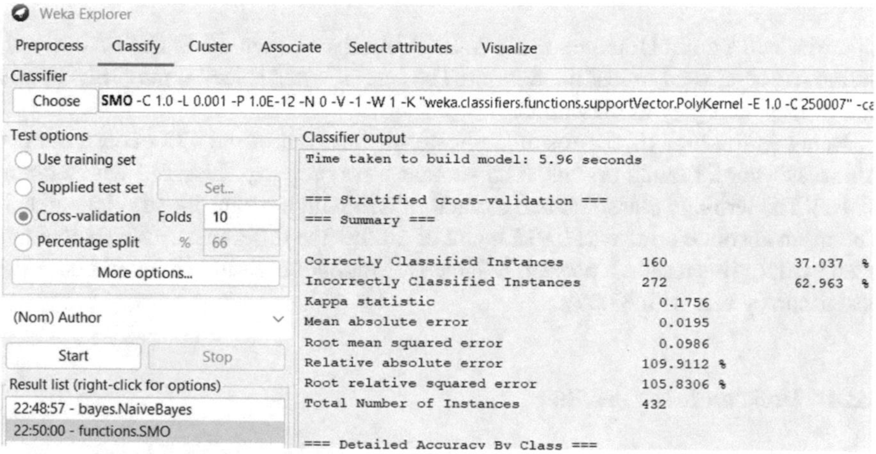

Fig. 9 Support vector machine classifier results

The mean absolute error was 0.0195. Out of all the five different classifiers that has been used for this research project, Support Vector Machine algorithm did not have the best accuracy rate with 37.04%.

4.2.3 K-Nearest Neighbour Classifier

The K-Nearest Neighbour classification approach has been used on the Training dataset for this level of research to forecast the 'Author' class for each cyberattack. In Sect. 2.4.4, the K-Nearest Neighbour classification approach was covered. The

Fig. 10 K-Nearest Neighbour classifier results

classifier will be implemented using WEKA Analyser, then 10 K-Fold cross validation has been used to test the dataset. The research has selected the 'IBk' option under 'lazy' category of classifiers.

Below in the Fig. 10, the results indicate the fact, out of the 432 occurrences of Chinese Cyber Criminal groups, 37.5 percent were properly classified with a count of 162. The wrongly classified had a count of 270 which meant it had 62.5 percent. The mean absolute error was 0.0153. Out of all the five different classifiers that has been used for this research project, K-Nearest Neighbour algorithm did not have the best accuracy rate with 37.5%.

4.2.4 Decision Tree Classifier

The Decision Tree classification approach has been used on the Training dataset for this level of research to forecast the 'Author' class for each cyberattack. In Sect. 2.4.3, the Decision Tree classification approach was covered. The classifier will be implemented using WEKA Analyser, then 10 K-Fold cross validation has been used to test the dataset. Two types of Decision Tree approaches are being used C4.5 and Random Forest [11]. Both of them has been implemented to decide out of the two which approach is best under Decision Tree algorithm.

C4.5 in detail depends on information entropy. The research has selected the 'J48' option to use C4.5 algorithm under 'trees' category of classifiers.

Below in the Fig. 11, the results indicate the fact, out of the 432 occurrences of Chinese Cyber Criminal groups, 35.6481 percent were properly classified with a count of 154. The wrongly classified had a count of 278 which meant it had 64.3519 percent. The mean absolute error was 0.0163. Out of all the five different classifiers

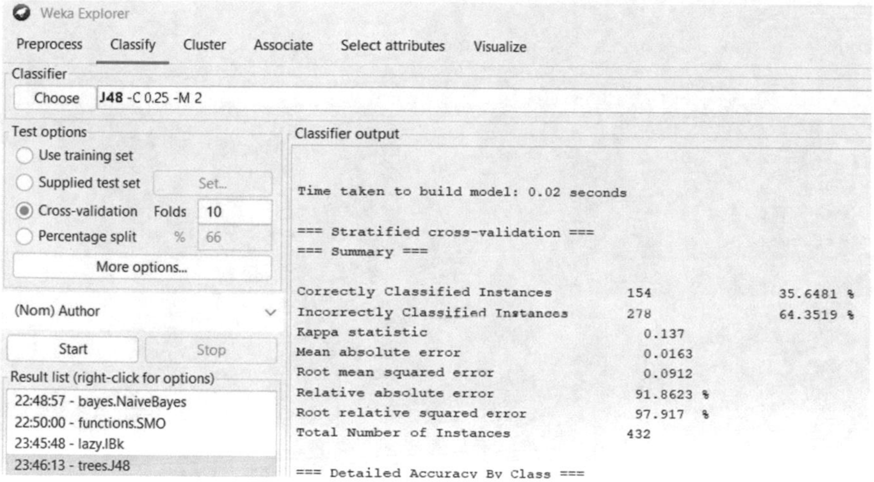

Fig. 11 C4.5 classifier results

that has been used for this research project, C4.5 algorithm did not have the best accuracy rate with 35.6%.

The categorization strategy used in the Decision Tree method includes Random Forest. That method uses a technique called bagging for creating decision trees. Bagging seems to be the technique of merging training networks to increase accuracy of classification approaches [11].

The Attacker groups from the Training set for the research model has been trained using Random Forest technique. Cross-validation is configured to the 10 K-Fold value for the training sample. Random Forest Approach is categorized under the name 'RandomForest' in the 'trees' section of WEKA.

Below in the Fig. 12, the results indicate the fact, out of the 432 occurrences of Chinese Cyber Criminal groups, 31.9444 percent were properly classified with a count of 138. The wrongly classified had a count of 294 which meant it had 68.0556 percent. The mean absolute error was 0.0153. Out of all the five different classifiers that has been used for this research project, Random Forest algorithm did not have the best accuracy rate with 31.9%. Also based on the comparison with C4.5 algorithm, the research has noted that C4.5 is the better of the two approaches taken under Decision Tree algorithm.

4.2.5 Artificial Neural Network Classifier

The Artificial Neural Network classification approach has been used on the Training dataset for this level of research to forecast the 'Author' class for each cyber-attack. In Sect. 2.3.1, the Artificial Neural Network classification approach was

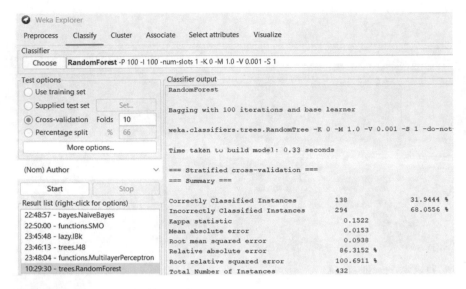

Fig. 12 Random forest classifier results

covered. The classifier will be implemented using WEKA Analyser, then 10 K-Fold cross validation has been used to test the dataset. The research has selected the 'MultilayerPerceptron' option under 'functions' category of classifiers.

Below in the Fig. 13, the results indicate the fact, out of the 432 occurrences of Chinese Cyber Criminal groups, 30.0926 percent were properly classified with a count of 130. The wrongly classified had a count of 302 which meant it had 69.9074 percent. The mean absolute error was 0.0155. Out of all the five different classifiers that has been used for this research project, Artificial Neural Network algorithm had the worst accuracy rate with 30.09%.

4.3 Section Conclusion

The dataset results from applying classification techniques was analysed in this segment. The dataset was processed using Support Vector Machine, Random Forest, Naïve Bayes, K-Nearest Neighbour, and Neural Networks, and a variety of evaluation criteria were reviewed and examined. The important findings from the study were discussed for every method throughout the discussions. According to the results, Naïve Bayes seems to have the best precision for identifying Chinese cyber criminals, compared to the other approaches being slightly lower. This algorithm properly identified 165 cases of Chinese cybercrime organisations with a 38.1944 percent accuracy rate, and mistakenly forecasted 267 occurrences with a 61.8056 percent accuracy rate. Next Section will have the detailed discussion with respect to the classifier with the best accuracy.

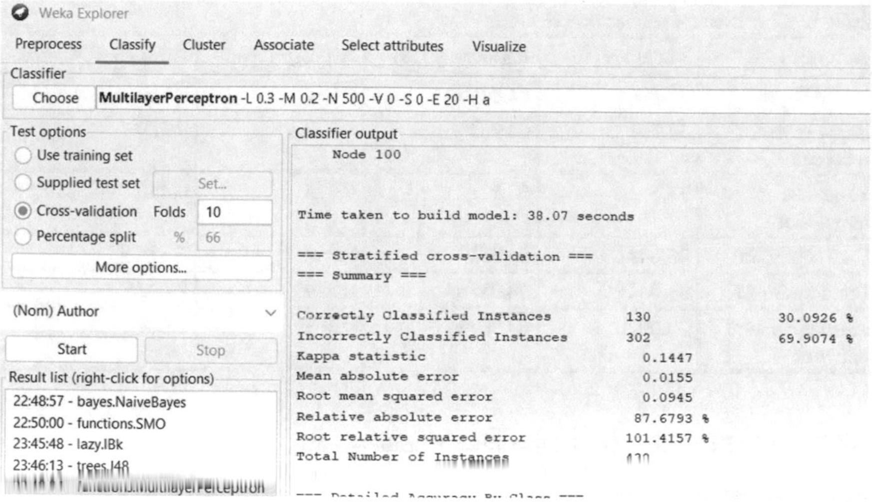

Fig. 13 Artificial neural network classifier results

5 Research Discussions

5.1 Introduction

In this Section, the research provides detailed discussions and information regarding the results from the five Classification algorithms and approaches that had been implemented in the previous Section. The best algorithm with the highest accuracy rate has been further discussed upon with the help of metrics used.

5.2 Evaluation of Research Results

Table 2 provides a summary of all results after the five classification techniques had been applied to the Training Dataset.

The research has compared the results to use measures that will be used to explore in further depth about the top 6 Chinese cybercriminal organizations which achieved the greatest results and had been accurately anticipated, along with data and metrics.

Precision: Accuracy also referred to as positive expected values, has been the ratio of properly categorised cases inside a category to the overall amount of classified examples within that group. The formula for Precision is as follows, TP/TP + FP [5].

Here, TP and FP are known as True Positive and False Positive respectively.

Table 2 Classifier accuracies overview

Approach	Accuracy (%)	Inaccuracy (%)	Average recall	Average ROC area
Naïve Bayes	38.1944	61.8056	0.382	0.590
Support vector machine	37.037	62.963	0.370	0.570
K-nearest neighbour	37.5	62.5	0.375	0.646
C4.5 algorithm	35.6481	64.3519	0.356	0.538
Random forest	31.9444	68.0556	0.319	0.647
Artificial neural network	30.0926	69.9074	0.301	0.664

ROC Area: ROC (Receiver Operating Characteristic) Area image shows a two-dimensional plot using False Positive ratings on the X axis as well as the True Positive rate on the Y axis of the graph. The effectiveness of classifiers depends critically upon the area immediately underneath the ROC curve [5].

F-Measure: Another classification metric used to classify and evaluate groups which is basically the average of Recall and Precision. Equation for F-Measure is: 2 * Recall * Precision/Recall + Precision [5].

Recall: The amount of cases accurately predicted inside one group divided using the total number of occurrences is known as the Recall also known as True Positive Rate. The equation is written as: Rate of TP = TP/(TP + FN) [5].

F-Measure: Another classification metric used to classify and evaluate groups which is basically the average of Recall and Precision. Equation for F-Measure is: 2 * Recall * Precision/Recall + Precision [5].

Given below Table 3 provides detailed measures with regard to the top six Chinese Cyber Criminal groups for each metric category only for Naïve Bayes Classifier because it had the best accuracy rate out of the five classification approaches taken.

Table 3 Overview of metrics for top six classified groups

Precision	Recall	F-measure	ROC area	Groups
0.336	0.777	0.469	0.595	Chinese hackers
0.100	0.056	0.071	0.740	APT10
0.400	0.250	0.308	0.660	BlackTech
0.333	0.125	0.182	0.577	APT27
0.500	0.500	0.500	0.998	RedFoxtrot
0.250	0.500	0.333	0.991	Gallium

The Accuracy rate for Naïve Bayes was 38.1944% as mentioned in Fig. 8. This doesn't get rid of the issue that the accuracy rate is not high enough to be considered as reliable method to cyber profile hackers using these datasets.

Based on Fig. 14, the following judgements have been made regarding the results to classify the top six Chinese cyber-criminal groups. Metrics such as Precision, Recall, F-Measure and ROC Area have been used for the 'Author' section of the Training Dataset.

The study will discuss each of the top 6 'Author' groups according to the respective metrics used.

Based on Precision rate results after implementing Classification Techniques on the Training Dataset. 'RedFoxtrot' hacker group has the highest Precision with 0.5, which is then followed closed by 'BlackTech' hacker group with 0.4 as the Precision. 'APT10' also known as RedApollo hacker group has the lowest Precision rate with 0.1. According to the Precision rates, 'RedFoxtrot' group is most likely to be best predicted for Precision from the unknown dataset.

Next the focus can be towards the Recall rate results after implementing Classification Techniques on the Training Dataset. 'Chinese Hackers' also known as groups who haven't been named but are known to be from China has the highest

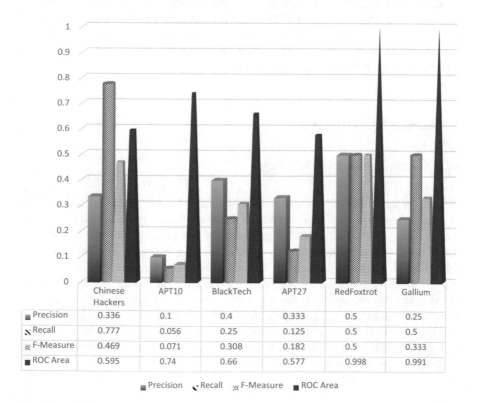

	Chinese Hackers	APT10	BlackTech	APT27	RedFoxtrot	Gallium
▪ Precision	0.336	0.1	0.4	0.333	0.5	0.25
◥ Recall	0.777	0.056	0.25	0.125	0.5	0.5
▨ F-Measure	0.469	0.071	0.308	0.182	0.5	0.333
▪ ROC Area	0.595	0.74	0.66	0.577	0.998	0.991

▪ Precision   ◥ Recall   ▨ F-Measure   ▪ ROC Area

Fig. 14 Evaluation of Chinese hacker groups

Recall rate with 0.777, which is then followed closely by two groups 'RedFoxtrot' and 'Gallium' with 0.5 as the Recall rate. The group with the lowest Recall rate is 'APT10'. According to the Recall rate, 'Chinese Hackers' are most likely to be best predicted for Recall from the unknown dataset.

Based on the F-Measure results after applying Classification techniques on the training dataset. The results for the top 2 are really close. But 'RedFoxtrot' group has the lead with 0.5 as the F-Measure followed closely by 'Chinese Hacker' groups with 0.469. 'APT10' group had the lowest F-Measure results.

Next the study focusses on ROC Area after conducting the Classification methods on the Training dataset. 'RedFoxtrot' and 'Gallium' had the highest ROC Area rates, they were 0.998 and 0.991 respectively. The group with the lowest ROC area results was 'APT27' with ROC rate of 0.577.

5.3   Section Conclusion

To summarise, this Section includes the discussion on the comparison of the best predictive method to the other classification techniques. The statistics and accuracy rates which led to the conclusion of the best one based on the various measurements of accuracy. The best algorithm with the highest accuracy rate has been further discussed upon with the help of metrics used. This has led to classify the top 6 Chinese Cyber Criminals groups based on various metrics.

6   Research Conclusion

6.1   Contribution to Knowledge

Day by day the global reliance on internet connectivity grows and so does various dangers or consequences of cybersecurity threats. China currently holds the top position for the number of hacker groups in the world. Security experts require even more data and tools to keep up with these cyber threats from Chinese Cyber-Criminals. They need to acquire top strategies to fight back in the warfare involving cybercriminals and defences. Cybersecurity monitoring is a technique which tries to fix cyber security strategic planning by studying cybersecurity risks. Since this rate of cyber threats has caused an expansion which substantially, machine learning and data-mining techniques have become an essential part in recognizing security risks. According to this research study, classification methods has been implemented to correctly categorise as well as forecast future cyber-attack vectors.

The research had aimed to cyber profile and improve the current strategies of cyber-attack prevention from Chinese attack groups. This has been done by evaluating previous studies on implementing data-mining techniques in the field of cyber-security. This research used the Classification Data-Mining framework on the dataset which has been gathered from an open-source data blog Hackmageddon. The dataset had been set to focus on the attacks that took place from 2016 to 2022. 5 of the major classification techniques has been used in order to find the best suitable method to predict and classify future cyber-crime instances from China. Out of the 5, the best one was found to be Naïve-Bayes algorithm.

Listed below are the report's contribution to knowledge in regard to the research topic, "Can Classification Data-Mining techniques or approaches be utilised to identify and anticipate prospective cybercrimes by Chinese Cyber Criminals?".

The utilisation of OSINT Dataset for the training phase has been one of the research's new characteristics. Compiled dataset began with 10,836 reports of cyber-crimes that had taken place from 2016 to 2022. This compiled dataset was then pre-processed into 432 reports of cybercrimes by Chinese Cyber Criminals. Open-Refine had been utilised to cleanse and conduct pre-processing on the dataset. Both automatic and manual pre-processing methods were used. After pre-processing the dataset, there were 8272 rows of data left. This can indeed be helpful for the future research projects. The five essential features of the dataset were: Author, Attack, Target Class, Attack Class and Country.

Additional contribution to knowledge would be that the study has sought out and proved that Naïve Bayes method predicted Chinese Cyber Criminals with the best accuracy. Although 38.1944% accuracy with 165 correctly classified occurrences could be deemed insufficient for academics and corporations. The top six cyber criminal groups mentioned in Sect. 5.2 of this paper had the best rating throughout every category of measurements. The cause of the insufficiency and low sufficient accuracy is due to the fact that the OSINT dataset had its shortcomings and errors. Thus, resulting in a bad accuracy rating for all classes.

In conclusion, the research utilised solely the Attacker or 'Author' class from the dataset to cyber-profile Chinese Cyber Criminals with the help of five main categorisation methodologies from classification. Though all these could not be utilised in isolation, numerous different class classes, such like 'Attack', 'Target Class' and 'Country' may also be examined in order to give investigators, crime control authorities, as well as institutions more information once they try to implement countermeasures or mitigation strategies for cyberattacks.

6.2 Limitations to the Study

Along with lots of similar studies, this one too contains its own set of limitations in forms of participants in the research. The dataset elements being the major concern due to its incompleteness as well as other aspects of the dataset is where the limits are most strongly concentrated. The datasets were gathered using OSINT and it has been said earlier that all material gathered via OSINT comprises a sizable quantity of distortion and unrelated knowledge. Inside the segment of database pre-processing including dataset purification, the process was mechanically performed through, that included eliminating extraneous cybercrime reports. Just one method to confirm cybercrimes would have been to analyse every report by reviewing the details as well as going through each link attached and related to the cyber-attack. If we were to do that, the procedure would take more time and would not be possible to complete it more quickly or efficiently.

Private and confidential sensitivity to each cyber-attack topic placed additional restrictions on accessing data, making it challenging to ensure that the dataset is accurate and full. To protect business reputations, several businesses and government agencies prefer to keep their security breaches a secret. Whenever it comes to predicting statistics, a precise and detailed dataset may produce findings that are more exact and dependable.

6.3   Recommendations for Future Research Ideas

Proposed experiments on this subject may employ yet another data-mining method to cyber-profile Chinese Cyber Criminals Following upon examining the published evidence on the issue, Classification is just one method to conduct profiling techniques. Another notable mention is the K-Means Clustering. They can be easily used for better datasets and would produce much efficient results to different categories and classes. Clustering as well as few commonly used data-mining techniques could be made useful for creating business cyber-security plan, investigating security occurrences, and working with investigators and infosec specialists. Future investigations might utilise time-series datasets or other data-mining methods to characterise Chinese Cyber Criminals. Even though time-series dataset analytics is great for forecasting and may make similarities making it easy to spot, applying this could be challenging for a few reasons.

To improve the quality of the dataset being used could also be another main goal of the next study to better understand and receive much better accuracy rates. This report's conclusions could be affected by the dataset's noise because it was gathered through Open-Source Intelligence. If at all feasible, collecting digital intrusions and incidents directly from the corporations might considerably increase the results' reliability. Key determinants could have been introduced as unique identifiers like Internet address & attack region. Much precise analysis may be produced if additional attributes were included in the dataset that had been collected for the purpose of the research.

References

1. Back S, LaPrade J, Shehadeh L, Kim M (2019) Youth hackers and adult hackers in South Korea: an application of cybercriminal profiling. IEEE Xplore. https://doi.org/10.1109/Eur oSPW.2019.00052
2. Burgess M (2022) How China hacked US phone networks. Wired. https://www.wired.com/ story/china-hacking-phone-network-security-roundup/
3. CBS (2022) Chinese hackers took trillions in intellectual property from about 30 multinational companies. https://www.cbsnews.com/news/chinese-hackers-took-trillions-in-intellectual-pro perty-from-about-30-multinational-companies/
4. Crowe S, Pournouri S, Ibbotson G (2021) Use of classification techniques to predict targets of cyber attacks for improving cyber situational awareness during the COVID-19 pandemic. In: Information security technologies for controlling pandemics, pp 231–268. https://doi.org/10. 1007/978-3-030-72120-6_9
5. Fawcett T (2006) An introduction to ROC analysis. Pattern Recognit Lett 27(8):861–874. https://doi.org/10.1016/j.patrec.2005.10.010
6. FBI (2022) Countering threats posed by the Chinese Government Inside the U.S. Federal Bureau of Investigation. https://www.fbi.gov/news/speeches/countering-threats-posed-by-the-chinese-government-inside-the-us-wray-013122
7. Marcot BG, Hanea AM (2020) What is an optimal value of k in k-fold cross-validation in discrete Bayesian network analysis? Comput Stat. https://doi.org/10.1007/s00180-020-009 99-9
8. MITRE (2015) MITRE ATT&CK^{TM}. Mitre.org. https://attack.mitre.org/
9. OpenRefine (2021) OpenRefine. Openrefine.org. https://openrefine.org/
10. Osman AS (2019) Data mining techniques: review. Int J Data Sci Res 2(1):1–5. http://ojs. mediu.edu.my/index.php/IJDSR/article/view/1841/717
11. Pournouri S, Zargari S, Akhgar B (2018) Predicting the cyber attackers; a comparison of different classification techniques. In: Advanced sciences and technologies for security applications, pp 169–181. https://doi.org/10.1007/978-3-319-97181-0_8
12. Rapid7. (2018). Common Types of Cybersecurity Attacks and Hacking Techniques | Rapid7. Rapid7. https://www.rapid7.com/fundamentals/types-of-attacks/
13. Rutgers (2021) What Is data mining? A Beginner's Guide (2022) Rutgers bootcamps. https:// bootcamp.rutgers.edu/blog/what-is-data-mining/
14. Shawky A, El-Bhrawy M, El N, Mohamed K (2016) Artificial neural networks in data mining. Int J Sci Eng Res 7(11). https://www.ijser.org/researchpaper/Artificial-Neural-Networks-in-Data-Mining.pdf
15. Waikato (2015) Weka 3—data mining with open source machine learning software in Java. https://www.cs.waikato.ac.nz/ml/weka/#:~:text=Weka%20is%20a%20collection%20of
16. Wass S, Pournouri S, Ibbotson G (2021) Prediction of cyber attacks during coronavirus pandemic by classification techniques and open source intelligence. IN: Cybersecurity, privacy and freedom protection in the connected world, pp 67–100. https://doi.org/10.1007/978-3-030-68534-8_6
17. Wolff R (2020) Classification algorithms in machine learning: how they work. MonkeyLearn Blog. https://monkeylearn.com/blog/classification-algorithms/
18. Yip M (2010) An investigation into Chinese cybercrime and the underground economy in comparison with the West. https://www.academia.edu/2728559/An_investigation_into_Chi nese_cybercrime_and_the_underground_economy_in_comparison_with_the_West
19. Zulfadhilah M, Prayudi Y, Riadi I (2016) Cyber Profiling using log analysis and k-means clustering. Int J Adv Comput Sci Appl 7(7). https://doi.org/10.14569/ijacsa.2016.070759

What Drives Generation Z to Behave Security Compliant? An Extended Analysis Using the Theory of Planned Behaviour

Fabrizia Vollenweider and Hamid Jahankhani

Abstract Cyber security remains a relevant topic for organisations. While companies invest in expensive security tools security awareness training often is neglected, even though human error still accounts for a large part of cyber incidents (Gartner, 2022). At the same time there is currently an important generational shift, as Generation Z (Gen Z) is starting to enter the workforce and is said to be soon overtaking millennials. However, Gen Z is said to experience a lot more security related issues compared to older generations. This paper aims to understand and analyse security behaviour for Gen Z by using the theory of planned behaviour to understand what influences the security behaviour of this generation. The theory of planned behaviour has been often utilised in previous research to understand how security behaviour is formed and how it can be influenced, however, it has not yet been researched in the context of Gen Z. The results conclude that Gen Z has a rather indifferent attitude towards security and generally values convenience much more than security. This can be led back to the lack of security awareness education as opposed to older generations that often experienced a training in the work environment. Gen Z is highly influenced by their environment; therefore, a company should strive to create a security culture. Lastly, Gen Z experiences a low perceived behavioural control, as they feel a lack of control regarding to the internet and lack education how to behave more secure. This gap could be closed by providing a suitable security awareness training.

Keywords Cyber security · Generation Z · Theory of planned behaviour · Digital · Vulnerabilities · Attack patterns

F. Vollenweider
Checkpoint, Spreitenbach, Switzerland
e-mail: fabriziav@checkpoint.com

H. Jahankhani (✉)
Northumbria University, London, UK
e-mail: Hamid.jahankhani@northumbria.ac.uk

1   Introduction

This project aims to understand and explain security behaviour of Generation Z (Gen Z) using the theory of planned behaviour to provide a proposal how companies can plan and adapt their security awareness programme to correctly target Gen Z.

Today technology is everywhere and even companies that are not in the information technology (IT) sector leverage IT and IT infrastructure for their core business and are highly dependent on it. However, simultaneously the number of attacks on this infrastructure is also increasing. For the first four months of 2022 the research team of Check Point, a renowned cyber security vendor, published fourteen blog entries on newly discovered vulnerabilities and attack patterns [1]. While there are many technologies around that can protect IT infrastructure on a technical level, human failure still accounts for the majority of security incidents [2]. At the same time, statistics have shown that there is a difference within generations in cyber security behaviour. The younger generations and especially Gen Z are statistically more prone to a phishing scam [3] and experience more security incidents [4]. While Gen Z was exposed with modern technology at a younger age this does not implicate a higher security awareness [5]. There are even some attacks reported that specifically target Gen Z, as they are conducted on the amongst Gen Z most popular platform TikTok [6]. Studies also show that awareness trainings after a cyber-attack will be too late [7].

2   Literature Review

Security awareness plays a fundamental role to improve security compliance for employees in a company. However, how can a company improve the security awareness of its employees? Which factors need to be considered to efficiently communicate and implement a security awareness strategy? Previous research has focused on behavioural economics to answer these questions. While the relation and influence between these two factors has been studied in-depth, other factors have not been considered. As more and more young people are entering the workforce the difference in attitude and behaviour between generations becomes more visible. Therefore, the questions above will be answered by analysing in-depth behavioural economics—in particular the theory of planned behavior—in the context of security awareness and Gen Z. Gen Z is gaining more importance for companies. Not only is Gen Z expected to soon overtake Millennials as the most populous generation in the world but also many of them are expected to enter the workforce in the next couple of years [8]. The theory of planned behaviour has been widely researched in the context of security awareness and it has been proven in various studies that security awareness can significantly influence the factors that influence an individual's intention to conduct a certain behaviour (for example proven in Safa et al. [9], Tsohou et al. [10], Tam et al. [11]).

This literature review first describes the importance of security awareness and security compliance for companies. Secondly, the topic of behavioural economics is analysed, and the theory of planned behaviour is presented in detail. The literature review provides an overview of the studies that have put the theory of planned behaviour in context of security awareness and security compliance and describe the results. Lastly, Gen Z and its characteristics are described. The review briefly touches upon existing studies on Gen Z and the theory of planned behaviour as well as research that investigates Gen Z and security awareness.

2.1 Security Awareness

Security awareness can be described as a sub-part of information security. Many cyber-attacks have been appointed to the human factor, therefore this variable has been gaining more attention. Countless papers have been published over the years on how to minimise the risk of a cyber-attack by increasing an individual's security awareness and simultaneously security compliance (for example proven in [10, 12]). Studies thereby focus on different approaches. While some research emphasizes the importance of communication of information security policies and transformational leadership [13] other studies research the dependency of the organisation form [14] or of the geographical area [15]. Furthermore, there are numerous studies that focus on gamification within the context of security awareness and the development of serious games. Examples include password guessing games [16], 2D puzzle adventures [17] and tabletop games where participants can both play the role of an attacker and defender [18]. Other studies compare the effectiveness of different elements of serious games and the general success of gamification [19].

The above-mentioned studies all look at examples of practical implementation of improving security awareness for individuals or in an organisation. At the same time other research aims to understand why individuals do not behave security compliant in the first place by trying to analyse and understand psychological factors. Stewart and Lacey [20] use the word "bounded rationality" to describe, that decisions are not conducted rationally but based on an individual's existing beliefs and attitude. Other studies look at personality traits based on the five factors model [21, 22] or nudging that describes how an individual's decision making can be influenced based on rewards for correct behaviour [23, 24]. While these studies all bring value to security awareness research in general, they do not explain how individuals form an intention to conduct a certain behaviour (in this case: behave security compliant).

There has been a lot of research conducted in the field of behavioural economics in correspondence with security awareness and security compliance. Xu and Guo [25] describe in their study that employees use procrastination and psychological detachment to avoid performing security tasks. They prioritise other business tasks and see security risk as not relevant to their person. This shows the importance on the behavioural aspect regarding security compliance.

2.2 Behavioural Economics and Theory of Planned Behaviour

To provide further context on the psychological aspects that influence and decide the behaviour of a human, the theory of planned behaviour is discussed in-depth. Other relevant theories in behavioural economics include the theory of reasoned action (in later models further expanded as the theory of planned behaviour), behaviourism theory and the behavioural intention model. The behaviourism theory and behavioural intention model are used to understand how an individual's intention can be influenced and changed. However, most research focuses on the theory of planned behaviour; thus, this theory will be further described [26]. The theory of planned behaviour states that an individual's behaviour can be predicted and explained with accuracy in a specific context using the individual's attitude towards the behaviour, the subjective norms, and the perceived behavioural control. Examples for an attitude can be something like a certain belief, profession, minority group etc. However, to properly consider attitude as a factor it is important to understand that attitude might be unique to a certain situation and therefore only shows relevance when using aggregated samples. The subjective norm describes if an individual thinks their surrounding environment such as their friends or peers approve or disapprove of the behaviour, meaning if there is a social pressure to perform a certain behaviour. Lastly, perceived behavioural control is discussed. This describes an individual's perception as to how easy or difficult it is to perform a behaviour of interest. Examples for that can be things such as time, money, skills, and motivation [27].

2.2.1 Theory of Planned Behaviour and Security Awareness

Theories in behavioural economics have been extensively researched in the context of security awareness. There have been various studies on the theory of planned behaviour and security awareness and compliance. While the theory of planned behaviour can only describe the intention to comply with a security behaviour, Sommestad et al. [28] determine that an intention to comply is the best predictor for an actual compliant security behaviour. Thus, the theory can be utilised to understand what drives individuals to behave security compliant.

One example for applying the theory of planned behaviour on security awareness is the study conducted by Tam et al. [11]. The authors research how the theory of planned behaviour can be used to explain why an employee will behave security compliant. The authors focus on the job satisfaction of individuals and how it impacts security behaviour by considering value congruence, meaning the alignment of an individual with the companies' values. Value congruence suggests that employees will give higher importance to perceived social pressure—in the theory of planned behaviour known subjective norms—and have a better attitude behaviour to keep the company safe as well perceive their behavioural control as higher as the employee has a strong intention to keep the manager happy. Running a statistical analysis, it was concluded that the working environment has a strong impact on the intention of behaving security compliant. That means that individuals are influenced strongly by an organisation with a security environment or colleagues that show a strong security attitude. Additionally, the job satisfaction and security awareness have also an impact [11]. These findings are highly relevant and used as a basis for the further analysis of this dissertation.

Other authors combine the theory of planned behaviour with additional theories, such as the protection motivation theory. This theory describes an individual's intention to perform a protective action based on appraisal and self-efficacy. The authors describe additional components to a conscious behaviour, such as information security awareness, information security organization policy and information security experience and involvement. The study concluded that information security awareness significantly influenced the attitude and subjective norms towards a secure behaviour. Additionally, with more experience and involvement the perceived behavioural control was said to be higher [9]. Kim and Kim [29] proved in their study that the active implementation of a compliance support system where participants are actively involved will lead to a higher overall security compliance. Additionally, knowledge of compliance for users was found to be important as it leads to more engagement in a platform.

Another example is presented by the research of Tsohou et al. [10] that analyses how information security behaviour is influenced by an individual's perception and beliefs on security, so called "behavioural biases". Taking Tsohou et al.'s points into consideration will further support the understanding of how behaviour in the context of information security can be influenced and the points will be reflected in the discussion in Chap. 4.

Tsohou et al. describe different biases that individuals often experience in the context of security awareness as visible in Table 1.

These biases will influence an employee's perception and behaviour of security. Additionally, Tsohou et al.'s study concluded that employees will not put a lot of effort in being compliant with an ISP if they do not perceive the benefits as significant. Based on loss aversion theories, people tend to overweight sure losses compared probable wins.

2.3 Generation Z

Security awareness in the context of the theory of planned behaviour has been researched and analysed in detail as visible in the previous sub-chapter. However, none of the studies have included how the age and generation of individuals would affect the results. Meanwhile, several other studies have shown that characteristics and traits between different generations differ greatly [30]. Theory suggests people experience similar things at similar ages, however this can vary greatly between the different generations, meaning that a person from the boomer generation has experienced different things, for example, at age 30 compared to a person from Gen Z

Table 1 Security awareness bias

Bias name	Description	Example
Affect bias	Decisions are taken based on impressions and images associated	Ex1: negative images of armed guards connected with security Ex2: sharing passwords is appreciated/"seen cool" by peers
Anchoring and confirmation bias	People do not like or believe information that contradict their current beliefs	Ex1: even though people are presented with facts that cyber-attacks are mostly conducted by organised crime organisations they still believe hackers are a misbehaved teenager
Availability heuristic	Current easy-to-remember information is weighted heavier in the risk assessment process than more complex information	Ex1: events covered in the media with strong narratives and videos tend to more influence people
Optimism	People see themselves to less likely experience a negative event compared to others	Ex1: people think that social engineering techniques will not happen to them

Source Tsohou et al. [10]

[31]. Furthermore, Gen Z has a high importance for the workforce. While currently only around 5% of Gen Z are part of the workforce this number is expected to rise to over 20% over the next years, meanwhile boomers are approaching the retirement age [8].

There is no common definition as to when Gen Z exactly starts. While some sources (for example [32]) state that Gen Z can be considered as people born between 2000 and after that, other sources see Gen Z as the generation born after 1995 (for example [33–35]). To understand how Gen Z is defined compared to the generations before, Andersen et al. [36] define a total of five relevant generations and their main characteristics as described in Table 2.

This interpretation of generations is used in the majority of studies; therefore, this will be adapted for this paper and Gen Z considered as people born in 1995 and later. Furthermore, it is important to note that the beliefs and values of each generation described are limited to the western world and might differ for other countries [36].

2.3.1   Gen Z and the Workplace

Gen Z are highly relevant to the workforce. It is estimated that by 2030 most of the entry-level roles in the US will be filled by people from Gen Z [37]. Several studies have researched work preferences and expectations by Gen Z. Understanding these will enable the analysis on how their behaviour can be influenced. Studies on

Table 2 Overview of the five different generations

Name	Year	Values
Traditionalists	Born between 1922 and 1944	Respect authorities and act obedient, diligent and dutiful
Baby boomers	Born between 1945 and 1964	Optimistic, collectivistic and embrace the free expression of opinion
Generation X	Born between 1965 and 1979	Independent, autonomous, have less faith in politics
Generation Y/millennials	Born between 1980 and 1994	Flexible life model, were partially introduced to new media, optimistic
Generation Z	Born from 1995 on and later	Born in information age with online and digital platforms, often named "digital natives", can easily operate on social media platforms, have a global connection

Source Andersen et al. [36]

work preferences have concluded that Gen Z is looking for a place that is "fun to work" that has a positive culture. It is important for them to find a sense of purpose in their role and feeling valued by their employer. They also expect regular feedback from their superiors. Climbing the career-ladder quickly is not their priority. However, their upbringing has been different from earlier generations. Gen Z is characterised of being less autonomous, meaning Gen Z will have more difficulties to take responsibilities and ownership. Twenge describes that Gen Z grew up protected and safe but at the same time prefers to avoid adult responsibilities [38]. Additionally, as they have started to work later compared to earlier generations, they are less used to things like handling calls and writing emails [37, 39].

Literature recommends for managers to show trust in Gen Z to make their own decisions in order to facilitate autonomy in the workplace. Furthermore, communication is a key aspect. As Gen Z is purpose-driven, it is important for them to understand why something is important and putting it in the context of their own growth and achievement [39].

2.3.2 Gen Z Learning Preferences

Analysing the learning process for Gen Z will help to further understand how security programmes should be designed to improve the learning process. Literature describes that Gen Z is used to spend over nine hours per day in a digital environment and therefore also prefers to learn with the use of technology, ideally cutting edge [40]. Social media is Gen Z's constant companion, and they rely on it also for studying by using YouTube and educational apps to get additional information. However, it is important to note that content cannot be simply transferred to these new learning platforms. The author takes the example of marketing, where the advertisement in offline formats cannot be simply copied to social media platforms but needs to be reshaped [41]. This should also be considered when coming up with suggestions for security awareness trainings.

Additionally, Gen Z prefers symbols and emojis over text messages. They are characterised to use their phones much more often. Experts recommend any communication to Gen Z to be a maximum of five words and content reflect three key-points: be eye catching, short, and easy to follow as the attention span is short and distraction high. Decision-making with Gen Z often happens instantaneous [42]. It is furthermore important to note that Gen Z is characterised by being exposed to a high amount of technology and being dependent on it [43] while spending approximately 11 h consuming material on online devices every day [44]. This further solidifies the importance of technology as part of Gen Z's everyday life.

2.3.3   Gen Z and Theory of Planned Behaviour

Overall, there has not been much research in behavioural economics in the context of Gen Z. However, a recently published theory by Djafarova & Foots [34] applies the theory of planned behaviour for Gen Z in the context of ethical purchases. While ethical purchases differentiate greatly from security awareness, information on attitude, subjective norm and perceived behavioural control for Gen Z can be considered and applied for this research as well.

Regarding the attitude, participants answered a strong attitude towards ethical purchase, even when they did not purchase ethically yet. A higher income and further technology would motivate a later ethical purchase [34]. This information is not applicable to security awareness and will not be further considered. For subjective norms the participants were discussing how their image is viewed and considered by others which was a strong factor why they would purchase ethically [34]. This aspect could also be considered for security awareness theory as a non-compliance could result in a damage of the personal reputation. The personal brand is very important to Gen Z, they want to give the best impression of themselves to others.

Lastly, the perceived behavioural control was assessed and factors such as financial position, accessibility, time, and inconvenience discussed. The financial position especially puts a burden to the participants, as well as the extra travel costs to purchase ethical products. Participants however also answered that knowledge brought to them by social media and influencers also lowered the control barriers [34]. From these answers it can be concluded that inconvenience can be a factor for Gen Z to lower their perceived behavioural control.

2.3.4   Gen Z and Security Awareness

Gen Z has been selected as the target group as they are more likely to experience security-related issues in the workplace compared to older generations. Generation

Z employees experience on average 4 security related issues per week, meanwhile employees older than 45 years old only experience on average 1 security issue per week [4].

White takes another aspect into account by analysing how Gen Z would behave and respond in the event of a cyber-attack. The experiment consisted of letting the participants read about a cyber-attack while measuring the factors optimism, self-efficacy, and general worry to see if these factors increase or decrease when being presented by possible countermeasures. The research concluded that reading about countermeasures after an attack has been already executed did not improve the factors of optimism and self-efficacy of Gen Z to mitigate the attack and further studies should thus focus on awareness on training of behaviour before a cyber-attack [7].

There are opposite opinions on the privacy and security awareness of Gen Z regarding of what to share and what not to share online. Fromm and Read [41] argue that Gen Z is extremely privacy and security aware and considers carefully what they post. Choong et al. [45] also conclude that Gen Z value their privacy and even learned from mistakes of the older generations by observing situations where information shared before on the internet damaged people's reputation. In contrast newer surveys conclude that Gen Z is much less privacy and security aware compared to older generations [5, 46].

This concludes that while Gen Z seems to be experiencing more security issues than other generations, there is insufficient knowledge on their privacy and security behaviour intentions, especially in the context of the corporate world. Thus, this research aims to close this gap by analysing the behavioural intention of Gen Z to behave security compliant using the theory of planned behaviour. Ultimately, companies should have an understanding how they can plan, conceptualise, and implement a security awareness strategy that would lead to a higher security compliance.

3 Research Methodology

To gain additional insights in understanding Gen Z's security behaviour the study is further analysing the theory of planned behaviour by using qualitative and quantitative research data collected through secondary datasets from existing studies and interviews in similar contexts.

As there has not been any research published that analyses the theory of planned behaviour in the context of security awareness with Gen Z, different research methods were considered to gain better insights. Primary research could have included conducting a quantitative survey or qualitative interviews with individuals from Gen Z. However, conducting primary research could lead to results that would be only relevant to a particular region and would require gaining a relevant participant number to ensure significant results. Furthermore, part of Gen Z is still minor of age which would force additional consent from their parents and increase complexity. Therefore, it was decided to use secondary datasets, as data has already been cleaned and properly anonymised by the relevant entities that conducted the research. Moreover, this also enabled the use of different data collected to provide a complete picture and different insights into the theory of planned behaviour. On the other hand, by using only secondary data some details might have been missed that were not reflected in the existing surveys. As the author did not have access to the original data sets due to privacy reasons, only already interpreted and filtered analysis could be included, leading to a potentially false conclusion.

The data was collected from various sources, as visible in Table 3. Most of these studies used quantitative surveys as a methodology. While these studies provide information on Gen Z and their technology and/or security views, they do not explicitly focus on behavioural theory and therefore do not provide a clear understanding why Gen Z on average experiences more security breaches [4] compared to older generations. As research has shown that the theory of planned behavior—and its three attributes attitude, subjective norms and perceived behaviour control—can predict security behaviour intention [11] and thus also security compliance of individuals. Therefore, different results provided in these studies are brought together here for further discussions.

The objective is to identify different characteristics of Gen Z using the theory of planned behaviour to understand how companies can plan, conceptualise, and implement a security awareness strategy that would lead to a higher security compliance.

To structure the data in the context of the three attributes of the theory of planned behaviour the approach of Djafavora and Foots [34] was found to be suitable, where each attribute was clustered in different themes that provided insights on several characteristics. Additionally, the defined variables from Tam et al. [11] were used to provide an idea of what information is relevant in the context of the theory of planned behaviour and security awareness.

3.1   First Attribute: Attitude

The first attribute attitude describes the motivation that an individual has towards a certain behaviour. There are some examples listed below, that were used in a study by Tam et al. [11] to gain an understanding how attitude in the context of security awareness can be interpreted:

ATT1   Information security conscious care behaviour is necessary.
ATT2   Information security conscious care behaviour is beneficial.
ATT3   Practicing information security behaviour is useful.
ATT4   I have a positive view about changing the user's information security behaviour to conscious care.
ATT5   My attitude toward information security care behaviour is favourable.
ATT6   I believe that the information security conscious care is something valuable to any organization.

Table 3 Overview of collected data

Research title	Author(s) and year	Research methodology	Number of participants	Research goal
Gen Z: the future has arrived	Dell Technologies [47]	Quantitative survey	12,000 participants, age 16–23	Capture data on current attitudes on technology and the workplace
What Generation Z can teach us about cybersecurity	Buitta and Johnson [46]	Qualitative interviews	9 participants, age 16–19	Understand experiences of Gen Z online and how it shapes understanding of cybersecurity
The future of digital experiences: how Gen Z is changing everything	WP Engine and CGK [48]	Quantitative survey	1014 participants, age 14–59; Gen Z was defined as age 14–21	Understand mindset, preferences and expectations of Gen Z compared to other generations
Are Gen Z-ers more security savvy online than millennials?	Cohen [5]	Quantitative survey	520 participants, age 18+; Gen Z was defind as age 18–22	Understand how Gen Z use the internet and think about cybersecurity and online privacy compared to other generations
Managing the narrative—young people's use of online safety tools	Family online safety institute [49]	Quantitative survey	1000 participants, age 13–24	Understand young people's experiences and attitudes towards online safety
How some millennials and Gen Zers are cybersecurity liabilities	Donegan [4]	Quantitative survey	1500 participants, age 18+; Gen Z was defined as age 18–24	Understand remote work and employee behaviour
'True Gen': Generation Z and its implications for companies	Francis and Hoefel [50]	Qualitative interviews	120 participants	Understand Gen Z and their characteristics
Coaching generation Z athletes	Gould et al. [51]	Qualitative interviews	12 participants	How Gen Z athletes' characteristics influence coaching practice

(continued)

Table 3 (continued)

Research title	Author(s) and year	Research methodology	Number of participants	Research goal
Welcome to generation Z	Mawhinney and Betts [8]	Quantitative survey	6000 participants	Analyse the factors that have been shaping Gen Z and the impact of Gen Z entering the workforce
Exploring ethical consumption of generation Z: theory of planned behaviour	Djafarova and Foots [34]	Qualitative semi-structured interviews	18 participants	Understand the factors that encourage Gen Z's behaviour to purchase ethically

3.1.1 Results

Theme 1: Motivation towards behaving security compliant

Gen Z highly values convenience, even when it comes at the cost of security. For example, they save passwords on websites to gain quicker and easier access [46]. They state that they prefer a digital world where algorithms predict and suggest what the user needs instead of being completely anonymous [48]. Individuals from Gen Z were also found the most likely to still visit a website, even after receiving a notification that the site is not secure [4]. The lack of privacy has been normalised amongst Gen Z and leads to Gen Z feeling indifferent regarding their personal security [46]. Gen Z does not necessarily strive for more privacy online. Furthermore, they only sometimes or rarely consider what they post online and the resulting impact and generally have a high perception that their information online is private [5]. In a study that analysed Gen Z athletes, it was determined that they are mostly extrinsically motivated to do well by receiving appreciation and physical goodies by their parents and companies [51].

Theme 2: Limitations today on Attitude

New technology provides new possibilities, however, most of the times no proper education is provided for technology and cybersecurity. Young girls perceive that it should be more accessible through in-school learning or dedicated programmes. Additionally, responsible technology such as multifactor authentication and password protection should be implemented and enforced by companies in the first place [46]. Moreover, younger people more often have an "optimism bias" which leads them to not take the needed safety precautions [3].

3.2 Second Attribute: Subjective Norms

The second attribute, subjective norms, describes how a behaviour is perceived by an individual's environment. There are some examples listed below, that were used in a study by Tam et al. [11] to gain an understanding how subjective norms in the context of security awareness can be interpreted:

SN1 The information security policies in the company where I work are important to my colleagues.
SN2 The information security behaviour of my colleagues influences my behaviour.
SN3 Information security behaviour culture of the company where I work influences my behaviour.
SN4 The information security behaviour of my manager influences my behaviour.

3.2.1 Results

Theme 1: How Gen Z's behaviour is influenced by their environment

Gen Z has generally high expectations towards themselves and feels pressure from people around that. They want to aim highly in what they do and will be disappointed if the result is not as they expected. Additionally, they struggle with negative feedback and quickly take it personally [51].

They highly value company culture and expect a company's values to align with their own values [8].

Gen Z put a lot of importance on their image online. They understand that mistakes on social media could result to them losing their jobs or influence the hiring of a future possible employment [34]. Another survey showed that Gen Z would even be willing to share more information in exchange for social media fame or receiving new followers or having access to features [52].

Lastly, Gen Z's parents also often lack training and learning as they did not grow up with the same access to technology and lack awareness of certain threats [46].

3.3 Third Attribute: Perceived Behavioural Control

The third attribute perceived behavioural control defines how an individual perceives their control over a behaviour, meaning how easy or difficult it is to conduct a certain behaviour. There are some examples listed below, that were used in a study by 2. (2022) to gain an understanding how perceived behavioural control in the context of security awareness can be interpreted:

PBC1 I believe that security behaviour isn't a hard practice.
PBC2 I believe that my experiences help me to have security behaviour about the data security.
PBC3 Following procedures and policies that lead to security behaviour is easy for me.
PBC4 Security behaviour is an achievable practice.

3.4 Results

Theme 1: Limitations of Gen Z to behave security compliant

Gen Z find the process of having to manage online security overwhelming and time-consuming [49]. They consider themselves as tech-savy [5] and consider data security a high priority but do not feel good enough on their efforts. 22% of the participants mentioned they would like to increase their security efforts but are unsure what to do [47].

Individuals of Gen Z feel that technology cannot be controlled. Some participants mentioned they feel like they have lost control on the internet. Other participants believe that laws should be more user-friendly for the users to have more trust in the technology again [46]. In general, Gen Z mention they prefer to be in control and understand what's going around them [50]. Individuals feel that it would be beneficial for them to get an easy way to identify malicious websites as well as having blocking viruses and spyware in place [48].

4 Results, Analysis and Evaluation

4.1 General Discussions

The previous section gathered data on the three main attributes of the theory of planned behaviour. This data will be analysed and discussed further. Additionally, some points raised in the literature review will be mentioned again as well. Furthermore, the results will be compared to the results in other security awareness studies as described in the literature review.

4.1.1 Attribute 1: Attitude

Gen Z in general does not seem to have a specific attitude towards secure behaviour. While some of the respondents answered that they would like to behave more securely but lack the knowledge to do so, many of the participants highly favour convenience and ease of access of programmes, much more than older generations. This perceived conflict between security and convenience also explains the different results of studies, where some of the research concludes that Gen Z value privacy much more compared to older generations [41, 45] while other studies determine that Gen Z is much less privacy aware compared to other generations [5, 46]. Generally, it can be concluded that convenience is an important factor and should be reflected in the security processes. The secure way should be simultaneously the convenient way and not require a lot of extra efforts. For example, it should be easy to report a phishing email.

Furthermore, Gen Z in general does not strive for more security as they believe that privacy in general is high on the internet and are affected by the optimism bias, where they believe that a security incident will not happen to them. Employers can break these wrong perceptions by working with statistics in their security awareness strategy. This is also reflected in the study by Tsohou et al. [10] which will be further discussed in the implementation proposal.

Gen Z is seemingly motivated by extrinsic motivational factors such as appreciation and physical goodies to perform a certain behaviour. While this also could be reflected in a security awareness strategy, the intrinsic motivation is much more fundamental. Managers should therefore seek to motivate employees intrinsically [53].

Education seems to be a key aspect to improve Gen Z's attitude towards a secure behaviour. While Gen Z was exposed with modern technology at a younger age this does not implicate a higher security awareness. Moreover, Gen Z has received less education on safety online compared to older generations. This can be explained through the fact that workplaces often offer these types of trainings, therefore the older generations had more possibilities to have experienced such a training [5]. As cyber security is not part of the mandatory education there is not any chance for them to gain the knowledge unless they are highly interested in the topic. An organisation should reflect this aspect in the planning of security awareness trainings.

4.1.2   Attribute 2: Subjective Norms

The literature review, that summarised research on Gen Z's expectations towards the workplace, concluded that Gen Z is looking for a fun place to work that has a positive culture and with values and beliefs that they can identify themselves with. This implies that it is even more important for Gen Z to have an environment and company culture of security, as this will heavily influence their own behaviour. The personal reputation is very important for Gen Z, they want to leave the best impression on others [54] and have high expectations towards themselves. This can also be adapted to the corporate world, meaning that Gen Z want to leave a positive picture of their behaviour. Therefore, if their colleagues and superiors express the importance of security behaviour this can positively influence them.

Moreover, Gen Z want to feel valued and have a purpose in what they do. Therefore, it is important for the employer to give them an understanding why they should behave security-compliant and what the effects would be if they do not. Gen Z is disappointed if they cannot perform as expected and quickly take negative feedback personally. This should be reflected in the messaging of the security awareness campaigns and considered by their colleagues and superiors.

Lastly, Gen Z is facing the challenge that their environment, especially older generations lack training and knowledge of the specific tools and programmes Gen Z is using in their everyday life. For a company this could still result in a data breach. Therefore, it is important for a company to incorporate Gen Z in the information security process to ensure knowledge of tools and programmes that Gen Z are regularly using and to avoid "Shadow IT".

4.1.3   Attribute 3: Perceived Behavioural Control

Lastly, perceived behavioural control describes the perceived ease to behave in a certain way. Compared to previous generations, Gen Z is characterised by being less autonomous and taking less responsibility in the workplace. Studies argue that their confidence can be influenced by superiors through putting trust into them and their actions. Additionally, inconvenience is named as a factor that can lower the perceived behavioural control of Gen Z. As already discussed in Sect. 4.1.1, companies should use security tools that are simultaneously convenient.

Gen Z likes to be in control but when it comes to the internet many of the respondents believe that they have lost control. They argue that managing security is overwhelming and time-consuming. While many of them mention that they feel like they do not put enough effort in security or would like to increase their efforts they are not aware how to do so. This leads to the implication that there is a lack of education and knowledge, which causes respondents to see their perceived behavioural control as low. To solve this, companies should invest in proper security education.

Furthermore, participants mention that it should be the role of the law to put up appropriate measures, some also mention that it should be easier to identify a malicious website. Companies should invest in user-friendly security programmes that block security incidents and provide some education to the user on the security incident after.

In general, the perceived behavioural control for Gen Z's security compliant behaviour is low.

4.2   Proposal for Implementation

After discussing the roles of behavioural economics and Gen Z in and outside of the context of security awareness, the practical work will focus on closing this gap by giving a proposal how this information can be leveraged and implemented.

To achieve that, security awareness implementation best practices by NIST and PCI-DSS will be used to organise and describe the different steps that are needed for a successful planning and implementation. As part of the recommendations, visuals will be presented that could be used to efficiently target Gen Z to improve security awareness. To ensure the effectiveness of the recommendations, appropriate measures of success will be discussed.

A security awareness programme is typically divided in three steps: first the programme needs to be planned, the material needs to be developed and prepared and lastly the programme needs to be implemented [10].

4.2.1   Planning Stage

The planning stage centres around designing the security awareness programme. Companies should focus at this stage on the priorities and assessing the current state. The assessment can then simplify the developing of the materials and understanding the priorities [12].

As part of the planning stage, it first should be determined who is part of the security awareness taskforce. As a best practice, it is recommended to use team members from different areas of the organisation [55]. This was also recognised as part of the analysis in 4.1, as Gen Z would understand their own generation and tools used in the best way. An organisation should therefore have someone from Gen Z as part of their security awareness team to properly address this generation. As another best practice in the planning stage, it is recommended to use different channels to reach all employees [55]. For Gen Z therefore an appropriate channel and media form should be used. In the literature review it was stated that Gen Z likes to have learning material on social media. Some of the social media providers even went as far as to provide their own security awareness trainings for their users. In October 2021 during cyber security awareness month the amongst Gen Z popular platform TikTok started a campaign with the name "#BeCyberSmart". The campaign featured several TikTok creators and employees, amongst them TikTok's Chief Security Officer [56].

Some of the Gen Z respondents mentioned a lack of general security education as a problem. As security education is not part of the mandatory school or university schedule, Gen Z did not have a chance to learn basic security best practices before joining the workforce. This should be considered in the onboarding process. A company could for example offer a voluntary security awareness basic training for any new joiner, where participants will have the chance to learn security best practices in an interactive way. This does not have to be necessarily only targeted to Gen Z but for any person that is interested in improving and further educating their security behaviour. As a relevant amount of Gen Z seems to be generally interested in improving their security knowledge a decent participant rate should be expected.

When considering educational experiences with Gen Z, literature suggests to use learning materials that are interactive and incorporate technology and social media. This could for example achieved by posting security awareness advice and recommendations on the company's social media. On the other hand, technology can be used to achieve more interaction. An example for that is the introduction of augmented reality and virtual reality. Knowledge is said to be transmitted faster through this technology [40].

Bartels et al. [57] argue that the messaging of information should be structured based on the perceived risk and potential benefits. Gain-framed messages describe messages that focus on the benefits an individual gains when they engage in a certain behaviour, while loss-framed messages focus on the possible costs if individuals do not engage in a certain behaviour. The authors conclude that gain-framed messaging is more effective if the probability to avoid the risk is high and there is a potential benefit. At the same time loss-framed messaging should be used when the probability to avoid the risk is low and there is a potential cost [57]. While the study was conducted in the context to promote health appeal, the findings could still be applied to security awareness.

4.2.2 Development of Material

In order to develop relevant awareness material, the target group should be considered. It should be understood who the material is developed for. PCI [55] recommends building a training based on the role and loosely split the groups in specialised roles, management, and everyone else. This grouping should be done based on the responsibilities the roles have in an organisation. In the case of Gen Z, it would not make sense to have a separate Gen Z role, as different members of Gen Z would take different roles and responsibilities within an organisation. Furthermore, a company should understand the security behaviour they would like to reinforce through awareness and the skills the target group should learn [12].

Additionally, the cultural theory of risk by Tsohou et al. [10] should be considered as well. The authors defined groupings that have a similar risk perception and should therefore be addressed in a similar way in a security awareness campaign as described in Table 4.

To identify in which category of the cultural theory of risk an employee falls, Tsohou et al. recommends conducting a questionnaire. Additionally, the materials should also reflect the consequences upon not complying with the security best practices and communicate them [10].

The development of the actual material eventually then can happen in-house or outsourced to a third-party vendor. This decision is dependent on the time and resources available. Examples for development of materials can include information on password usage, phishing, social engineering, shoulder surfing, incident response and more [12].

While the content of the material should be defined based on the role and responsibility of the individual, the communication material and channel can be highly targeted towards the recipient. Separate communication channels should be used to properly reach Gen Z. In the literature review it was concluded that Gen Z likes to learn with technology and having learning material on social media available. There are already some content creators available on TikTok, the most popular social media

Table 4 Cultural theory of risk

Name	Risk perception	Recommendation for security awareness programme
Hierarchy	Risks perception low when justified by governmental orders but high if this structure is threatened	Are more concerned with cybercrime/terrorism, they trust expert opinions
Egalitarianism	Risk perception high when backed up by government or experts, overestimate risks that will bring dangers to future generations	Emphasize equality aspect, let them take an active role and participate, they don't trust expert opinions
Individualism	Risks are seen as opportunities, high risk perception if potential limit of their freedom	Very afraid of personal freedom → create awareness by focusing on security attacks that threatens that → e.g. identify theft, phishing etc.
Fatalism	Prefer to not be aware of any risks as they perceive they cannot do anything about them (low perceived behavioural control)	Strongly influenced by other's opinion, therefore emphasize on commitment to others and the organisation by complying to the security policy

Source Tsohou et al. [10]

form for Gen Z [58] that create videos on security awareness. A short TikTok video, where a young girl is seemingly disgusted by a comment someone is making about how they are handling passwords. The whole video is only 23 s long and ends with a statement by the creator to the audience, soliciting to take control of their data.

A second example for a suitable security awareness video for Gen Z is 18-s long video brings awareness on most used passwords that are easy to hack for attackers.

Additionally, Gen Z prefers symbols and emojis over text messages. A marketing consultancy specialised on Gen Z argues that communication to Gen Z should not use more than five words and a big picture in order to reach the individuals [54]. Tsohou et al. [10] advice to use a story that is easy to remember and therefore take advantage of the availability bias. Figure 1 presents a visual communication that would be suitable to use for Gen Z as it uses only a small number of words and many symbols (Fig. 1).

4.2.3 Programme Implementation

In order to implement the security awareness programme, a company should first build on top of an existing programme if available by adopting the points reflected above. Security awareness should also be reflected as part of the onboarding process. Based on the six onboarding best practices by Chillakuri [60] security awareness could be implemented in the onboarding as described in Table 5.

A key point in the implementation is the initial communication of the security awareness efforts. NIST [12] recommends that this should be done centrally by the Chief Information Officer or the IT security programme manager to show the importance. In 4.1 it was concluded that a company culture of security plays a key role and Gen Z are influenced by how their colleagues and superiors see the importance of a certain behaviour. While the idea and importance of the programme should be enforced by a hierarchically relevant person, managers should also stress the importance to their teams. While Gen Z often prefers texting as communication channel over face-to-face the research of [51] concluded that important conversations should happen face-to-face, and texting should be used only for basic communication.

Furthermore, PCI-DSS [55] mentions the importance of an on-going process. Security awareness should not be an annual activity or a one-time affair but instead part of the company culture. Initiatives such as the cyber security awareness month can be leveraged to create company-wide activities and create awareness.

To ensure a successful programme, appropriate KPIs should be defined, and success of security awareness efforts measured. Examples for suitable KPIs could for example be an increase in reports of attempted e-mail or phone scams or an increase in reporting of security concerns.

Lastly, independent of the security awareness programme, security tools that have any interaction with company employees should be re-evaluated regarding the user friendliness. The importance of convenience was largely discussed in previous sections. While this cannot be realised with security awareness, it still heavily influ-

Fig. 1 Example for phishing awareness. *Source* Stanford University [59]

ences Gen Z's security behaviour. Companies should therefore consider the user interaction and friendliness when evaluating a security tool and not only technical features or price.

Table 5 Onboarding best practices

Name	Description	Example
Meaningful work	Let them understand how their work is meaningful will motivate people	Explain how security best practices can protect the company
Performance management	Instant and in-person feedback, understand how they are measured	Explain consequences of not complying with security policies
Work-life balance	Want to be flexible at work, connect seamlessly	Let them understand what chances security can provide for them, such as working from home
Personal connect	Human connection especially in the first years, in-person	Have in-person security training
Understanding the bigger picture	Prefer transparency and honesty, are ambitious, are not satisfied with low-work value	Explain the value and threat for the company dependent on security
Learning and development	Self-learning, own pace, equip them with necessary skills	Provide necessary tools to gain the relevant knowledge

Source Chillakuri [60]

5   Research Evaluation and Limitations

While overall the project output is viewed as successful by the author, this section should evaluate alternatives and limitations.

Firstly, the literature review focused in detail on the theory of planned behaviour as a psychological model of behavioural economics to predict and individual's behaviour. This model was chosen as it has been used several times to understand security behaviour. However, only focusing on the theory of planned behaviour might insufficiently predict the actual behaviour of individuals, as there are many other theories in the field of behavioural economics. Other studies focus on personality traits and research how messaging should be conducted to make individuals more receptive to security awareness messages (for example [21, 30, 61]). Another challenge presented in the literature review was the research on Gen Z being very current, therefore new research gets published every day and it is difficult to keep up to date.

Regarding the practical work and discussion, it is assumed that the theory of planned behaviour also accurately predicts Gen Z's security behaviour. While a correlation of the theory of planned behaviour and security intention has been proven in other studies, there has not been any distinct research that proves that this theory also applies to Gen Z.

Additionally, the datasets described in the methodology part all mainly focus on Gen Z in the western world. The results might differ greatly for other geographical locations and development levels. This is partially visible in the study of Francis and Hoefel [50] conducted with Gen Z in Brazil or the research in Sriprom et al.

[30] that analyses Gen Z in Thailand. Using the theory of planned behaviour, the characteristics would differ and therefore lead to different results. Additionally, it becomes more and more difficult to generalise a whole generation [51] as there are limitations such as the increasing gap in wealth, even within the same geographical location.

Many of the surveys and interviews included here are regarding Gen Z as a consumer or an individual but not in the role of an employee of a company. In general, there has not been a lot of research on Gen Z's behaviour in the workplace, as Gen Z is only starting to enter the job market [62]. This could lead to potentially different results of attitude and subjective norms if Gen Z would be questioned within a different setting.

Conclusion and Recommendations

This project presents an in-depth overview on how security awareness for Gen Z can be improved by understanding and explaining the behaviour using the theory of planned behaviour by Ajzen [27]. Over the course of the paper, the literature review provided an insight into the current state of security awareness and presented in detail the theory of planned behaviour, a renowned theory from the field of behavioural economics that has been used in past research to research and explain security compliant behaviour of individuals. Next, papers that researched the theory of planned behaviour in the context of security behaviour were presented and critically discussed. Lastly, as part of the literature review, Gen Z was presented and compared to earlier generations. Current research on Gen Z's characteristics was summarised to provide insight in preferences in learning and in the workplace. A recent study that researched Gen Z and the theory of planned behaviour within another context was analysed in detail to understand how the methodology and findings could be leveraged for this dissertation. The literature review concluded that while Gen Z is facing a lot more security issues than other generations before them, there has not been any study that analysed Gen Z's security behaviour.

By analysing different surveys and interviews conducted with Gen Z participants, important insights on Gen Z's attitude, subjective norms and perceived behavioural control were gained. For the attribute attitude the literature review presented conflicting statements regarding Gen Z's privacy. While some authors mentioned that Gen Z values privacy much more compared to other generations, other authors claimed that Gen Z is much less aware of privacy. The analysis of the surveys concluded that Gen Z feels indifferent towards personal security. They value convenience much more than anything else. Some of the interviewed Gen Z participants mentioned that the indifference of attitude results from the lack of education that Gen Z has received on this topic, as security awareness is not part of the mandatory school schedule and as individuals from Gen Z are just starting to enter the workforce, they have also not experienced any trainings at the workplace compared to older generations.

The analysis of the second attribute subjective norms presented that Gen Z are highly influenced by subjective norms in many ways. They put a lot of importance on the work culture and identifying themselves with the values of the company, have a feeling of purpose and understand their role. At the same time their personal reputation online is very important to them. This leads to the conclusion that a company should put an effort to have a security culture throughout the company to ensure Gen Z will align with those values. As their image is very important to them it will positively influence Gen Z to behave security compliant if their superiors and colleagues express the importance of it.

Lastly, the third attribute perceived behavioural control discussed that inconvenient tools that make it difficult to report security events lower the perceived behavioural control for Gen Z. Therefore, when evaluation a security tool that provides user interaction the user-friendliness and ease to understand should be considered by companies. Perceived behavioural control seems to be in general low for Gen Z, as some of the respondents of surveys mentioned they feel like they have lost control on the internet or would like to improve their security efforts but do not know why. This leads to the conclusion that security awareness programmes and training could improve perceived behavioural control and thus overall security compliance.

Besides discussing the results, a proposal was presented for companies to create or adapt their current security awareness programme to ensure that Gen Z is properly addressed. As part of that, advice on planning, content creation and implementation were provided. This should enable companies to reflect and improve their current programme.

As it was noted previously, a limitation of the study is that the considered secondary data—and particularly the questions on security behaviour to Gen Z—was not surveyed in the context of a company environment. This can lead to inaccurate results, as the participants might have answered differently in another context. Therefore, future research could focus to conduct direct quantitative or qualitative surveys with Gen Z and their views on security behaviour and security awareness at the workplace. Furthermore, the presented research focused on providing insights and a proposal for implementation. While some examples for content were discussed in the dissertation this could be developed further. One could focus on creating awareness content such as videos or images or even using gamification elements to create a serious game targeted at Gen Z participants based on the insights provided.

Finally, as the generation is constantly evolving and starting to enter the workforce it will be interesting to see how the behavioural patterns might change with maturity and age of this generation.

References

1. Check Point Research (2022) Threat research. https://www.research.checkpoint.com/category/threat-research/
2. Gartner (2022) Gartner identifies top security and risk management trends for 2022. https://www.gartner.com/en/newsroom/press-releases/2022-03-07-gartner-identifies-top-security-and-risk-management-trends-for-2022

3. Bisson D (2021) More generation Zs are falling for online scams. https://www.securityintelli gence.com/news/more-generation-zs-falling-online-scams/

4. Donegan J (2020) The kids are not alright: How some Millennials and Gen Zers are cybersecurity liabilities. https://www.securitymagazine.com/articles/94205-the-kids-are-not-alright-how-some-millennials-and-gen-zers-are-cybersecurity-liabilities

5. Cohen R (2020) Are gen Z-ers more security savvy online than millennials? https://www.f5.com/labs/articles/threat-intelligence/are-gen-z-ers-more-security-savvy-online-than-millen nials

6. Sjouwerman S (2021) Phishing campaign targets TikTok influencers. https://www.blog.kno wbe4.com/phishing-campaign-targets-tiktok-influencers

7. White G (2021) Generation Z: cyber-attack awareness training effectiveness. J Comput Inf Syst 62(3):560–571

8. Mawhinney T, Betts K (2019) Understanding Generation Z in the workplace—new employee engagement tactics for changing demographics. https://www2.deloitte.com/us/en/pages/con sumer-business/articles/understanding-generation-z-in-the-workplace.html

9. Safa NS, Sookhak M, Von Solms R, Furnell S, Ghani NA, Herawan T (2015) Information security conscious care behaviour formation in organizations. Comput Secur 53:65–78

10. Tsohou A, Karyda M, Kokolakis S (2015) Analyzing the role of cognitive and cultural biases in the internalization of information security policies: recommendations for information security awareness programs. Comput Secur 52:128–141

11. Tam C, de Matos C, Oliveira T (2022) What influences employees to follow security policies? Saf Sci 147:105595

12. NIST (2003) Building an information technology security awareness and training program. https://www.nvlpubs.nist.gov/nistpubs/legacy/sp/nistspecialpublication800-50.pdf

13. Rocha Flores W, Ekstedt M (2016) Shaping intention to resist social engineering through transformational leadership, information security culture and awareness. Comput Secur 59:26–44

14. Khando K, Gao S, Islam SM, Salman A (2021) Enhancing employees information security awareness in private and public organisations: a systematic literature review. Comput Secur 106:102267

15. Scholl MC (2019) Raising information security awareness in the field of urban and regional planning. Int J E-Planning Res 8(3):62–86

16. Francia G, Thornton D, Trifas M, Bowden T (2014) Gamification of information security awareness training, emerging trends in ICT security, pp 85–97

17. Jaffray A, Finn C, Nurse JRC (2021) SherLOCKED: a detective-themed serious game for cyber security education. Springer International Publishing, Manhatten, New York City

18. Hart S et al (2020) Riskio: a serious game for cyber security awareness and education. Comput Secur 95:101827

19. Hamari J, Koivisto J (2014) Measuring flow in gamification: dispositional flow scale-2. Comput Hum Behav 40:133–143

20. Stewart G, Lacey D (2012) Death by a thousand facts: criticising the technocratic approach to information security awareness. Inf Manage Comput Secur 20(1):29–38

21. Kajzer M, D'Arcy J, Crowell CR, Stiegel A, Van Bruggen D (2014) An exploratory investigation of message-person congruence in information security awareness campaigns. Comput Secur 43:64–76

22. Ong LP, Chong CF (2014) Information security awareness: an application of psychological factors—a study in Malaysia. In: Proceedings of the 2014 international conference on computer, communications and information technology

23. Acquisti A, Adjerid I, Balebako R, Brandimarte L, Cranor L, Komanduri S, Leon P, Sadeh N, Schaub F, Sleeper M, Wang Y, Wilson S (2017) Nudges for privacy and security: understanding and assisting users' choices online. ACM Comput Survey 50(3):1–41

24. Petrykina Y, Schwartz-Chassidim H, Toch E (2021) Nudging users towards online safety using gamified environments. Comput Secur 108:102270

25. Xu Z, Guo K (2019) It ain't my business: a coping perspective on employee effortful security behavior. J Enterp Inf Manage 32(5):824–842
26. Gundu T, Flowerday SV (2013) Ignorance to awareness: towards an information security awareness process. SAIEE Africa Res J 104(2):69–79
27. Ajzen I (1991) The theory of planned behavior. Organ Behav Hum Decis Process 50(2):179–211
28. Sommestad T, Hallberg J, Lundholm K, Bengtsson J (2014) Variables influencing information security policy compliance: a systematic review of quantitative studies. Inf Manage Comput Secur 22(1):42–75
29. Kim S, Kim Y (2021) Augmented compliance intention through the appropriation of compliance support systems. Behaviour & information technology, pp 1–17
30. Sriprom C, Rungswang A, Sukwitthayakul C, Chansri N (2019) Personality traits of Thai gen Z undergraduates: challenges in the efl classroom? PASAA 57:165–190
31. Wagner LS, Luger TM (2021) Generation to generation: effects of intergenerational interactions on attitudes. Educ Gerontol 47(1):1–12
32. Bennet J, Pitt M, Price S (2012) Understanding the impact of generational issues in the workplace. Facilities 30(7/8):278–288
33. Dimock M (2019) Defining generations: where millennials end and generation Z begins. Pew Res Center 17(1):1–7
34. Djafarova E, Foots S (2022) Exploring ethical consumption of generation Z: theory of planned behavior. Young Consumers, NA
35. Ismail AR, Nguyen B, Chen J, Melewar TC, Mohamad B (2021) Brand engagement in self-concept (BESC), value consciousness and brand loyalty: a study of generation Z consumers in Malaysia. Young Consum 22(1):112–130
36. Andersen K, Ohme J, Bjarnøe C, Bordacconi MJ, Albæk E, de Vreese C (2020) Generational gaps in political media use and civic engagement: from baby boomers to generation Z, 1st ed. Routledge
37. Gabrielova K, Buchko AA (2021) Here comes Generation Z: millennials as managers. Bus Horiz 64(4):489–499
38. Twenge JM (2017) IGen: why today's super-connected kids are growing up less rebellious, more tolerant, less happy–and completely unprepared for adulthood–and what that means for the rest of us. Simon and Schuster, New York City, NY
39. Schroth H (2019) Are you ready for gen Z in the workplace? Calif Manage Rev 61(3):5–18
40. Hernandez-de-Menendez M, Escobar Diaz CA, Morales-Menendez R (2020) Educational experiences with generation Z. Int J Interact Des Manuf (IJIDeM) 14:847–859
41. Fromm J, Read A (2015) Marketing to Gen Z : the rules for reaching this vast, and very different, generation of influencers. Amacom, New York
42. Fromm J, Read A (2018) Marketing to Gen Z: the rules for reaching this vast–and very different-generation of influencers. Amacom, New York
43. Johnston R (2018) Who is generation Z and how will they impact the future of associations. http://www.naylor.com/associationadviser/generation-z-future-associations/
44. Adobe (2019) 15 mind-blowing stats about generation Z. https://www.business.adobe.com/blog/the-latest/15-mind-blowing-stats-about-generation-z
45. Choong Y, Theofanos MF, Renaud K, Prior S (2019) Passwords protect my stuff—a study of children's password practices. J Cybersecur 5(1)
46. Buitta L, Johnson A (2022) What generation Z can teach us about cyberse-curity. https://www.microsoft.com/security/blog/2022/03/15/what-generation-z-can-teach-us-about-cybersecurity/
47. Dell Technologies (2019) Gen Z: the future has arrived. https://www.delltechnologies.com/asset/en-us/solutions/industry-solutions/briefs-summaries/gen-z-the-future-has-arrived-complete-findings.pdf
48. WP Engine & CGK (2017) The future of digital experiences: how gen Z is changing everything. https://www.wpengine.com/wp-content/uploads/2017/12/WPE-EBK-LT-GenZ-AUS_v04.pdf
49. Family Online Safety Institute (2021) Managing the narrative: young people's use of online safety tools. https://www.global-uploads.webflow.com/5f47b99bcd1b0e76b7a78b88/618d32fb1c370900fcd08ab0_FOSI%20Research%20Report%202021.pdf

50. Francis T, Hoefel F (2018) 'True Gen': generation Z and its implications for companies. https://www.mckinsey.com/industries/consumer-packaged-goods/our-insights/true-gen-generation-z-and-its-implications-for-companies
51. Gould D, Nalepa J, Mignano M (2019) Coaching generation Z athletes. J Appl Sport Psychol 32(1):104–120
52. Brady S (2021) 78% of generation Z would sacrifice online security for social media fame. https://www.valuepenguin.com/news/generation-z-sacrifices-online-security-over-social-media-fame
53. Zhang Y, Liu S (2022) Balancing employees' extrinsic requirements and intrinsic motivation: a paradoxical leader behaviour perspective. Eur Manage J 40(1):127–136
54. Williams A (2015) Move over millennials, here comes generation Z. https://www.nytimes.com/2015/09/20/fashion/move-over-millennials-here-comes-generation-z.html
55. PCI DSS (2014) Information supplement: best practices for implementing a security awareness program. https://www.pcisecuritystandards.org/documents/PCI_DSS_V1.0_Best_Practices_for_Implementing_Security_Awareness_Program.pdf
56. Wu L (2021) #BeCyberSmart: cybersecurity awareness champions. https://www.newsroom.tiktok.com/en-us/becybersmart-cybersecurity-awareness-champions
57. Bartels RD, Kelly KM, Rothman AJ (2010) Moving beyond the function of the health behaviour: the effect of message frame on behavioural decision-making. Psychol Health 25(7):821–838
58. Kalupski K (2021) TikTok is the reigning champ amongst gen Z. https://www.investisdigital.com/blog/news/tiktok-gen-z
59. Stanford University (2022) Phishing awareness program—learn to recognize malicious emails. https://www.uit.stanford.edu/service/phishingawareness
60. Chillakuri B (2020) Understanding generation Z expectations for effective onboarding. J Organ Chang Manage 33(7):1277–1296
61. Jaeger L, Eckhardt A (2020) Eyes wide open: the role of situational information security awareness for security-related behaviour. Inf Syst J 31(3):429–472
62. Dobrowolski Z, Drozdowski G, Panait M (2022) Understanding the impact of generation Z on risk management—a preliminary views on values, competencies, and ethics of the generation Z in public administration. Int J Environ Res Public Health 19(7):3868

Cyber Security Compliance Among Remote Workers

Diana Adjei Nyarko and Rose Cheuk-wai Fong

Abstract Remote working has become an important part of keeping operations up and running and managing the security risks associated with it is essential for organizational development. This study seeks to find out how remote employees comply with their organizations' cybersecurity regulations and policies. The objectives of this research are to investigate strategies organizations use to keep positive cyber security compliance, understand the challenges remote workers face, and how they are reducing their cyber risk exposure through compliance. Furthermore, it also aims at gathering the best compliance practices used for maintaining trust, safeguarding, and reducing insider risk/human error which will help in recommending ways that employees can adopt to improve upon their cybersecurity compliance A mixed-method research approach is used to investigate remote workers' awareness, commitment, and motivation to cyber security compliance. The findings show that although most organizations have strategies to keep positive cyber security compliance for their remote workers, more than half of the interviewees are ignorant about it or lack the requisite training needed to comply.

Keywords Cyber security · Compliance · Remote

1 Introduction

Remote work is growing more prevalent, with surveys indicating that 70% of people globally work remotely at least once a week [1]. Organizations that wish to tap into this growing pool of remote workers must be prepared to deal with a whole new level of cybersecurity and compliance vulnerabilities. Managing the security risks associated with remote working is a crucial part of helping businesses run safely

D. A. Nyarko (✉) · R. C. Fong
Northumbria University, London, UK
e-mail: diana.nyarko@northumbria.ac.uk

R. C. Fong
e-mail: rose.fong@northumbria.ac.uk

© The Author(s), under exclusive license to Springer Nature Switzerland AG 2023   343
H. Jahankhani (ed.), *Cybersecurity in the Age of Smart Societies*,
Advanced Sciences and Technologies for Security Applications,
https://doi.org/10.1007/978-3-031-20160-8_18

and securely. This is possible if employees can efficiently manage their cybersecurity compliance obligations. The purpose of this research is to analyze how remote workers are complying with cybersecurity policies and regulations remotely.

1.1 Background

Workplaces are changing across the world, as many people have found themselves in new remote firms. While individuals must make significant changes in their personal lives to accommodate these variations, businesses must also make rapid and significant changes in their policies and practices to keep up with the newly established remote workforce and ensure that it is not only functional but also secure and compliant. How organizations can develop a secure and compliant workforce outside of their own physical offices is a great challenge [2]. When remote working became the order of the day around the world, most employees were forced to adapt without the knowledge or training needed to meet cybersecurity standards. According to Wright and Fancourt [3], some remote employees are still catching up on their compliance adherence measures. Employees accessing sensitive data of an organization from non-corporate networks gives room for cybersecurity concerns, including data theft, breaches, and, as a result, significant financial loss.

According to Pałęga and Knapiński [4], organizations must consider how to increase physical and technical security for their employees at home, as well as the management of confidential information, while encouraging employees to follow security procedures. There are hundreds of controls and dozens of acronyms that overwhelm some employees as they have a responsibility to play in ensuring that their companies adhere to the established standards.

1.2 Problem Description

Individual security compliance behaviour has been the focus of recent information and cybersecurity studies as employees fail to adhere to best security practices for improving security performance. Remote employees are confronted with an enormous task in terms of cybersecurity compliance as they try to figure things out on their own [5]. When one is not working in a conventional workplace, it's easy to forget about data security best practices. Some employees use their personal email accounts to submit company documents. Compounding the problem, many remote employees use public Wi-Fi, pooled internet connections, or personal hotspots.

Hudson et al. [6], are of the view that compliance is a difficult task, let alone doing so while implementing remote working rules and processes. As a result, even in a remote working environment, it is necessary to invest in employees to preserve sensitive data and comply with requirements.

Some solutions, such as remote monitoring of staff usage of devices and logins, communication channels, electronic signatures, and virtual meetings for papers and authorizations, may be provided by technology, but they will necessitate a re-evaluation of regulation, risk, and trust. People will find solutions if security hinders them from completing their jobs, [7] and non-compliant employees are just as damaging to a company's cybersecurity as anonymous outside hackers. Employees forsaking standard security procedures when working from home are typically blamed for organizational security sanctions.

Employees want companies to prioritize security and compliance while also providing the tools required for safe remote work. Although it is impossible to anticipate how the workplace will evolve in the future, taking steps to increase security standards now will ensure that firms and people are prepared for whatever the future holds.

1.3 Significance of the Study

Cybersecurity is no longer only an issue that must be fixed but it is a constant threat that businesses must deal with daily. The complexity of data flows outside the confines of the office is multiplied. Employees normally only use designated office devices that are secured with physical and technological levels of protection at work, but as more gadgets are incorporated into firms' data systems, vulnerabilities increased. With increased cybercrime, remote working, and the growing need for data security, ensuring compliance has never been more important [8]. It is easy for the protocol to fall by the wayside, as personal devices and increasingly shared workspaces provide greater opportunities for data leakage. This study will raise security awareness, sensitize remote workers, and educate them about potential threats and best practices for prevention.

Data breaches, identity theft, and a myriad of other unfavourable events can all result from working from home [8]. Furthermore, the remote working environment has the potential to blur the lines between work and personal life, leading to bad cybersecurity practices such as the use of work equipment for personal purposes. As a result, this study will add to the body of knowledge on cybersecurity compliance and provide new information on how to close the gap between compliance and remote working.

1.4 Objectives

This research aims to access and analyse how employees are managing cybersecurity compliance instituted by their organizations.

- To explore cybersecurity strategies used by organizations to promote compliance among remote workers.
- To understand the challenges remote workers face and how they are reducing their cyber risk exposure through compliance.
- To gather best practices for compliance used to maintain trust, safeguard, and reduce insider risk/human error.
- To recommend ways that employees can adopt to improve upon their cybersecurity compliance.

2 Literature Review

Due to the dramatic surge in digitization that swept the world in the aftermath of the COVID 19 pandemic, cybersecurity compliance has progressively gained a position in public discourse in recent years. Cybersecurity Compliance involves adhering to various measures (normally implemented by governments, regulatory bodies, or industry associations) required to protect data confidentiality, integrity, and availability [9]. Compliance requirements vary by industry and business, but they often comprise several specific organizational policies and technologies to safeguard data. Some sources of control are the NIST Cybersecurity Framework, the CIS, and ISO 27001. According to Richter and Hauff [10], business leaders are still uncertain simply due to the sheer diversity in business conditions and data types that companies end up processing. Different industries and countries impose sets of controls on their business entities that collectively represent a fully functional cybersecurity program. These controls constitute a cybersecurity compliance framework which is why Hudnall [11] believes that there is no one-size-fits-all method for cybersecurity compliance frameworks, each industry is unique and handles its information differently. There is evidence that compliance issues at the organizational level are demonstrated in some recent cybersecurity research but there is also the need to research deep into how cybersecurity security compliance among remote workers is managed as data risk increases with remote access.

2.1 Cybersecurity Compliance Framework

Using a code of practice to manage cybersecurity compliance allows businesses to validate their security procedures through independent attestations, demonstrating the appropriateness and soundness of security measures to customers, clients, and industry authorities. Depending on the application domain, several versions of a framework for cybersecurity best practices exist. International standards, on the other hand, are universal [12].

The European Parliament's adoption of the General Data Protection Regulation (GDPR) brought in a new season for cybersecurity compliance, as technology businesses sought to gain the trust of their customers and authorities [13]. Compliance with state, industry and international cybersecurity legislation is now recognized as a strategic priority for organizations that are serious about conducting business. It is often essential, or at the very least strongly urged, for companies that want to comply with these regulations. For this reason alone, there are multiple frameworks, and most businesses end up using more than one framework depending on the kind of industries and regions they operate in.

2.1.1 International Organization for Standardization (ISO)

ISO 27001 which is an information security management model, provides a framework for securing and protecting sensitive and private data in businesses of all sizes [14]. It is a collection of standards designed to guide organizations looking to implement strong cybersecurity. ISO/IEC 27001:2013 is the most well-known of these, providing companies with guidance to develop an information security management system (ISMS).

Whiles complying with ISO 27001 is not mandatory for any organization, companies may choose to achieve and maintain ISO 27001 compliance to demonstrate that they have implemented the necessary security controls and processes to protect their systems and the sensitive data in their possession [15]. The ISO 27001 regulation's principal purpose is to assist enterprises in developing, implementing, and enforcing information security management system (ISMS). This ISMS outlines the controls, processes, and procedures in place at the company to protect the confidentiality, integrity, and availability of the information it holds.

To achieve ISO 27001 compliance, an organization must also document the steps that were taken in the process of developing the ISMS. Key documentation includes information security risk assessment process and plan, evidence of competence of people working in information security, information security policy, information security objectives, ISMS internal audit program and results of audits conducted, results of the information security risk assessment and treatment, evidence of nonconformities identified and corrective action results, and evidence of leadership reviews of the ISMS [14].

According to Lim [15], achieving ISO 27001 compliance is important as a differentiator in the marketplace and as a foundation for complying with other mandatory requirements and standards. An organization with ISO 27001 compliance is likely more secured than one without it, and the standard provides a solid framework for building many of the security controls required by other regulations.

2.1.2   Control Objectives for Information and Related Technology (COBIT)

COBIT is an IT governance framework that helps organizations bridge the gap between control requirements, technological obstacles, business risks, and security concerns [16]. It presents a globally recognized IT control architecture that enables companies to build an IT governance framework that covers their entire organization. The COBIT framework demonstrates how IT operations provide the data that the company requires to meet its goals.

This delivery is tracked and evaluated using 34 high-level control objectives, one for each IT process, and divided into four domains: planning and organization, acquisition and implementation, delivery and support, and monitoring and evaluation [17].

Management Guidelines, Control Objectives, Implementation Guide, IT Assurance Guide, Control Framework and Executive Summary, and Governance are all included in the current edition of COBIT, which is Version 5.

2.1.3   National Institute of Standards and Technology (NIST)

The NIST technique is divided into three sections, each of which builds on the preceding to assist a company in evaluating its current systems and developing a strategy. The five functional areas to evaluate are Identify, Protect, Detect, Respond, and Recover [18].

The National Institute of Standards and Technology (NIST), is a globally recognized standard that should be shared with all employees and vendors. Rather than leaving cyber security to the IT department, the framework works with top management to ensure data protection for all.

Organizations will be able to examine their current IT infrastructure and plug any gaps, lowering the risk of data compromise, by following the approach. It also implies that they will be prepared to put their backup plan into action if something goes wrong.

Apart from covering Data protection (through GDPR), it also covers risk management, cloud services, malware protection, vulnerability scanning, incident management, firewalls, security policy, and business continuity, amongst other crucial topics.

2.2   Organizational Cybersecurity Culture

The shared norms, values, and qualities of an organization are defined as the foundation for a feeling of shared purpose and the maintenance of links among people, processes, and policies [19]. It indicates that security management and governance are most effective when they are integrated into the organization's culture of behaviour and actions.

Cybersecurity culture is the ability to collectively harmonize the norms, values, and attributes of individuals or groups within the organization and operate subconsciously to identify, understand, and act on current and future risks [20]. Several studies have identified the need of cultivating a security culture in which employees have the attitude, talent, and knowledge to support cybersecurity compliance goals. Many debates on the subject suggest ways to make a security-conscious culture second nature to people while they work.

Cybersecurity culture should be all encompassing, such that it compliments technical security measures put in place by organizations while remaining ubiquitous to the daily activities of employees [21]. Studies suggest that there is a strong emphasis on security culture as a factor that encourages secure behaviour, it is described as the stage between the influencing factor and the sustaining factor. The intention to violate security protocols transcends some progressive stages before the intention translates to either secure or insecure behaviour.

2.3 Cybersecurity Compliance Practices

According to Kuhlman [22], a Cybersecurity program must include compliance management practices. However, ensuring that one is meeting regulatory obligations is getting increasingly challenging as complex extraterritorial rules, industry-specific restrictions, and general data protection laws are driving this trend.

To manage compliance, Lie et al. [21], suggest that it is required to develop a compliance monitoring plan capable of continuously reviewing the organization's compliance actions in real-time. To create a successful compliance monitoring program, businesses must first determine which rules and regulations apply to them, as well as what it takes to comply with them [23]. This will enable firms to do a gap analysis to determine what compliance controls and business processes they already have in place, as well as what new security measures are required.

This risk assessment approach will identify risk areas and should be used to develop the information security policy of the organization in addition to maintaining compliance with ever-changing privacy and security standards, every employee must have a role in controlling cyber threats. These standards and laws differ depending on the area and sector, making it difficult for businesses to stay compliant [22].

The implementation of best practices to maintain the confidentiality, accessibility, and security of information stored, transferred, or processed is critical. According to Donalds and Osei-Bryson [24], cybersecurity standards and regulations differ by industry, so there's nothing like a one-size-fits-all strategy for compliance management. As a result, establishing complete compliance systems necessitates constant risk management to identify and mitigate all potential dangers.

Working with the IT department to identify cybersecurity vulnerabilities will determine where compliance needs to be enhanced [16]. Client information, banking data, and technology infrastructure must all be kept protected from outside sources as well as insider threats and accounted for in the organization's risk management strategy. The firm's information systems and networks that they access must be identified to establish a thorough risk assessment plan. Santini et al. [25], are of the view that the level of risk associated with each category of data and where high-risk data is stored, transported, and collected must be determined. Once a risk has been recognized, security controls such as network firewalls, data encryption, an incident response strategy, a patch management schedule, and network access control should be implemented to manage it.

Developing an effective risk assessment plan allows an organization's compliance team to make changes to existing policies and procedures or create new ones [26]. Several regulatory agencies have requested that the compliance department clarify how the firm's rules and processes interact with its installed cybersecurity programs.

2.4 Work Design and Remote Working

According to Kessler [27], organizations are faced with numerous security challenges as a result of the global remote working transition, and conventional security solutions are inadequate in addressing accidental data loss and insider threats. Both internal (workers) and external threats have increased because of remote working. Since the COVID-19 pandemic, the number of cyber and hacking attempts has increased by 300% (FBI IC3 Report, April 2020).

The hunt for a globally accepted definition of remote employment has aroused debate and contention [2]. According to Allen et al. [28], the lack of a uniform definition of remote working has limited knowledge of this type of employment, highlighting the fact that the conclusions of diverse studies are sometimes conflicting. Remote working, according to Gleason [29], is defined as employing technology to do work tasks at home or in a location other than the workplace. Correa [30] also defines the practice as a working arrangement that allows workers to carry out their job responsibilities from a remote location while communicating with the organization via technology and internet connectivity. Remote working, on the other hand, is defined by Wheatley [31] as paid work done outside of one's typical office, such as at home or while actively traveling using ICT. Workers that work from a location other than their office are known as remote or mobile workers [32–34]. Most previous research, according to Chandra et al. [35], has concentrated on the broad concept of flexible working, while remote working has gotten less attention.

According to Vivekananth [36], technology improvements, societal trends, and cultural transformations are driving remote work. Amaddeo [37], in a similar vein, outlines five factors that are impacting remote working today, including demographic shifts, globalization, technical breakthroughs, sociocultural trends, and an emphasis on low-carbon technologies. In relation to that, Onakoya [38] predicts that remote working will continue to grow in the future as a result of individual differences in attitudes and beliefs, as well as significant developments in ICT.

Previous studies on cybersecurity compliance have identified elements that boost employee compliance [39, 40]. These studies have investigated what motivates remote workers to comply. Traditional models and constructs of cybersecurity organizational culture were investigated by Alvesson and Sveningsson [41], but they did not clarify, identify, or describe the construct of remote work cultures that affect cybersecurity compliance. As a result, it's vital to identify the situations and solutions that are critical for preserving, strengthening, or eroding remote workers' workplace non-compliance.

3 Methodology

Cybersecurity compliance is a diverse topic, and the nature of the study conducted tends to be complex, making choosing an effective research method difficult. These issues have long preoccupied cybersecurity experts, and they have played an important part in the discipline's development, resulting in a broad discussion of various approaches [36]. Experts believe that no single methodology will work for all studies, instead, several research methodologies, methods, and procedures can be applied in a variety of scenarios. The philosophical assumptions that drive this research, as well as the research methodology and procedures used, are defined and presented. Explaining the research approach is a key tactic for boosting the validity of social research, according to Luo and Creswell [42].

The nature of the research, the study's target, and the researcher's experience determine the strategy to be used. To aid in data collection, a convergent parallel mixed approach is used. This method allows for the simultaneous collection of quantitative and qualitative data, with the results being combined during the overall interpretation to detect convergence, divergence, inconsistencies, or correlations between the two sets of data. The purpose of using this strategy is to give the researcher a complete picture of the problem by collecting a variety of complementing data.

Furthermore, employing many methodologies strengthens the findings since triangulation or creating an explanation for all the data when they diverge will strengthen the findings [43]. Critical multiplism was coined by Creswell [44] to convey the idea that research topics may be studied from numerous perspectives and that integrating different methodologies with different biases is a good strategy.

This study focuses on mostly remote workers and a concurrent mixed-method strategy to collect codes and analyse data at the same time, utilizing random sampling procedures for the quantitative strand and purposive data techniques for the qualitative.

In most research, the sample size is effectively decided by two factors: (1) the nature of the planned data analysis and (2) the projected response rate. Abbott et al. [45] recommend a rule of thumb of 10 samples per measurement variable for quantitative studies. If your questionnaire has 25 measurement variables, for example, your sample size should be 250. Based on this, a sample size of 130 was chosen with a measurement instrument of 13.

To achieve data saturation in qualitative investigations, it has previously been advised that a minimum sample size of 12 be used [46–48]. As a result, a sample size of 13 was found adequate for qualitative analysis and study scale.

4 The Conceptual Framework and Hypothesis Development

The conceptual framework is based on the independent variables of cybersecurity, compliance, and remote employees, as well as the dependent variable of cybersecurity compliance. The goal of the conceptual framework is to demonstrate the link between remote work and cybersecurity compliance. Gauging information security performance, according to Barone [49], aids an organization in determining the degree to which its security needs are met. Information security efficacy can be measured in a variety of ways, including compliance with best practices, how a company has formed and enforced a data security policy, and the extent to which employees are provided with the appropriate tools and technologies.

Within the model, we look at the impact of these beliefs on the hazards of remote work, changes in cyber-attacks, the capacity to respond to cyber-attacks, significant problems in creating an acceptable response to cyber-attacks, and perceptions of important cyber-security challenges. Accordingly, the formulated Hypothesis is.

5 Results

Most respondents were UK residents (67.8%), with some responses from other countries (non-UK respondents 32.2%). Although the focus was on the United Kingdom, the international participants provided a broader perspective.

Although 130 questionnaires were sent out, the total number of respondents was 121 with 90 being men and 31 being women representing 74.38% and 25.62% respectively. The researcher had more men working remotely than women (Fig. 1).

The respondents' age ranged from 20 to 60 and over, with the age group of 20–29 having the highest number of respondents with 63.64% and the least age group of 40–49 with 16 respondents representing 13%. Also, 30–39 were the second-highest respondents with 27.27%, and the 50–59 had 21 respondents representing 17% with 60 and above having no respondents (Fig. 2).

When asked about the period they have worked remotely, 52% have worked remotely for more than a year but less than five years, 42% have worked within a year remotely compared to 6% who have worked more than five years but less than 10 years. No respondent was recorded for having worked more than ten years remotely.

Fig. 1 Gender distribution

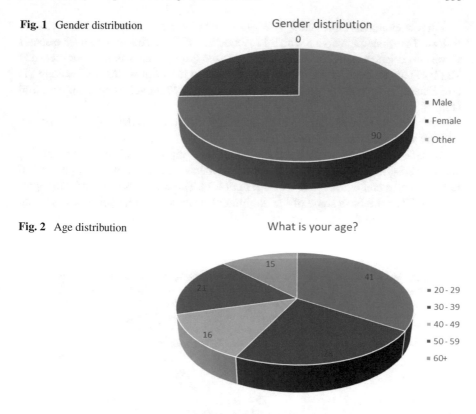

Fig. 2 Age distribution

This suggests that remote working became popular within the last five years, and it is declining as some have either returned to face-to-face working or are now doing hybrid (Fig. 3).

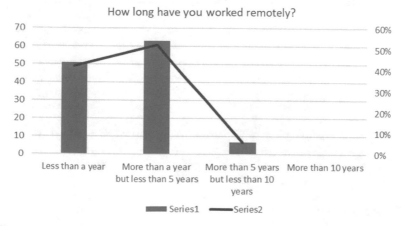

Fig. 3 Years of working remotely

Forty respondents representing (33%) are in the educational sector. This is closely followed by customer service with 34 respondents (28%). The administrative sector in various organizations also had 23 respondents (19%). The health sector recorded 12 (10%) respondents and a few respondents from the Information Technology (7) and Economic sector (5) representing 6% and 4% respectively were also captured (Fig. 4).

Respondents have similar job titles but different sectors. Below is the breakdown of their job titles (Fig. 5).

When asked if they have the relevant working equipment and dedicated working space, sixty-four respondents representing 53% have all the needed equipment and have created a conducive space to work remotely. There were 48 respondents representing 40% who have some of the equipment needed to work from home and not all too suitable workspace remotely. A small number (9) representing 7% have no

Fig. 4 Occupation or profession

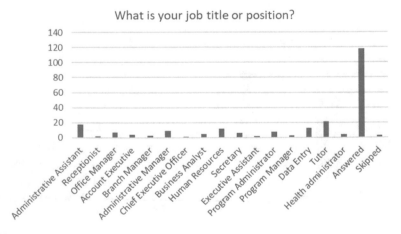

Fig. 5 Positions or titles of respondents

Fig. 6 Dedicated workspace
and requisite equipment

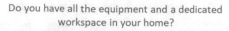

Do you have all the equipment and a dedicated
workspace in your home?

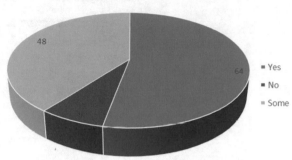

IT equipment or space to work but they are forced to do so because of current
circumstances (Fig. 6).

On the issue of their organizations having a cybersecurity compliance framework
policy in place, fifty-eight of the respondents representing 48% answered in affirma-
tive whilst twenty-nine representing 24% stated that their organizations do not have
it and 34 representing 28% said they are not aware whether their organizations have
it or not (Fig. 7).

When respondents were asked about the cybersecurity policy framework their
organizations were using the below breakdown was given (Fig. 8).

On the issue of familiarity with their organization's cybersecurity compliance
policy, respondents gave the below response (Fig. 9).

Comparing complying remotely to face to face, 21% of respondents strongly agree
that it is difficult to comply remotely, 42% agree, 9% neither agree nor disagree, 28%
strongly disagree and 7% disagree as they think it is the same as working face to face
(Fig. 10).

On formal training on cybersecurity compliance for remote working, 34%
responded yes to having received formal training whiles 66% have not (Fig. 11).

Fig. 7 Availability of
cybersecurity compliance
framework

Does your organization have
cybersecurity framework
policy?

Which cybersecurity framework and policy does your organization use?

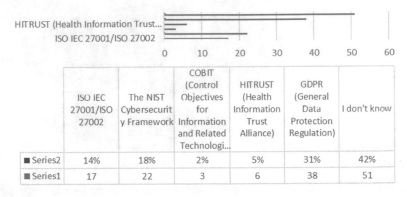

	ISO IEC 27001/ISO 27002	The NIST Cybersecurit y Framework	COBIT (Control Objectives for Information and Related Technologi...	HITRUST (Health Information Trust Alliance)	GDPR (General Data Protection Regulation)	I don't know
■ Series2	14%	18%	2%	5%	31%	42%
■ Series1	17	22	3	6	38	51

■ Series2 ■ Series1

Fig. 8 Existing cybersecurity compliance Framework

Fig. 9 Familiarity with a cybersecurity compliance framework

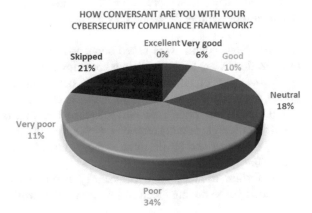

HOW CONVERSANT ARE YOU WITH YOUR CYBERSECURITY COMPLIANCE FRAMEWORK?

The below chart shows the strategies respondents' organization uses to keep positive cyber security compliance for their remote workers (Fig. 12).

On whether their organization's cybersecurity compliance practices maintain trust, nurture teamwork, safeguard, and reduce insider risk/human error, below are their responses (Fig. 13).

The results from the quantitative phase of the study expands the overall understanding of how remote workers comply with cybersecurity framework policies, and hence the study extends knowledge into cybersecurity culture and remote workplace practices.

It is difficult to comply with cybersecurity
regulations when working remotely

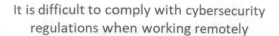

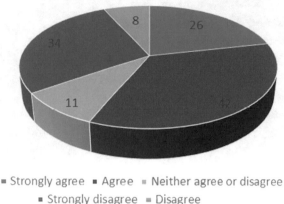

■ Strongly agree ■ Agree ■ Neither agree or disagree
■ Strongly disagree ■ Disagree

Fig. 10 Cybersecurity compliance when working remotely

Has your organization given you any
cybersecurity training for remote
working?

■ Yes ■ No

Fig. 11 Cybersecurity compliance training for remote employees

Profile of the Interviewees

This section summarises the profiles of the thirteen interviewees in order to provide a clear picture of their backgrounds, which will help to make sense of the subsequent discussion. No real names are used throughout the study as part of the agreement reached before it began. Seven males and six females were interviewed for the study, totalling thirteen distant employees. The majority of them work from home full-time, while some have flexible, part-time, or hybrid schedules. Their remote working experience span between two to five years. They were between the ages of 28 to 50. Nine of them are in the UK whiles the other four are from Ghana. Five of the respondents were from various administrative sectors whilst four were from customer care and four from the educational sector (Table 1).

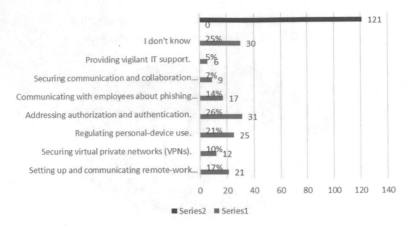

Fig. 12 Positive cybersecurity compliance strategies

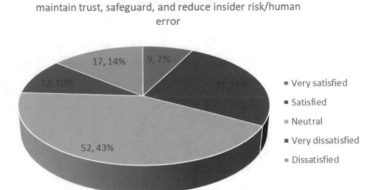

Fig. 13 Cybersecurity culture

Thematic analysis was used to examine the information gathered from the interviews. These discussions were taped and later transcribed. For the analysis, the shared data was highlighted, categorized, and sorted into themes. The data was cross-referenced with other transcripts after each transcript was reviewed individually to discover the most common themes. During the inquiry, 12 themes were revealed after meticulous analysis and coding of the material. Because several of the 12 themes overlapped, the researcher decided to organise them into five main themes based on the data. The researcher looked at each of these themes and divided them into five categories:

1. Cybersecurity compliance challenges
2. Awareness and Training
3. Monitoring

Table 1 Breakdown of sample participants

Interviewees	Gender	Age	Length of remote working	Type of remote work	Sector	Role
A	Male	47	5	Full time	Education	Teaching
B	Male	30	2	Flexible	Health	Administrator
C	Female	28	2	Full time	Banking	Customer care
D	Male	50	4	Full time	Health	Administrator
E	Female	42	3	Full time	Tourism	Customer care
F	Male	35	3	Hybrid	Education	Teaching
G	Male	31	2	Hybrid	Education	Administrator
H	Male	44	4	Full time	E-commerce	Customer care
I	Female	27	3	Hybrid	Retail	Customer care
J	Female	49	2	Part-time	Education	Teaching
K	Male	42	5	Full time	Telecom	Administrator
L	Female	33	5	Part-time	Retail	Customer care
M	Female	28	3	Hybrid	Banking	Customer care

4. Boundary Management
5. Work environment.

Cybersecurity Compliance Challenges

Organizations need to maintain security standards regardless of where individuals work so that they can comply with their regulatory body's requirements. However, this becomes challenging when data is transferred across offices and between remote personnel and devices. Employees themselves attest to the fact that it is difficult to comply with all the security protocols put in place by their company when working remotely. Respondents A, C, D, G, K, and M stated that they occasionally undertake work using non-corporate assets. Respondent G stated that:

> The IT staff had complete control over ensuring we don't violate the cybersecurity standards, but now since we work from home with no direct supervision, it's a little more difficult to follow the regulations, especially for those of us who had to improvise with our own equipment.

Respondents B and E Respondents B and E believe that when working remotely, they may get away with riskier security practices, which is why their cybersecurity behaviour at home differs from what they do at work. Respondent F acknowledged taking up bad cybersecurity habits and adopting security workarounds while working from home. Some responders use unsecured WI-FI, personal devices, and other illegal applications, which might increase the threat level and the number of loopholes available to cybercriminals. They also allow household members to use unprotected Wi-Fi, personal devices, and corporate equipment for personal usage. Respondent H stated that:

Ironically, you should inquire about security and privacy. Last week, I received three odd emails on my personal computer informing me that I needed to pay a little ransom because someone knew my password. Each email came from a different person, but they all contained the same few letters in the middle of my password. So, I changed the majority of my passwords, but I'm not sure I caught them all. Because the emails were sent to my personal email address, I didn't report it to IT, but I was concerned for a few days. You've made me think about it once again!

This is obvious because email is the major communication channel used in remote working, so it is easy for hackers to target them.

Awareness and Training

Cyber security awareness and training are primarily intended to influence the adoption of secure behaviour, particularly when working remotely. It requires more than simply educating people what they should and shouldn't do to have an effective impact: they must first agree that the knowledge is important, then understand how they should respond, and then be willing to do so in the face of multiple other demands.

This research found out that some people are aware of their cybersecurity policies, and some have even acquired training, but they do not act accordingly. According to Respondent H and I, users will eventually make mistakes and avoid security completely if the system is too complex to use. Respondent B said that:

I'm tempted to give up on any attempts at safety and overlook any threat when met with so many useless warnings and imprecise advice. Threatening security messages are ineffectual because they increase stress levels to the point where the individual becomes disgusted or denies the need for any security decision.

M also stated that:

I don't blame the remote worker altogether; when I was told that I could work from home and use my own device, I had to figure out how to be productive while still being extremely careful and secure for the good of the firm.

To further understand why remote workers are ignorant about their organization's cybersecurity framework the qualitative phase of the study tried to find out the reason behind it and respondent C stated that

Before the pandemic, firms may not have given cybersecurity high priority, thus employees were unconcerned about it.

Another respondent: B also said

As a result of the rise in remote work owing to the pandemic, we realized that some considerations, such as risk assessment, ongoing staff education, compliance difficulties, and top management level assistance, may have been missed.

The widespread talk about cybersecurity compliance awareness, remote worker training, and the attempt to secure the human element in many circumstances needs significant amount of effort and expertise from both the organization and its employees.

Monitoring

Employee monitoring is focused on preventing scenarios that jeopardize the organization's security, such as employees visiting shady websites or fiddling with sensitive information, detecting distracted employees who play computer games at work, as well as unapproved programmes that may compromise sensitive data.

Most employees aren't focused on cybersecurity, according to the findings of this study, they believe they are there to execute their jobs, not to prevent attacks or breaches. Phishing schemes and employee irresponsibility, unsurprisingly, remain two of the most common causes of cybersecurity threats.

Respondents F and H, think that they must know what they are being monitored for. They are also of the view that rather than secretly installing corporate spyware on their devices, organizations should allow them to download monitoring software themselves so that they will be responsible for it. They also want to know what exactly will be monitored (for example, web browsing and app activity) and when. Respondent J believes that using remote monitoring to pick out employees can have a negative impact on morale and lead to a cultural crisis.

Employee monitoring solutions may have a role to play in a business when other ways of gaining visibility into workforce behaviour or managing compliance are not possible. Even when the reasons for doing so are apparent and justifiable, any company that monitors employees must be honest about the process, cognizant of the hazards involved, and have a clear plan for what to do with the data acquired.

Boundary Management

Respondents were asked to describe how they keep work and family life separate while conforming to cybersecurity regulations. All the participants admitted to having difficulty defining boundaries and that the borders were frequently blurred. Respondent I said:

> For me, boundaries are hazy. I might be preparing dinner while checking emails on my laptop.

Respondents J and G made similar statements, saying, *"I might be doing homework with my kids while responding to emails."*

Other findings indicate that technology may contribute to blurred boundaries experience,

> "I don't feel like I separate the boundaries because I have my emails on my phone," remarked participant H.

> "It's always problematic because I usually find myself bouncing between work and social media when working from home," participant E said.

Having emails on your personal phone, on the other hand, may be both an advantage and a disadvantage, according to participant K, he said that: *"It's beneficial in the sense that my emails are quickly available, but it's also problematic in the sense that I could be lying in bed checking my emails,"*

The findings imply that self-discipline is key to regulating boundaries. *"It's about discipline,"* said participant D, *"and you have to convince yourself to stop at a specific time."* Similarly, participant A stated, *"It boils down to the individual and how they manage their time."* *"You have to be strict on yourself and have discipline, but it is incredibly difficult,"* said participant E.

Working Environment

The findings of this study imply that the participants' work environment may influence their compliance level. Participants were asked to comment on their work environment while working remotely. *"Having a good setting and the necessary equipment is important for yourself since it allows you to focus on exactly what you are told to do,"* said participant J. Participant K also asserts that: *"It's critical to create an environment with lots of space, enough light, and no distractions."*

Describing their working space, Participant C said: *"I work anywhere I can find myself: in coffee shops and hotels because of my frequent travelling"* Participant G said: *"I work on the move even when driving and especially where I will get a free Wi-Fi".*

When working remotely, four of the thirteen participants said they work from any available space in their homes whiles six participants said they have their own workplace at home and the remaining three have no space, so they work from any available space outside.

Participant B said that: *"an IT officer came to my house to ensure that I have a suitable place to work and furnished the place with a chair, desk, and laptop,"* in reference to due diligence made by the organization in setting up the work environment for remote working. Also participant, J emphasized that: *"My working environment had to be assessed to ensure that it was safe and that I had the necessary equipment."* Participants F and H reported that the organisation maintains security by enabling remote access for security reasons.

6 Discussion and Recommendations

6.1 *Organizational Cyber Security Compliance Strategies Awareness*

The quantitative phase of the study affirms claims that organizations have strategies in place to ensure positive cyber security compliance for their remote workers [2, 50], but the challenge is that most respondents are unaware of the kind of strategies and policies their organizations are using. Also, quite a significant number of respondents are unsure whether their organization has a strategy in place or not. However, according to Amaddeo [37], and Felstead and Henseke [1], remote workers must implement the same policy compliance and intellectual property rights controls as face-to-face workers in remote locations. This demonstrates that remote workers must be made aware of the importance of familiarizing themselves with and adhering to their company's cybersecurity policy.

Gleason [29] assertion that because of the shift to remote work, businesses must examine their compliance posture and ensure that security protections are in place will not achieve its purpose if employees who work remotely think that their company's compliance is inadequate in protecting information assets, so they have no desire to learn more.

The current trend, as well as the rapidly expanding popularity of remote working, necessitates a review of compliance procedures. This is not a one-time occurrence; the organization must continue to monitor its control environment to ensure that it is still effective. Although most firms have a compliance framework in place, the study findings show that security practices and compliance levels are low, which can have a substantial influence on compliance. Because the devices would be managed by the IT team, Whitley [31] proposes that access to the organization's internal network should be limited to employer-provided devices. He also stressed the need for remote workers to conduct a vulnerability evaluation of their home surroundings.

Employees must be made aware of and taught about any new behaviours or practices that may be required since cyber security measures at home differ from those at work. Remote employees exhibit different levels of security and privacy policy comprehension, self-efficacy, and compliance intents than office employees, according to Felstead and Henseke [1]. These findings suggest that lack of support, training, and monitoring hinders distant employees' ability to follow policies.

Kessler [27] proposes the E-Awareness Model (E-AM) as a solution to the problem of cybersecurity awareness when working remotely. Home users might be forced to learn about the existing cyber threats and according to this researcher, remote workers must be forced to take cybersecurity programs online. However, because awareness does not always lead to behavioural change, this model does not guarantee that users will follow organizational regulations. Employees may also react poorly to mandatory training, lowering their trust in the organization.

Researchers have argued that cybersecurity knowledge and awareness are transferred from the workplace to the home setting [30], but this study found that remote workers seek what is convenient to them always, even if it means non-compliance. This suggests organizations should move away from awareness initiatives towards awareness-raising tactics that will establish an all-encompassing individual security culture for users, regardless of whether they work in the office or remotely.

Establishing and fostering a cybersecurity and cyber awareness culture will start the process of assisting employees in adhering to applicable policy and understanding their role in the overall security of the firm. Failure to comply with security policy is a failure to meet one's work commitments and this should be emphasized.

6.2 Cybersecurity Compliance Challenges of Remote Workers

Employees who rely on their home networks and, in some circumstances, their own equipment to complete tasks with little or no control or monitoring run the danger of a variety of problems. Firms are substantially more vulnerable to cyber-attacks when they increasingly rely on technology without the security protections that office systems provide, such as firewalls and blacklisted IP addresses.

According to McGrath [2], the blurred lines between personal and professional life increase the risk of sensitive data falling into an insecure environment. Many employees use their personal devices for two-factor authentication, and they may have mobile app versions of IM clients like Teams and Zoom. Some respondents claimed that mobile devices are a serious challenge when it comes to cyber security. Organizations must ensure that their personnel are aware of security principles, plans, and procedures. Companies should review their plans to ensure that they are prepared to handle a data breach or security issue.

Management directions on cybersecurity, according to Onakoya [38], should be recorded in the security policy, and standards, processes, and guidelines should be developed to support these directives. These directives, however, will be ineffective if workers refused or are not motivated to comply. Employees need to understand the importance of cybersecurity and the specific security requirements that are expected of them. The nature of the data being handled should be reflected in the management of the cybersecurity program. It should be designed in a way that makes sense to everyone in the company.

6.3 Maintaining Trust, Safeguarding, and Reducing Insider Risk/Human Error Through Compliance

According to a global poll of Information Security Forum (ISF) members, remote workers were responsible for a number of vulnerabilities, including taking data home to work on in their leisure time, unwittingly opening a phishing email, or clicking on a harmful link. However, people who are looking for ways to get around policies that they believe are impeding their work may engage in careless behaviour. While most people are aware of security issues and understand the value of compliance, their workarounds can be risky.

This study agrees with Donalds, and Osei-Bryson's [24] assertion that an organization's compliance status is not always a genuine indicator that targeted changes in employee behaviour have been achieved. However, if compliance is engrained in the organization's culture and daily routine, personnel are unlikely to be persuaded to disobey the law in objective situations. The compliance program must incorporate security into the organization's culture. Embedding security into organizational culture is an important aspect of the compliance program. In the context of security culture, this means that the top management hierarchy is accountable for enacting policies that have a significant impact on staff attitudes, behaviour, and motivation. The big picture in the context of secure behaviour will begin to take shape by paying attention to small security issues within an organization. Before there can be any success in incorporating security procedures into remote working, management must demonstrate their commitment.

A committed employee may gain the motive and ability to retaliate when they feel mistreated, disrespected, or abused. As a result, avoiding putting employees in situations that are likely to weaken their trust and breed animosity is a key element of the solution. That is why Wright and Skitter [19] believe that cultivating a culture of trust is necessary. The trust-building is a necessity and must be a continuous process even when people are working remotely. This goes hand in hand beginning with onboarding procedures and training them with the knowledge and skills expected of trusted insiders.

Expectations of trustworthy behaviour as well as the repercussions of non-compliance should be stated upfront. Over time, trust should be a major consideration infrequent performance evaluation. Cultivating a culture of trust based on shared values, ethical behaviour, and the truth is essential. Cyber hygiene and security awareness must be addressed regularly in communications, training, and rules.

6.4 Recommendations

Because its surroundings do not always have the same safeguards as the office, remote work poses a distinct cybersecurity risk. When data leaves the protected confines of its storage, new threats emerge, necessitating the implementation of extra security policies. As a result, the following suggestions are required:

Public Wi-Fi Usage

Public Wi-Fi poses substantial security risks and should be avoided if at all feasible, because other individuals have access to the network, and threat actors can access and destroy data without a firewall.

VPNs allow a flexible connection to connect to multiple services (web pages, email, a SQL server) and can safeguard traffic for many remote access applications. It is possible to set up encrypted remote connections to a remote desktop or another server. Many of these connection types (RDP, HTTPS, SSH) offer encryption as part of their service direction, so they don't need a VPN or another encryption provider to protect data in transit.

Keeping Work Data on Work on Computers

A firm with a good IT team will install frequent updates and run antivirus scans that block harmful websites on workplace computers but not on personal computers. Personal computers are not safe for work information without those running in the background since they could be hacked by a third party, resulting in the possible liability of large business losses due to poor cybersecurity practices.

Work must be done online when firms provide access to a portal or remote access environment such as Office 365 to prevent downloading or syncing files or emails to a personal device. Personal business should always be kept on personal equipment, and work-issued laptops should only be used for work-related business.

Sensitive Data Encryption in Emails and on Devices

Sending emails containing sensitive information is always a risk. A third party could intercept it or see it. When data is encrypted and attached to an email, it is impossible for an unwanted receiver to see it. In the event that a device is stolen, all data will remain safe if all stored data are encrypted. Remote working is made safer with strong passwords, VPNs, and proper email habits. With end-to-end encryption and the guidelines outlined above, this process can be readily safeguarded, ensuring successful remote working during these pandemic times.

Formalizing Cybersecurity Policies of Remote Work

While good technologies and regulations are beneficial, the individuals who help the company expand are the main source of security risk. Policies on computer and internet use for work from home and remote workers can help, and these policies can be enforced with both technological and administrative controls.

These are the security awareness elements that will assist employees use business devices and information safely regardless of where they work. These work-from-home security measures can be readily included into official employee and cybersecurity policies by CISOs and IT administrators. When onboarding new employees, during regular security awareness training, and especially when organisations change their security policies, companies should train their employees on these policies.

As discussed in the literature review, using a code of practice to manage cybersecurity compliance allows businesses to validate their security procedures through independent attestations. Depending on the application domain, several versions of a framework for cybersecurity best practices exist. International standards, on the other hand, are universal [12]. The findings of this study affirm the first hypothesis that organizations have strategies to keep positive cyber security compliance for their remote workers but although they have adopted many cybersecurity compliance frameworks, their implementation is not well felt by their remote workers.

Report from ISACA and CMMI Institute shows remote workers are given extra security checks to comply with which helps in protecting companies' information, but this shows that there is much progress to be made, as participants identify gaps between their present and desired cybersecurity organizational culture. This research has shown that most especially remote workers lack the requisite training whiles some show deliberate attempts not to conform to the cybersecurity culture that promotes compliance. This necessitates additional research that prioritises investment in employee training and motivation, which can be a significant driver of strong cybersecurity culture, as well as measuring and assessing employee perspectives on cybersecurity, which can lead to increased awareness and improved culture.

According to Piatt and Woodruff [26], developing an effective risk assessment plan allows an organization's compliance team to make changes to existing policies and procedures or create new ones, however, the findings of this research suggest that without getting the employees themselves involve and making them play a vital role in the risk assessment planning its implementation will be difficult especially now that they do not have direct supervision.

According to Hunter [6], technical improvements, societal trends, and cultural transformations are driving remote working, as evidenced by the findings of this study. As a result, firms must re-evaluate their procedures and close important gaps using compliance to mitigate all types of cyber threats.

7 Conclusion

The primary goal of this study was to see how remote workers comply to cybersecurity regulations set by their organizations. The foundation of compliance is demonstrated security awareness and readiness in all scenarios, which evolves and adapts over time. The findings show considerable differences among participants' compliance levels, demonstrating that, while cybersecurity compliance is slowly evolving, it has a long and meticulous path to become an unbreakable element of organizational operations and workplace realities.

However, whether compliance is a positive or negative component in cybersecurity is frequently a question of a number of circumstances that may either decrease or increase compliance's impact on cybersecurity. By analysing compliance as a vital aspect in the organization's cybersecurity strategy, this study provides a better understanding of these factors.

More study is needed to better understand the changing cyber security landscape and employee involvement with cyber security policies, procedures, and training plans. The practical implications of not being able to get rapid in-person help with technology and cyber security should also be considered, as well as whether such challenges exist.

References

1. Felstead A, Henseke G (2017) Assessing the growth of remote working and its consequences for effort, well-being and work-life balance. N Technol Work Employ 32(3):195–212
2. McGrath A (2020) Working remotely. ITNOW 52(6):25–25
3. Wright L, Steptoe A, Fancourt D (2021) Patterns of compliance with COVID-19 preventive behaviours: a latent class analysis of 20 000 UK adults. J Epidemiol Community Health 76(3):247–253
4. Pałęga M, Knapiński M (2019) Assessment of employees level of awareness in the aspect of information security. Syst Saf Human Tech Facility Environ 1(1):132–140
5. Furnell S, Heyburn H, Whitehead A, Shah J (2020) Understanding the full cost of cyber security breaches. Comput Fraud Secur 2020(12):6–12
6. Hudson B, Hunter D, Peckham S (2019) Policy failure and the policy-implementation gap: can policy support programs help? Policy Des Pract 2(1):1–14

7. Henrichsen JR (2020) Breaking through the ambivalence: journalistic responses to information security technologies, Digit Journal 8(3):328–346. https://doi.org/10.1080/21670811.2019.1653207

8. Collier J (2018) Cyber security assemblages: a framework for understanding the dynamic and contested nature of security provision. Polit Gover 6(2):13–21

9. Marotta A, Madnick S (2020) Analyzing the interplay between regulatory compliance and cybersecurity (Revised). SSRN Electron J

10. Richter N, Hauff S (2022) Necessary conditions in international business research—advancing the field with a new perspective on causality and data analysis. J World Bus 57(5):101310

11. Hudnall M (2019) Educational and workforce cybersecurity frameworks: comparing, contrasting, and mapping. Computer 52(3):18–28

12. Karie N, Sahri N, Yang W, Valli C, Kebande V (2021) A review of security standards and frameworks for IoT-based smart environments. IEEE Access 9:121975–121995

13. Markopoulou D, Papakonstantinou V, de Hert P (2019) The new EU cybersecurity framework: the NIS directive, ENISA's role and the general data protection regulation. Comput Law Secur Rev 35(6):105336

14. Lopes I, Guarda T, Oliveira P (2019) Implementation of ISO 27001 standards as GDPR compliance facilitator. J Inf Syst Eng Manage 4(2)

15. Lim H (2021) Development of requirements for information security management system (ISO 27001) with CPTED in account. J Inf Secur 21(1):19–24

16. Mulgund P, Pahwa P, Chaudhari G (2019) Strengthening IT governance and controls using COBIT. Int J Risk Conting Manage 8(4):66–90

17. Setiawan A, Andry J (2019) Information technology governance performance measurement at national library using cobit framework 5. Jurnal Terapan Teknologi Informasi 3(1):53–63

18. Frayssinet Delgado M, Esenarro D, Juárez Regalado F, Díaz Reátegui M (2021) Methodology based on the NIST cybersecurity framework as a proposal for cybersecurity management in government organizations. 3C TIC: Cuadernos de desarrollo aplicados a las TIC 10(2):123–141

19. Skiter I (2021) Cyber security culture level assessment model in the information system. Cybersecur Educ Sci Tech 1(13):158–169

20. Gavin M, Poorhosseinzadeh M, Arrowsmith J (2022) The transformation of work and employment relations: COVID-19 and beyond. Labour Ind 32(1):1–9

21. Lie L, Utomo P, Winarno P (2021) Investigating the impact of cybersecurity culture on employees' cybersecurity protection behaviours: a conceptual paper. Conf Ser 3(2):295–305

22. Kuhlman R, Kempf J (2015) FINRA publishes its 2015 "Report on cybersecurity practices". J Invest Compl 16(2):47–51

23. Costigan S, Tagarev T (2015) Good practices and challenges in organizing for cybersecurity. Inf Secur Int J 32:5–8

24. Donalds C, Osei-Bryson K (2020) Cybersecurity compliance behavior: exploring the influences of individual decision style and other antecedents. Int J Inf Manage 51:102056

25. Santini P, Gottardi G, Baldi M, Chiaraluce F (2019) A data-driven approach to cyber risk assessment. Secur Commun Netw 2019:1–8

26. Piatt K, Woodruff T (2016) Developing a comprehensive assessment plan. New Direct Stud Leadership 2016(151):19–34

27. Kessler G (2019) Book review: cyber security and global information assurance: threat analysis and response solutions. J Dig Forensics Secur Law

28. Allen T, Golden T, Shockley K (2015) How effective is telecommuting? Assessing the Status of our scientific findings. Psychol Sci Public Interest 16(2):40–68

29. Gleason A (2021) Remote monitoring of a work-from-home employee to identify stress: a case report. Workplace Health Saf 69(9):419–422

30. Correa M (2022) Working from a new home? Exposure to remote work and out-migration from large cities. SSRN Electron J

31. Wheatley D (2016) Employee satisfaction and use of flexible working arrangements. Work Employ Soc 31(4):567–585

32. Kelliher C, Anderson D (2010) Doing more with less flexible working practices and the intensification of work. Hum Relat 63(1):83–106. https://doi.org/10.1177/0018726709349199

33. Crawford JO, Mac Calman L, Jackson CA (2011) The health and well-being of remote and mobile workers. Occup Med 61:385–394

34. Keeling T, Clements-Croome D, Roesch E (2015) The effect of agile workspace and remote working on experiences of privacy, crowding and satisfaction. Buildings (5):880–898

35. Chandra Putra K, Aris Pratama T, Aureri Linggautama R, Wulan Prasetyaningtyas S (2020) The impact of flexible working hours, remote working, and work life balance to employee satisfaction in banking industry during Covid-19 pandemic period. J Bus Manage Rev 1(5):341–353

36. Vivekananth P (2022) Cybersecurity risks in remote working environment and strategies to mitigate them. Int J Eng Manage Res 12(1):108–111

37. Amaddeo F (2022) Commuters and remote working. Jusletter (1107)

38. Onakoya S (2018) The paper investigates the role of remote working in export management, determining how virtual organizations can manage remote working effectively. Texila Int J Manage 4(2):12–18

39. De Menezes LM. Kelliher C (2017) Flexible working, individual performance and employee attitudes: comparing formal and informal arrangements. Hum Resour Manage 56(6):1051–1070

40. Eddleston KA, Mulki J (2017) Toward understanding remote workers' management of work–family boundaries: the complexity of workplace embeddedness. Group Organ Manag 42(3:346–387

41. Alvesson M, Sveningsson S (2015) Changing organizational culture: cultural change work in progress (2nd ed.). Routledge. https://doi.org/10.4324/9781315688404

42. Luo S, Creswell J (2016) Designing and developing an app for a mixed methods research design approach. Int J Desi Learn 7(3)

43. Stoecker R, Avila E (2020) From mixed methods to strategic research design. Int J Soc Res Methodol 24(6):627–640

44. Creswell J (2013) Achieving integration in mixed methods designs-principles and practices. Health Serv Res 48(6pt2):2134–2156

45. Abbott P, DiGiacomo M, Magin P, Hu W (2018) A scoping review of qualitative research methods used with people in prison. Int J Qual Meth 17(1):160940691880382

46. Clarke V, Braun V (2013) Successful qualitative research: a practical guide for beginners

47. Fugard AJB, Potts HWW (2015) Supporting thinking on sample sizes for thematic analyses: a quantitative tool. Int J Soc Res Methodol 18:(6):669–684. https://doi.org/10.1080/13645579.2015.1005453

48. Guest G, Bunce A, Johnson L (2006) How many interviews are enough?: an experiment with data saturation and variability. Field Methods 18(1):59–82. https://doi.org/10.1177/1525822X05279903

49. Barone D (2021) Exploring lenses used in case study research in literacy over time. The Qualitative Report

50. Holbeche LS (2018) Organisational effectiveness and agility. J Organ Eff: People and Performance 5(4):302–313. https://doi.org/10.1108/JOEPP-07-2018-0044

An Analysis of the Dark Web Challenges to Digital Policing

Reza Montasari and Abigail Boon

Abstract The Dark Web is the hidden group of Internet sites that can only be accessed through specific software. The Dark Web enables private computer networks to communicate anonymously without revealing identifying information. Keeping Internet activity anonymous and private can be beneficial for both legal and illegal applications. Although it is used to evade government censorship, it is also deployed for highly illegal activity. The aim of this paper is to provide a critical analysis of the technical, legal, and ethical challenges to policing the Dark Web. The most significant recommendation identified in this paper is the need for stronger national cyber security strategies, increased awareness and use of the UN Cybercrime Repository, and greater support from intergovernmental organisations. This would help to contribute towards addressing many of the technical, legal and ethical challenges concerning the multi-jurisdictional nature of Dark Web investigations and lack of reliable data and resources while ensuring transparency and accountability. The recommendations proposed in this paper are restricted by certain limitations, therefore, further research is recommended into the field of digital policing and the Dark Web.

Keywords The Dark Web · Digital policing · Digital forensics · The internet · Cyberspace · Cybercrime

1 Introduction

The development and growth of the Internet has transformed global communication, becoming one of the driving forces of social change and evolution [1]. The United Nations International Telecommunication Union "…estimates that approximately 4.9 billion people are using the Internet in 2021". At 63% of the world's population,

R. Montasari · A. Boon (✉)
Department of Criminology, Sociology and Social Policy, School of Social Sciences, Swansea University, Richard Price Building, Singleton Park, Swansea SA2 8PP, UK
e-mail: Abigailcboon@gmail.com

R. Montasari
e-mail: Reza.Montasari@Swansea.ac.uk

Advanced Sciences and Technologies for Security Applications,
https://doi.org/10.1007/978-3-031-20160-8_19

that represents a significant increase over the 1 billion people—or 16% of the world's population—using the Internet in 2005 [2]. However, the growth of the Internet and the emergence of new technologies have led to increased opportunities for criminal activity [3]. The Internet, often referred to as a network of networks, is a telecommunications system that uses a suite of communication protocols, such as the World Wide Web, to connect many smaller networks and systems around the globe [4]. Often confused for being synonymous with the Internet, the World Wide Web is a standardised system for accessing and navigating the Internet and can be accessed by web-based search engines, such as Google Chrome and Microsoft Edge [5]. The information accessible on the World Wide Web can be categorised into two sections, known as the surface web and the deep web. Only content that is indexable and searchable can be accessed through the surface web, while the deep web accounts for all other content that cannot be accessed via a search engine [6].

However, most of the criminal activity that takes place on the Internet occurs on the Dark Web, a subset of the deep web, which can only be accessed using an anonymous browser. The Dark Web provides the "...ability to traverse the Internet with complete anonymity [and] nurtures a platform ripe for...illegal activities", such as credit card fraud, arms and drug trafficking, and the leaking of sensitive information [7, p. 3]. The use of the Dark Web and the ready availability of hacking and cyber tools has also been highlighted as factors contributing to the rapid rise in cybercrime in recent years [8]. Meanwhile, strong encryption and anonymity protocols make it increasingly difficult for law enforcement to police and regulate the Dark Web [8]. Against this backdrop, this paper aims to provide a critical analysis of the Dark Web challenges to digital policing and, accordingly, recommend a number of solutions that could be adopted by law enforcement to address these challenges. The topic of the paper is notably relevant to current research because there are clear "...legal and technological gaps that exist in law enforcement's ability to cope with and respond to...cyber-crime", especially when dealing with matters on the Dark Web [9, para. 3].

The remainder of the paper is structured as follow. Section 2 will cover the necessary background information, including the development and technical aspects of the Internet. It will also provide a further in-depth review of the uses and users of the surface and the Dark Web. Section 3 will examine technical, legal and ethical challenges while Sect. 4 offers a number of recommended solutions to address these challenges. Section 5 provides a discussion, and finally, the paper will be concluded in Sect. 6.

2 Background and Context

The creation of the Internet, whose origins are generally traced back to the 1960s, facilitated a societal and technological change in how information is created, distributed and stored [5, 10]. The Internet operates by using a protocol suite, named after the Transmission Control Protocol (TCP) and Internet Protocol (IP), that sets a standard of rules for message formats and procedures that allow computers to systematically transfer data [11]. Within the Internet protocol suite (TCP/IP), there are other protocols that work on top of TCP/IP protocol to provide the necessary networking infrastructure for Internet-based information systems, such as the World Wide Web [12]. The World Wide Web provides a "...uniform, user-friendly interface to the net" and uses a series of protocols to create, transmit, and retrieve data [4, p. 1552]. These protocols, such as Uniform Resource Locators (URLs) and Hypertext Transfer Protocol (HTTP), operate at different layers of the TCP/IP networking model and have remained fundamental to the functioning of the Internet and subsequent applications [4]. Nevertheless, recent advancements in technology have allowed for a new level of connectivity facilitated by the "...Internet of Things, an evolution of connectivity expanded from the Internet [and World Wide Web]..." [13, p. 107]. The Internet of Things (IoT) simply refers to a collection of digital technologies, such as smartphones and fitness trackers, that can connect, generate, and share data over a network [14]. However, the evolution of technology and the IoT have brought about new security challenges in cyberspace [15].

At present, the Internet and subsequent information and communication technologies play a significant and vital role in all aspects of how we communicate and acquire needed information [16]. Yet only a small fraction of the information on the Internet is readily available to Internet users [7]. The surface web, one of the most accessible layers of cyberspace, refers to the network of indexable public web pages that use HTTP [17]. This content can be accessed using a basic search engine and includes public information on sites such as Facebook and YouTube. Although the surface web is the most easily accessible to the general public, it only represents a small portion of the information available on the Internet. The deep web, which is responsible for 90% of all traffic on the Internet, is almost 500 times bigger than the surface web and includes all content that is not indexed or searchable [18, 19]. This data set includes legitimate content such as digital material blocked by a paywall and personal accounts for email and banking websites, along with more illicit content that can be found on the Dark Web.

The Dark Web, a subsection of the deep web, provides more anonymity for Internet users than the surface web but is inaccessible through a standard web browser [7]. As the Internet and World Wide Web became more popular in the 1990s, concerns were raised regarding the lack of privacy and the Internet's ability to be used for tracking and surveillance [20]. These concerns spurred the development of an anonymizing browser, known as The Onion Router (Tor), by researchers at the U.S. Naval Research Laboratory [20]. The Tor browser offers anonymity by guiding users' data traffic through different servers, which are located around the world, and ensuring that the user's IP-address changes at every server, this helps preserve anonymity [21]. Averaging almost two million users per day in 2021, Tor has quickly become a popular tool for Internet users whose ability to remain anonymous is paramount, such as political activists and whistleblowers [22].

Although there are legitimate reasons for using Tor, it has become a relatively well-known resource for accessing illicit content that can be found on the Dark Web [7]. Malicious actors use the Dark Web to buy, sell, and distribute illegal drugs, weapons, and ransomware, with child pornography accounting for the most popular type of content on the Dark Web [23]. In recent years, the use of the Dark Web to commit crimes has grown exponentially due to factors such as the use of cryptocurrencies, such as Bitcoin, and the increasing profitability of Dark Web markets [24]. The growing threats of the Dark Web and dark marketplaces have elicited a global response from law enforcement agencies, with many governments designating dedicated Dark Web investigation teams [9]. However, even with recent technological advancements in digital policing, many of these investigatory powers still face technical, legal, and ethical challenges to policing the Dark Web, as discussed in the following section.

3 Challenges

3.1 Technical

One of the main technical challenges faced by law enforcement is the anonymity that is associated with strong encryption techniques on the Dark Web, which makes it "…much more difficult for investigators to assemble the evidence puzzle and prove that a crime has been committed" [25, p. 2]. Policing anonymity-granting technologies is difficult because these systems, such as Tor, are decentralised, use globally located servers, and do not retain data [6]. Even if these anonymising browsers did record data traffic, the global and volunteer nature of these servers makes it virtually impossible for law enforcement in any one country to locate it. This anonymity, aided by the use of cryptocurrencies (Bitcoin, Litecoin, Monero, etc.), enable buyers and vendors on the Dark Web to interact "…instantly, directly, freely and safely, without requiring any form of introduction or 'vetting'" [26, p. 69]. However, the increasing popularity of cryptocurrency and the large number of legitimate users make it difficult for law enforcement to differentiate between the legal and illegal trade of goods and services [26, 27].

Another technical challenge encountered by law enforcement agencies is their inability to accurately track and estimate increases in crimes committed on the Dark Web [25]. While part of this issue stems from the anonymous nature of the Dark Web, another significant contribution is the underreporting of crimes instigated by Dark Web users [28]. In recent years, the Dark Web has become a popular location for hacker forums and the distribution of malware, exploits, distributed denial of service attacks (DDoS) and hacking services [24] (Fig. 1).

However, many individuals and organisations who are victims of these digital attacks do not report it to local law enforcement. The Federal Bureau of Investigation (FBI) estimates that only one in every seven Internet-enabled crimes is reported to law enforcement in the US, and Europol also notes that victim organisations "…appear to be reluctant to come forward to law enforcement authorities or the public when they have been victimised" [30 p. 7; 31, p. 3]. Research has shown that

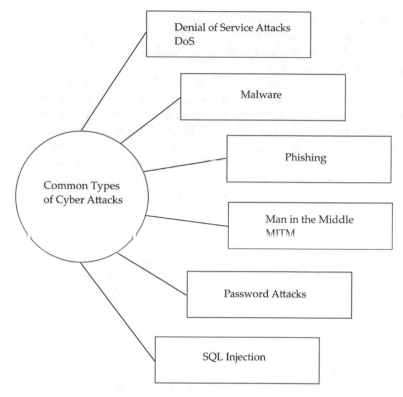

Fig. 1 Common types of cyber-attacks based on the UND's [29] classification

there are many factors that contribute to the underreporting of these crimes, including reputational risks that come with publicising data breaches, low confidence in the effectiveness of law enforcement, lack of awareness on where to report crimes, and being unaware that victimisation has even occurred [32–34]. However, the lack of access to sufficient and reliable data poses a major challenge for investigators who are responsible for monitoring and policing crimes committed on the Dark Web. Without the necessary information to provide an accurate representation of the threat landscape, law enforcement agencies lack the understanding and resources to handle the crimes that are reported. The anonymising encryption technologies of the Dark Web and the lack of reliable data prove to be significant technical challenges that investigators face when policing the Dark Web.

3.2 Legal

The multi-jurisdictional nature of Dark Web-enabled crimes is one of the many legal challenges faced by investigators. It can be very difficult for law enforcement agencies to investigate criminals when the offences have impacted victims in multiple jurisdictions [6]. Currently, most national law enforcement agencies rely on the Mutual Legal Assistance Treaties (MLATs), which are agreements between two or more countries with the purpose of collecting and sharing information in an effort to aid criminal investigations [35]. However, the current MLAT process is in need of significant reform, with calls for increased emphasis on the protection of human rights and heightened efficiency, if it is to become an effective police tool in aiding the prosecution of malicious actors on the Dark Web [36]. Another issue is that not all countries share the same fundamental values and legal principles, which can make international cooperation extremely difficult. Coordination failures also occur at the local level, with knowledge and resource deficiencies between local and national police posing great challenges to investigations.

Another legal challenge faced by law enforcement is the lack of universal definitions, guidelines, training, and technical support for Dark Web investigations. In the US, research has shown that "…there is limited familiarity in many, if not most, state and local law enforcement agencies regarding the Dark Web and how it is used to facilitate criminal activity" [25, p. 9]. This relative lack of experience with investigatory techniques can make it almost impossible to build a strong criminal case, and police can often overlook physical evidence, such as cryptocurrency wallets and encryption keys, when conducting an investigation into the Dark Web [25]. Lack of appropriate training can also threaten the legality of undercover operations and collection of intelligence on the Dark Web and increase the risk of violating national and international criminal law while conducting investigations [8]. Many factors can lessen the risk of victimisation such as undertaking the appropriate risk assessments, instituting rigorous security safeguards, and adhering to relevant cybersecurity practices [37].

Legal questions about the public and private aspect of the information on the Internet also raise some fundamental concerns. One example is the debate regarding the public nature of IP addresses, and their potential to be used as evidence in legal proceedings. Under the General Data Protection Regulation (GDPR), the European Union (EU) classifies IP addresses as "Personally Identifiable Information", which then could be subject to data privacy laws. However, in the US, the view that IP addresses are public remains contested, despite a federal court ruling IP addresses are considered public information in *United States v. Forrester (2007)* [38]. These legal and policy discrepancies pose a significant challenge and "…greater harmonisation between jurisdictions and law enforcement agencies operating practises is desirable in order to retain democratic accountability and legitimacy" and increase the success of investigations on the Dark Web [39, p. 6].

3.3 Ethical

One of the biggest ethical challenges faced by law enforcement when policing the Dark Web is the concern regarding digital surveillance and the protection of civil liberties. In recent years, with the "…growing threat of terrorism, far-right radicalization and extremist groups emerging in western societies, surveillance is viewed not only politically necessary, but also electorally popular" [40, p. 1]. In Dark Web investigations, mass surveillance techniques are frequently used when law enforcement have been unable to clearly identify suspects. However, when these types of techniques are used, "…it is often difficult to preserve the privacy of non-suspects, especially in situations where a machine has been used by multiple individuals", such as someone accessing the Dark Web using a machine belonging to a relative or in a public library [25, p. 7]. As illegal activity continues to grow on the Dark Web, law enforcement are more likely to face dilemmas around the use of otherwise ethically harmful methods with increasing frequency [39]. The encouragement, normalisation, and facilitation of harm by law enforcement in these types of operations is another legal grey area [37]. These issues are more likely to occur when operations require investigators to go undercover and work closely with criminals on the Dark Web. These scenarios frequently place law enforcement personnel into difficult situations, often having direct knowledge of the occurrence of serious crimes but not taking opportunities to intervene because of the potential effect on the overall aims of the operation [39].

Another ethical challenge is preventing harm to the welfare of law enforcement officers carrying out investigations on the Dark Web. Oftentimes, officers are exposed to very traumatic and illicit content, such as the torture and abuse of children, and repeated exposure can cause psychological distress and posttraumatic stress disorder (PTSD) [41]. Reports have also shown that officers involved in covert operations struggle to maintain their moral identity when having to interact and collaborate with criminals on the Dark Web [39]. While the online policing of crime on the Dark Web has serious implications for the welfare of law enforcement officers, it also becomes an issue of preserving valuable expertise and ensuring the success of operations. Investigators who "…have breakdowns or who can no longer perceive the distinction between morally acceptable and unacceptable tactics can become a risk rather than an asset to covert units" [39, p. 15]. These are just some of the significant ethical challenges faced by law enforcement when policing the Dark Web.

4 Recommendations

4.1 Technical

As noted earlier, one of the main technical challenges facing law enforcement is the use of strong encryption techniques in the Dark Web. Although most law enforcement agencies remain "…understandably secretive in relation to the tools and techniques used to unmask Dark Web criminals…", there are a wide array of policing techniques that can be used to identify criminals hidden under the veil of technological anonymity [20, para. 5]. One of the most well-known techniques is the use of Open-Source

Fig. 2 Common types of OSINT based on [43] work

Intelligence (OSINT), which is the collection of information legally available in the public domain and is increasingly driven by Internet and ICT based technological developments [20, 42]. OSINT can be collected from a wide range of sources as represented in Fig. 2.

OSINT allows law enforcement to legally gather information on suspects, usually left through human error. However, the vast amount of OSINT can cause difficulties for investigators when attempting to filter out relevant and valuable information. Therefore, to leverage OSINT techniques successfully, law enforcement personnel must be given the necessary training and resources, particularly with regards to handling voluminous data analyses. The second technical challenge highlighted was the lack of reliable data, in part due to the anonymous nature of the Dark Web, but also resulting from the underreporting of crimes. At the individual level, this could be addressed by increasing public knowledge through the use of social media campaigns about online threats, such as malware attacks, along with information for victims on where to report incidents and seek support. The strengthening of cyber security strategies at the national level could also allow for stricter cybersecurity breach disclosure laws, which would help address underreporting of crimes by organisations and businesses. It could also provide the necessary policy infrastructure to increase law enforcement resources and training on matters pertaining to the Dark Web at the local and national levels.

4.2 Legal

The first two legal challenges identified in the previous section of this paper were issues arising from the multi-jurisdictional nature of Dark Web-enabled crimes, and the lack of universal definitions, guidelines, and technical support for Dark Web operations. As previously highlighted in the technical recommendations section, a comprehensive cybersecurity strategy, along with the fostering of a strong international cybersecurity community, would greatly benefit local and international Dark Web investigations. The convening of an "…international court for cyber events and passing [of] international laws concerning cyberspace and cyber weapons", as well as requiring countries to report any vulnerabilities, would enable law enforcement

Fig. 3 A graphical representation of the interpol operating model

agencies to increase their general knowledge and understand the best techniques for ensuring successful operations [44, p. 1].

The third legal challenge discussed focused on the fundamental concerns regarding the public and private aspects of an individual's information on the Internet. This challenge also highlights the resulting issues posed by legal and policy discrepancies between different countries, and the need for international collaboration and cooperation. A partial solution to addressing these challenges would be for countries to ensure that their local and national law enforcement agencies have knowledge of and access to the UN Cybercrime Repository, "…a central database of legislation, case law and lessons-learned on cybercrime and electronic evidence" [45]. Along with support from Interpol, an inter-governmental organisation that facilitates police cooperation and crime control, this would allow for a more unified approach to addressing dark-web enabled crime while addressing the resource and skill gap between international agencies [46]. Figure 3 is a graphical representation of the Interpol Operating Model.

4.3 Ethical

The first two ethical challenges identified in the previous section were regarding the concerns about digital surveillance and the protection of civil liberties along with the encouragement, normalisation, and facilitation of harm by law enforcement in Dark Web investigations. Although it is difficult to police the Dark Web, law enforcement must find a balance between safeguarding the benefits of anonymity for non-malicious users such as activists and whistleblowers, while ensuring that the Dark Web does not empower child abusers and arms traffickers [23]. To address these concerns, law enforcement agencies should systematically monitor the effectiveness of tactics and outcomes of all Dark Web investigations. This should be similar to the UN Cybercrime Repository. However, more focus should be placed on practises tailored to a country's political and social culture.

The third ethical challenge identified was the risk of harm to the welfare of law enforcement officers carrying out investigations on the Dark Web. To mitigate the harm to the welfare of personnel, law enforcement agencies must ensure that there is "…strategic and long-term provision of psychological, moral, and professional support for officers involved in operations that bring them close to disturbing activities…" [39, p. 15]. However, it is important to note that law enforcement officers who experience mental health problems often do not seek psychological help due to the stigmatisation of mental health issues [47]. Therefore, any type of psychological, moral, and/or professional support provided to law enforcement would need to address this concern to be beneficial.

5 Discussion

The most significant recommendation identified in this paper is the strengthening of international cooperation and resources in relation to the policing of the Dark Web. At the international and national level, this could be achieved through stronger national cyber security strategies, increased awareness and use of the UN Cybercrime Repository, and greater support from intergovernmental organisations such as Interpol. This would help to combat several of the technical and legal challenges that come with the multi-jurisdictional nature of policing the Dark Web and the lack of relevant methods, data, and resources. At the national and local level, this could be achieved through encouraging countries to develop their own cybercrime databases. This could be similar to the UN Cybercrime Repository to systematically monitor the effectiveness of their own practises used in investigations and retain greater democratic accountability and legitimacy. This would help tackle some of the legal and ethical challenges regarding the legality of certain techniques which in turn would ensure better protection of civil liberties and reduction of the normalisation and facilitation of harm on the Dark Web. The development of social media campaigns to educate the public about cyberattacks and where to report victimisation, along with stricter breach disclosure laws, would help expand the information available for these databases and would increase the overall visibility of the threat landscape.

It should also be noted that there are several significant limitations with regards to these findings. The space constraint for this paper is a critical limitation, and further opportunities to develop this paper would add to its significance in current research. Due to the space constraint, only the most prevalent technical, legal, and ethical challenges were selected based on previous research in the field of Digital Policing and the Dark Web. However, many of these sources were written in English and therefore focused on the experiences of law enforcement agencies, Digital Policing, and Dark Web threats in western societies. As a result, many of the recommendations, if not all of them, may only be applicable to law enforcement agencies who share the same fundamental values about human rights, and other social, cultural, and political values deemed important in many western democracies. These limitations identify a need for further research, especially with regards to the challenges faced by law enforcement agencies in 'non-western' democracies such as Russia or China. However, it also must be noted that the anonymous and secretive nature of Dark Web investigations, regardless of the investigating country, poses the biggest limitation to research in this field. This could be attributed to the dangerous repercussions of publicly identifying investigation techniques and tools to criminals on the Dark Web.

6 Conclusion

As the Internet and emerging technologies allow for increased communication and connectivity, they also provide criminals with new opportunities to conduct illicit activities on a global scale [25]. The Dark Web, a subset of the deep web and part of the overall World Wide Web, presents numerous challenges to online policing. The aim of this paper was to provide a critical analysis of the technical, legal, and ethical challenges faced by law enforcement when policing the Dark Web. First, the technical challenges were identified as a lack of reliable threat landscape due to the underreporting of crimes and the anonymity of using specialised browsers, such as Tor to access the Dark Web. Then, the legal challenges were addressed in relation to the multi-jurisdictional nature of Dark Web-enabled crimes, lack of universal methods and guidelines, and legality of Dark Web operations. Finally, the ethical challenges were explored in regard to digital surveillance for operations and the protection of civil liberties, the encouragement and facilitation of harm in operations, and the harm to law enforcement's welfare.

The most significant recommendation identified in this research paper is the need for stronger national cyber security strategies, increased awareness and use of the UN Cybercrime Repository, and greater support from intergovernmental organisations. This would help combat many of the technical, legal, and ethical challenges, including some of the most significant challenges, such as the multi-jurisdictional nature of policing the Dark Web and lack of reliable data, methods, guidelines, and resources that many law enforcement agencies face when policing the Dark Web. However, as previously mentioned in the Discussion section, it should be noted that the space constraint for this paper was a significant limitation to reviewing all challenges and providing related recommendations. Further research would prove beneficial in identifying challenges unique to individual countries. As the Dark Web becomes more popular with malicious and non-malicious users, law enforcement will need to continue to address these challenges to ensure the physical and digital safety of individuals around the world.

References

1. Hilbert M (2020) Digital technology and social change: the digital transformation of society from a historical perspective. Dialog Clin Neurosci 22(2):189–194. https://doi.org/10.31887/dcns.2020.22.2/mhilbert
2. International Telecommunication Union. (2021). World telecommunication/ICT indicators database 2021, 25th edn. United Nations

3. Diamond B, Bachmann M (2015) Out of the beta phase: obstacles, challenges, and promising paths in the study of cyber criminology. Int J Cyber Criminol 9(1):24–34. https://doi.org/10.5281/zenodo.22196

4. Pallen M (1995) The world wide web. BMJ: Brit Med J 311(7019):1552–1556. http://www.jstor.org/stable/29729797

5. Van Sluyters RC (1997) Introduction to the internet and world wide web. ILAR J 38(4):162–167. https://doi.org/10.1093/ilar.38.4.162

6. Jardine E (2017) The dark web dilemma: Tor, anonymity and online policing. In: Cyber security in a volatile world. Centre for International Governance Innovation, pp 37–50. http://www.jstor.org/stable/resrep05239.8

7. Chertoff M, Simon T (2017) Cyber security in a volatile world. In: The impact of the dark web on internet governance and cyber security. Centre for International Governance Innovation, pp 29–36

8. Omand D (2015) The dark net: policing the internet's underworld. World Policy J 32(4):75–82. http://www.jstor.org/stable/44214265

9. Merchant S (2014) How the web presents new challenges for law enforcement agencies. The Office of Community Oriented Policing Services. U.S. Department of Justice

10. Naughton J (1999) A brief history of the future: the origins of the internet. Weidenfeld & Nicholson

11. Goralski W (2017) The illustrated network: how TCP/IP works in a modern network, 2nd edn. Morgan Kaufmann. https://doi.org/10.1016/B978-0-12-811027-0.00001-1

12. Berners-Lee T (1992) The world-wide web. Comput Netw ISDN Syst 25(4–5):454–459. https://doi.org/10.1016/0169-7552(92)90039-s

13. Chou S-Y (2018) The fourth industrial revolution: digital fusion with internet of things. J Int Aff 72(1):107–120. https://www.jstor.org/stable/26588346

14. Madakam S, Ramaswamy R, Tripathi S (2015) Internet of things (IoT): a literature review. J Comput Commun 03(05):164–173. https://doi.org/10.4236/jcc.2015.35021

15. Dulhare UN, Rasool S (2019) IoT evolution and security challenges in cyber space. Count Cyber Attacks Preserv Integr Availab Crit Syst 99–127. https://doi.org/10.4018/978-1-5225-8241-0.ch005

16. Roztocki N, Soja P, Weistroffer HR (2019) The role of information and communication technologies in socioeconomic development: towards a multi-dimensional framework. Inf Technol Dev 25(2):171–183. https://doi.org/10.1080/02681102.2019.1596654

17. Hai-Jew S (2014) Conducting surface web-based research with Maltego carbon. Kanas State University

18. Greenberg A (2014) What is the dark web? Wired. https://www.wired.com/2014/11/hacker-lexicon-whats-dark-web/

19. Barker D, Barker M (2013) Internet research illustrated. Cengage Learning, Independence, KY, C-4

20. Davies G (2020) Shining a light on policing of the dark web: an analysis of UK investigatory powers. J Crim Law 84(5):407–426. https://doi.org/10.1177/0022018320952557

21. Monk B, Mitchell J, Frank R, Davies G (2018) Uncovering Tor: an examination of the network structure. Secur Commun Netw 2018:1–12. https://doi.org/10.1155/2018/4231326

22. The Tor Project (2021) Users—Tor metrics. https://www.metrics.torproject.org/userstats-relay-country.html

23. Owen G, Savage N (2017) The Tor dark net. In: Cyber security in a volatile world. Centre for International Governance Innovation, pp 51–62. http://www.jstor.org/stable/resrep05239.9

24. Kethineni S, Cao Y (2020) The rise in popularity of cryptocurrency and associated criminal activity. Int Crim Justice Rev 30(3):325–344. https://doi.org/10.1177/1057567719827051

25. Goodison S, Woods D, Barnum J, Kemerer A, Jackson B (2019) Identifying law enforcement needs for conducting criminal investigations involving evidence on the dark web. RAND Corporation. https://doi.org/10.7249/RR2704

26. Paoli G, Aldrige J, Ryan N, Warnes R (2017) Behind the curtain: the illicit trade of firearms. RAND Corporation, Explosives and ammunition on the dark web

27. Martin K (2016) Data aggregators, consumer data, and responsibility online: who is tracking consumers online and should they stop? Inf Soc 32(1):51–63. https://doi.org/10.1080/019 72243.2015.1107166
28. National Crime Agency (2018) National strategic assessment of serious and organised crime 2018. https://www.nationalcrimeagency.gov.uk/who-we-are/publications/173-national-strategic-assessment-of-serious-and-organised-crime-2018/file
29. University of North Dakota (2020) 7 types of cyber security threats. University of North Dakota Online. https://www.onlinedegrees.und.edu/blog/types-of-cyber-security-threats/
30. Europol (2020) Internet organised crime threat assessment. https://www.europol.europa.eu/cms/sites/default/files/documents/Internet_organised_crime_threat_assessment_iocta_2020.pdf
31. Federal bureau of investigation internet crime complaint center (2016) Internet crime report. https://www.pdf.ic3.gov/2016_IC3Report.pdf
32. Maras M (2016) Cybercriminology, 1st edn. Oxford University Press
33. McGuire M, Dowling S (2013) Cyber crime: a review of the evidence. Home Office. https://www.assets.publishing.service.gov.uk/government/uploads/system/uploads/attachment_data/file/246754/horr75-chap3.pdf
34. Tcherni M, Davies A, Lopes G, Lizotte A (2015) The dark figure of online property crime: is cyberspace hiding a crime wave? Justice Q 33(5):890–911. https://doi.org/10.1080/07418825.2014.994658
35. Kendall M, Funk T (2014) The role of mutual legal assistance treaties in obtaining foreign evidence. Litig J 40(2):1–3
36. Woods A (2017) Mutual legal assistance in the digital age. Cambridge University Press, In The Cambridge Handbook of Surveillance Law
37. U.S. Department of Justice (2020) Legal considerations when gathering online cyber threat intelligence and purchasing data from illicit sources. Computer crime & intellectual property section criminal division. https://www.justice.gov/criminal-ccips/page/file/1252341/download
38. Chertoff M, Jardine E (2021) Policing the dark web: legal challenges in the 2015 playpen case (CIGI Papers No 259). Centre for International Governance Innovation
39. Hadjimatheou K (2017) Policing the dark web: ethical and legal issues. University of Warwick and TNO
40. Ünver HA (2018) Politics of digital surveillance, national security and privacy. Centre for Economics and Foreign Policy Studies. http://www.jstor.org/stable/resrep17009
41. Brewin C, Miller J, Soffia M, Peart A, Burchell B (2020) Posttraumatic stress disorder and complex posttraumatic stress disorder in UK police officers. Psychol Med 1–9. https://doi.org/10.1017/S0033291720003025
42. Ünver HA (2018) Digital open source intelligence and international security: a primer. Centre for Economics and Foreign Policy Studies. http://www.jstor.org/stable/resrep21048
43. Böhm I, Lolagar S (2021) Open source intelligence. Int Cybersecur Law Rev 2(2):317–337. https://doi.org/10.1365/s43439-021-00042-7
44. Chernenko E, Demidov O, Lukyanov F (2018) Increasing international cooperation in cyber-security and adapting cyber norms. Council on Foreign Relations. https://www.cfr.org/report/increasing-international-cooperation-cybersecurity-and-adapting-cyber-norms
45. United Nations Office on Drugs and Crime (n.d.) Cybercrime repository. https://www.unodc.org/unodc/en/cybercrime/cybercrime-repository.html
46. Interpol (n.d.) What is Interpol? https://www.interpol.int/en/Who-we-are/What-is-INTERPOL
47. Sharp ML, Fear NT, Rona RJ, Wessely S, Greenberg N, Jones N, Goodwin L (2015) Stigma as a barrier to seeking health care among military personnel with mental health problems. Epidemiol Rev 37(1):144–162. https://doi.org/10.1093/epirev/mxu012

Are Small Medium Enterprises Cyber Aware?

Homan Forouzan, Amin Hosseinian-Far, and Dilshad Sarwar

Abstract Technology has become a pivotal point in our society, this dependency is becoming increasingly more critical on a daily basis. This ranges from people to businesses and on a larger scale government organisations who are now increasingly focusing on becoming more cyber resilient. This paper intends to provide an overview as to why a comprehensive knowledge management framework is necessity for SMEs on tackling cyber and cyber-enabled crimes. The paper explores new sources of data to reliably understand the importance as to why such a framework is required. This type of system can pave the way for SME's to devise their cyber strategy and to be able to respond efficiently to cyber-related incidents. One of the cyber weakness and vulnerabilities for the SME's are through their interactivity and or engagement with their suppliers and customers. Namely the interactions which take place via their respective internet sites, email communications, ports (using external devices, USB, CD drive, SD cards etc....) or the router (The use of their WIFI systems). The benefits of this framework model will be primarily to educate SME's in becoming more cyber resilient and provide them with the knowledge, awareness and techniques to identify weaknesses and vulnerabilities in their computer networks, devices and internet usage.

Keywords Cybercrime · Cyber · Crime · Conceptual framework · Design science research · Small and medium sized enterprises

H. Forouzan · A. Hosseinian-Far (✉) · D. Sarwar
Department of Business Systems and Operations, University of Northampton, Northampton NN1 5PH, UK
e-mail: Amin.Hosseinianfar@northampton.ac.uk

H. Forouzan
e-mail: Homan.Forouzan@northampton.ac.uk

D. Sarwar
e-mail: Dilshad.Sarwar@northampton.ac.uk

© The Author(s), under exclusive license to Springer Nature Switzerland AG 2023
H. Jahankhani (ed.), *Cybersecurity in the Age of Smart Societies*,
Advanced Sciences and Technologies for Security Applications,
https://doi.org/10.1007/978-3-031-20160-8_20

1 Introduction

The term SME's can be defined as any business being a small and or medium sized enterprises (SMEs) with fewer than 250 employees and a turnover of less than £50 million. Three different categories come under the SME umbrella; micro, small and medium size businesses [1]. Micro businesses are those employed with nine or less employees, according to the statistics from figure four that accounts for ninety-six percent of all the operating businesses within the UK. Small businesses employ just under forty-nine employees and medium businesses employ just under two hundred and forty nine employees, which makes up three percent of the total businesses operating in the UK. In total, there are only 8000 large businesses that employee over the 250 employees and this in comparison to all the businesses operating in the UK is only 0.1% [2].

There are millions of people within the UK employed by the SMEs, and as such the SMEs are one of the main driving factors of the UK economy in creating more employments then larger organizations. As the SMEs dependency of technology is rapidly advancing so are the opportunities of cyber criminals to carry out increasingly complex and sophisticated crimes. This could also have a ripple effect on larger businesses and organisations who are in business relationship with those SMEs. As cyber criminals could either attack the intended SMEs or even use those as proxies to launch an attack on the lager businesses or organizations.

According to Federation of Small Businesses [2] there are two main vulnerability factors why businesses increasingly face cyber threat challenges they are technological and organizational vulnerabilities.

- Technological vulnerabilities: the weaknesses of networks, devices, hardware, software and programmes, which may be exploited for breaches and attacks.
- Organisational vulnerabilities: the weaknesses within the staff (who either knowingly or unknowingly divulge information), the processes and procedures within that SME. Limited staff knowledge and understanding of the technological vulnerabilities.

For SMEs, such vulnerabilities could be the supplier's data, customer's information, and financial details held on their devices. One could argue that the Internet is one of the most important assets for SMEs as through the Internet they will build upon their reputation, advertise and sale their products/services. If the integrity of such assets are breached that would mean significant financial and reputational consequences to the SME's. If SME's are not cyber resilient then such attacks and breaches could push the company towards liquidation.

In conjunction with these vulnerabilities, most SME's also function with limited resource and knowledge constrains, for example lesser assets base, due to their seize the limitations of the bargaining power, limited access to finance capital, and limited internal resource capacity [3]. Consequently, most SME's will have limited access to resources and knowledge to be able to grow their cyber resilience. As a result of these limitations alongside the organisational and technological vulnerabilities, generally

SME's are not well positioned to be able to tackle their own internal cyber-attacks and exposures [2].

There are currently a number of different products, guidance and approaches available on the market for SME's to be able to increase their cyber resilience, however due to the limited resources SME's have the capacity and their lack of awareness allows for a wider intelligence gap around cyber resilience. One of the key arguments is that larger organisations are more robust when encountering cyber-attacks as they are more adequately equipped and prepared with more financial and human resources; on the other hand, SME's are more vulnerable in the approach to attacks due to their poorer resilience capability [4]. What is apparent is that the need for SME's who form ninety nine percent of all the UK businesses to have cyber resilience embedded as part of their core business functions. In order to raise the profile and the importance of embedding cyber security within SMEs it is imperative that larger organisations, government, and law enforcement agencies share the burden of resilience to support SME's for a successful economy.

This paper is organised as follows: Sect. 2 provides a succinct overview of the methods used for the development of a conceptual framework that can inform a pertinent knowledge management system for preparing small and medium sized enterprises additionally improving their cyber resilience. Section 3 outlines the developed conceptual models. A critical discussion of results is provided in Sect. 4. The paper is concluded in Sect. 5.

2 Methods

Methodology is at the heart of any research and plays a pivotal point in providing a structure for that research. Within this research the researcher intends to use Design Science Research (DSR) qualitative approach to capture the information through iterative interviews.

Design Science Research can be described as a road map that focuses on a set of procedures or guidelines for evaluation and repetition within research development. Design Science Research Methodology results in the development of a system and/or an artefact and has gained pace in the information technology and engineering subject areas. However, the methodology has been applied in other subject disciplines where there is no tangible artefact. For instance, [5] has applied the methodology to develop a framework for magistrate and crown court judges to assess and evaluate cybercrime cases.

Such an approach emphasises the performance and development of the implemented artefacts with the clear purpose in mind to improve the well-designed performance of that artefact. DSR is aimed at developing a new approach in tackling the said problem through understanding the behavioral aspects attributed to that research by improving or considering new approaches, techniques and methods. Van Aken [6] describes DSR as an instrument in developing knowledge to assist experts of that particular field to design and implement a solution to tackle the said problem. Such an approach works on three main factors which are to describe, explain and forecast the problem with an element of providing a solution for that problem. According to Hevner [7] the purpose of DSR is accomplishing knowledge and obtaining a bigger understanding of the said problem by developing a designed artefact.

DSR is applied to a number of artefacts namely human/computer interfaces, algorithms, languages and design methodology. There are numerous variants to the Design Science Methodology (DSRM) for example computer science [8] and engineering [9].

According to Peffers [10] research in such fields are seen as a mission in understanding the concept of the current problem and ways to improve and develop human performance. Design science research has had its footprints in many fields and disciplines, more notably in information technology and engineering. What is apparent is that DSR is not confined to one set of guidelines or rules; there are a number of different methods, approaches and techniques used by Design Science Research [10, 11].

2.1 Interviewees Profiles

One of the main aspects of this research is heavily reliant on a range of different interviews being completed, in order to gain a better understanding of the issues and to be able to work towards implementing a knowledge management framework to support SMEs. Interviews will provide a platform to the researcher to support in building the framework of this project. In order to get a detailed understanding of the problem the researcher intends to conduct ten interviews from different sources namely:

- SMEs and start-ups—their understanding of cybercrime and the adequate provision provided to them by support agencies, insurance companies and the law enforcement agencies.
- The law enforcement agencies—to understand the police and government approach towards cybercrime and the current measures in place in supporting SMEs. In addition, to understanding the current crime trends, methods, and approaches that affect SMEs.
- Companies Supporting SMEs—It is essential for the researcher to be aware on products currently available in supporting SMEs, in order to gain an in depth understanding of what they provide and to avoid duplications.
- Academics—liaising with subject matter experts to understanding the areas of vulnerabilities and weaknesses that SMEs encounter.

The purpose of the interviews was to obtain comments and feedbacks from the interviewers around their experience and knowledge on the developed framework. Although there will be ten participants within this research, the number of interviews conducted were far higher than the number of participants. During this research thirty interviews were carried out this resulted to three interviews per participant.

The comments and feedback provided by the interviewee enables the research team to identify any shortcomings or missed variables within the developed framework. Such feedback proved to be a crucial contributor to the process of developing a fit for purpose knowledge management framework.

3 Results

After conducting thirty interviews the output from the comments and feedback provided on the proposed conceptual modelling framework model was embraced within Figs. 1, 2, 3, 4, 5 and 6. The modelling tool used to develop this framework is called a Systemigram. Systemigram was used to create a structured approach as such visualisation would be more effective in demonstrating what the framework is actually trying to illustrate.

Figure 1 illustrates the conceptual model framework cycle that forms the foundation of the knowledge management framework to support SMEs in protecting their business from cyber-attack. The conceptual model consist of five stages, the preparation stage, implementation stage, operational stage, incident reaction stage and post incident recovery stage.

The preparation stage is the most important stage of the conceptual framework as this will alert SMEs to identify and explore their businesses weaknesses and vulnerabilities. As highlighted within Fig. 2, SMEs shall navigate through each of the strands (staff, ecommerce infrastructure planning, compliances, device system

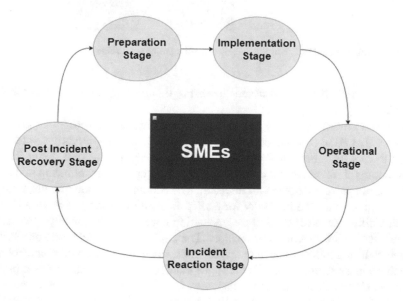

Fig. 1 The conceptual framework cycle iteration three

Fig. 2 Illustrates the first stage of the framework—the preparation stage

security and risk assessment strand) to ensure that each processes and actions are considered.

The staff strand highlights periodical staff learning and development processes to ensure that SME staff are adequately trained and more importantly aware of the recent threats, risk, harm and social engineering methods that staff could fall foul of as victims. If these are not addressed there could be significant business damage both in terms of reputational and financial. The ecommerce infrastructure planning strand demonstrates the necessity for SMEs to ensure that they have a transaction and due diligence process in place preventing spear phishing or any other forms of breaches on their invoices or receipts. The compliance strand alerts the SME to contemplate and follow some of the regulations, standards and requirements relevant to their business. Complying with the appropriate social and ethical factors to prevent the latter, significantly impacts the business. Consideration needs to be given to the legality of the business and the work that is intended to be carried out to prevent any criminal proceedings attributed to the business.

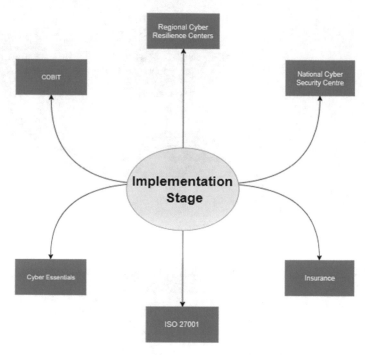

Fig. 3 Illustrates the second stage of the framework—the implementation stage

Fig. 4 Illustrates the third stage of the framework—the operational stage

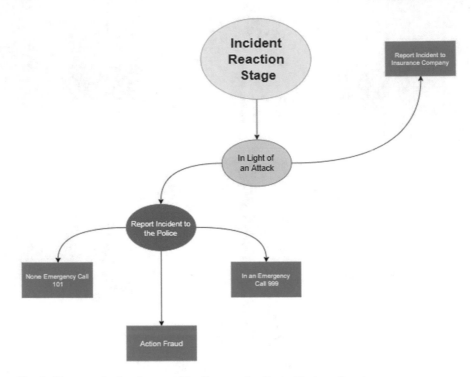

Fig. 5 Illustrates the fourth stage of the framework—the incident reaction stage

The device system security strand highlights the importance of periodical system security measures to be considered for the business. The following considerations are provided;

- Having firewalls and Anti-virus installed on networks and devices to monitor and control incoming and outgoing network traffics. Such barriers will work in preventing malware, breaches or other forms of attack penetrating the network and devices within the business.
- Having mandatory passwords for all devices and networks within the business, creating an administration account to restrict employee's access to certain important accounts, folders and files.
- The use of a two-factor authentication (2FA) system in place for those more important accounts, folders and files.

The risk assessment strand highlights the importance to businesses in having secure systems in place to protect their ICT (information communication technologies). Therefore the need to identify and address their ICT vulnerabilities and weaknesses. These could consist of communications from suppliers, customers and staff interactions with the ICT [2]. Fundamentally it is imperative that these potential risks are identified before a business is started as such risks if not mitigated could cause financial, reputational damage and can lead the businesses becoming bankrupt.

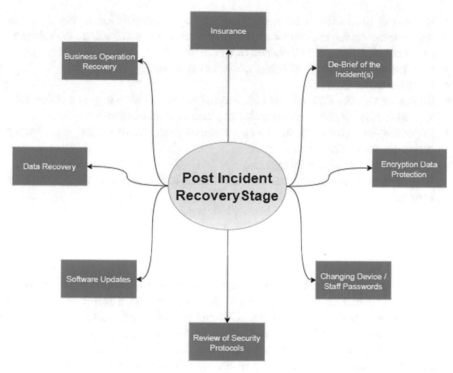

Fig. 6 Illustrates the last stage of the framework—the post incident recovery stage

The implementation stage as illustrated in Fig. 3 allows business to consider the use of available business support services (such as NSCS, RSRC, Cyber Essentials, COBIT, ISO 27001, and insurances companies) to be up-to-date with topical cyber trends, threats, methods and techniques which will assist SMEs in raising their cyber security awareness.

The operational stage as illustrated in Fig. 4 highlights the necessity for businesses to have periodical software and Hardware device updates and frequent data backups. Data security is extremely critical to any business or organisation. As such, leakages or breaches of information could have significant impact on the business financially, the functionality and reputation.

The incident recovery stage as illustrated in Figure five highlights the consideration that needs to be given In light of a breach, hacking and or a cyber-attack. To make contact with Action Fraud, insurance company, in an emergency to call 999 and none emergency to call 101.

The Post Incident Recovery Stage as illustrated in Fig. 6 highlights what SMEs should do after they have been victims of a cyber-attack. Considerations needs to be given to the following:

- Providing encrypted data protection for devices and staff to protect loss of data.

- Reviewing the business recovery plan, having contingencies in place to run the business on the backup when the Information communication technologies, systems and devices are all cleared of any malware.
- Recovering data through utilising accredited agencies,
- Software updates,
- Review of security Protocols and Identifying where the weakness and vulnerabilities where and putting contingencies in place to address them,
- Reporting the attack to insurance company for compensation and further assistance for business recovery,
- Changing Device / Staff Passwords,
- It is extremely important that debrief is held to identify knowledge gap and learning.

4 Discussion

Over previous decades, the UK Economy has moved towards a very complex communications infrastructure within the digital environment. As a result of this, there are tremendous opportunities now available for small to medium sized enterprises. This digital economy information age started quite innocently with the introduction of the personal computer, mobile telephone, the Internet and email. Technology has advanced considerably resulting in the introduction of tablets, smart phones and social media. Although the emergence of such technological advancements has helped to reduce cost considerably for businesses, this has helped to increase the overall efficiency and provide global reach for small to medium enterprises [12, 13]. The Internet thus far has offered phenomenal opportunities for SMEs, with these opportunities there are also additional risks.

With the adoption of the Internet and development of various digital business models, small to medium enterprises must have a web presence to remain competitive. Most people are now using the Internet to order and purchase their desirable products and services therefore business are more internet dependent than ever [14]. Although, many benefits are made available to these types of businesses through their cyber presence, there have also been numerous consequential challenges. Businesses are now facing complex dilemmas such as the ensuring privacy and security of consumer data, as well as enhancing the customers' trust. The digital and online business models have been evolving since the first Business to Consumer (B2C) instance in 1997, and so as adversaries' techniques targeting such online platforms [15, 16]. The rise in the adoption of cyber platforms has also seen a larger increase in related cyber, and cyber-enabled crimes. This research paper has focused on how to support SMEs from falling fowl of cyber criminals and more importantly highlights how organisations can be resilient to cyber attacks.

The findings of the results identified within this research highlighted that despite the increases in cyber-attack year on year on SMEs there is a lack of investment on cyber security within the business infrastructures. This was supported with one of the interview participants who stated that cybersecurity was not at the forefront of their business strategy, as their main focus was on making revenue. Such false beliefs highlighted not just through the literature review but also through interviews conducted with a number of SMEs during this research that their business would not be targeted as they were of an opinion that they had nothing of value for cyber criminals was rather concerning that such lack of knowledge and awareness was still apparent in the current ever dependant digital society [17]. There may be a number of factors as to why such misconceptions are still apparent within SMEs;

- Lack of awareness campaigns from government sectors, organisations, their party providers,
- Lack of a one stop shop platforms so that SMEs could be sign posted in identifying cyber related advice when setting up or running a business,
- Lack of mandatory enforcements by governing bodies on making cyber compliance compulsory for SMEs.

There are at times misconceptions by SME's that staff are adequately trained and educated relating to basic cyber-security awareness. Clearly as highlighted within this paper such assumptions could be costly to a business or an organisation. One of the interview participant's touched upon the fact that that some of the elementary process were not being carried out for example staff were not aware that passwords needed to be changed periodically. Such findings were alarming as assumptions of this nature would be detrimental to SMEs whom without the adequate cyber security measures could be left with no option but to terminate their business in light of an attack as there could be exposures of significant financial and reputational damage to the business.

The results from the interviews further highlighted that there are increasing needs for a knowledge management framework for SMEs to consider implementing as the purpose of this framework would be essential in raising understanding, awareness and knowledge of the significance of cyber security within SMEs. The benefits of such a framework would be invaluable in forewarning SME's in considering risk factors that need to be well-thought-out before a business is started, whilst the business is functioning and in light of an attack being resilient in keeping the business operational. This does not mean that the framework is the 'be-all and end-all' but more of a tool to support SMEs in preventing cyber-attacks or in light of an attack to be resilient.

The proposed conceptual model intends to reduce technological vulnerability, organisational vulnerability, improve policing of cybercrime and cyber criminals. Where there is a reduced level of pliability across digital networks, cost imposed on businesses can be phenomenal. These costs can be detrimental to SMEs both reputationally and financially. The costs of these vulnerabilities are not limited to the attacks directly influencing the SME but rather the day to day functioning of the business. For instance, the SME's ability to meet customer needs, meet order requests, obtain new customers and the digital reputation of the SME. As SME's main concern is to focus on developing and sustaining their businesses, SMEs tend not to have the skills base or the resources to adequately match cyber threats and vulnerabilities as highlighted within this paper. Therefore, it is imperative that cyber resilience is adequately and appropriately directed to support SMEs.

Further benefits of such a framework were highlighted by another interview participant, that the conceptual model mitigates start-ups and SMEs falling fowl of becoming victims of a cyber-attack as through early involvement in raising awareness. Furthermore, being mindful of the information contained within the proposed conceptual model can help mitigate risks and identify digital vulnerabilities.

5 Conclusion

The findings of the results identified within this paper highlighted that despite the increases in cyber-attack year on year on SMEs there is a lack of investment on cyber security within the business infrastructures. This was supported by one of the interviewee participants who stated that cybersecurity was not at the forefront of their business strategy, as their main focus was on making revenue. Such false beliefs highlighted not just through the literature review but also through the interviews conducted that some SMEs believed that their businesses would not be targeted as they had nothing of value for cyber criminals, was rather concerning that such lack of knowledge and awareness was still apparent in the current ever depended digital society. There could be a number of factors as to why such misconceptions are still apparent within SMEs;

- Lack of awareness campaigns from government sectors, organisations, third party providers,
- Lack of a one stop shop platforms so that SMEs could be sign posted in identifying cyber related advice when setting up or running a business,
- Lack of mandatory enforcements by governing bodies on making cyber compliance compulsory for SMEs.

The conceptual framework has many benefits one in particular is that it mitigates start-ups and SMEs from falling fowl of becoming victims of a cyber-attack as through early involvement in raising awareness, being mindful of the information contained within the proposed conceptual model can help mitigate risks and identify digital vulnerabilities.

References

1. Sainidis E, Robson A (2016) Environmental turbulence: impact on UK SMEs' manufacturing priorities. Manage Res Rev 39(10):1239–1264
2. FSB (2018) Cyber resilience: how to protect small firms in the digital economy, FSB, UK. https://www.fsb.org.uk/docs/default-source/fsb-org-uk/FSB-Cyber-Resilience-rep ort-2016.pdf?sfvrsn=0. Accessed 13 July 2019

3. Hayes J, Bodhani A (2013) Cyber security: small firms under fire [Information Technology Professionalism]. Eng Technol 8(6):80–83
4. Valli C, Martinus IC, Johnstone MN (2014) Small to medium enterprise cyber security awareness: an initial survey of Western Australian business
5. Montasari R (2021) The comprehensive digital forensic investigation process model (CDFIPM) for digital forensic practice. University of Derby (United Kingdom)
6. Van Aken JE (2005) Management research as a design science: articulating the research products of mode 2 knowledge production in management. Br J Manage 16(1):19–36
7. Hevner AR, March ST, Park J, Ram S (2004) Design science in information systems research. MIS Q 28:75–106. http://www.citeseerx.ist.psu.edu/viewdoc/download?doi=10.1.1.103.1725&rep=rep1&type=pdf
8. Takeda H, Veerkamp P, Yoshikawa H (1990) Modelling design process. AI Mag 11(4):37–37
9. Fulcher AJ, Hills P (1996) Towards a strategic framework for design research. J Eng Des 7(2):183–193
10. Peffers K, Tuunanen T, Niehaves B (2018) Design science research genres: introduction to the special issue on exemplars and criteria for applicable design science research. Eur J Inf Syst 27(2):129–139
11. Peffers K, Tuunanen T, Rothenberger MA, Chatterjee S (2007) A design science research methodology for information systems research. J Manage Inf Syst 24(3):45–77
12. Attaran M, Woods J (2019) Cloud computing technology: improving small business performance using the Internet. J Small Bus Entrep 31(6):495–519
13. Porter ME, Kramer MR (2019) Creating shared value. In: Managing sustainable business. Springer, Dordrecht, pp 323–346
14. Okamoto T, Yatsuhashi J, Mizutani N (2017) Priorities of smartphone online shopping applications for young people
15. Timmers P (1998) Business models for electronic markets. Electron Mark 8(2):3–8
16. Webb H, Webb L (2001) Business to consumer electronic commerce Website quality: integrating information and service dimensions. AMCIS 2001 proceedings, p 111
17. Forouzan H, Jahankhani H, McCarthy J (2018) An examination into the level of training, education and awareness among frontline police officers in tackling cybercrime within the metropolitan police service. In: Cyber criminology. Springer, Cham, pp 307–323

'HOAXIMETER'—An Effective Framework for Fake News Detection on the World Wide Web

Ishrat Zaheer Chowdhary and Umair B. Chaudhry ⓘ

Abstract Fake news and misinformation have become a serious predicament, especially, in recent times as it has become easier to spread and harder to recognize. Not only has it created an environment of distrust around the world and misled people, but also incited violence and resulted in people losing their lives. State of the art fake news detection includes the use of verifying news from a trustable dataset, BERT filtering or using cues from Lexical Structure, Simplicity and Emotion and more. In addition, several frameworks have also been proposed to deal with this issue e.g., SpotFake and FR-detect to name a few. However, these frameworks strongly rely on the users to verify the news, modification of information, categorization of the news, and reliability of the dataset (if a dataset is used). This paper proposes 'HOAXIMETER', a framework to detect fake news on the World Wide Web covering the weakness of the aforementioned existing ones. It does so by putting forward an in-depth study and literature review on the existing frameworks by analyzing and evaluating them in the context of how well they work with each step, thereby highlighting their strengths and weaknesses. Ultimately, 'HOAXIMETER' is proposed which is meant to be the most effective fake news detection framework free from the issues of the existing one.

Keywords Misinformation · Fake news · BERT filtering · Long short-term memory (LSTM) · Recurrent neural network (RNN)

I. Z. Chowdhary · U. B. Chaudhry (✉)
Queen Mary University of London, EECS, 10 Godward Square, Mile End Rd, London E1 4FZ, UK
e-mail: u.b.chaudhry@qmul.ac.uk

U. B. Chaudhry
Northumbria University, 110 Middlesex Street, London E1 7HT, UK

399

1 Introduction

News should be newsworthy but more than that it needs to be true. The purpose of the media is to give information on various topics of interest, however, what started as a kind initiative to inform and empower people has gone astray and lost most of its purpose. Media in today's day and age is spreading more misinformation than information. While some newsmakers want to report true and reliable news, sometimes the desire to be the first to report a piece of news makes people forget to verify the source and the credibility of a report. Another and more common reason for the spread of misinformation is income generated from sales of newspapers and Television Rating Points of news channels and clicks on social media which generate income from AD revenue.

Now, journalism is all about titillating the reader and making them tune in or click. However, this has resulted in lots of misinformation and damage. This misinformation is also known as Fake news. Fake news at its core is conjecture, speculation or rumours mainly started to either attain monetary benefits or spreading of propaganda. This allure for money has not only has it created a hostile environment of mistrust and misinformed people but also incited hate, violence and death of many.

Technology in recent times has made great advances it has specially made communication easy but as we see in the iconic book 'Frankenstein' by Mary Shelley, if dealt with carelessly with great advancement come their own set of problems, just like with easy and fast communication spreading fake news and rumours have also become easy and fast and before we know it the damage is done. Fake news is not a recent problem and has been around for quite some time but has become a bigger predicament in recent times, this is because Fake news is now spread the most via the internet. The Reason for this is that spreading fake news on the internet/online is easier to post, faster to spread, and cheaper compared to other ways of news communication. Conformity also plays a huge role in the fast-spreading of fake news as due to social media echo chamber fake news is considered legitimate.

There have been numerous efforts to tackle this problem most common and reliable one is Expert-based Fact-checkers, Expert-based Fact-checking Websites and Crowd-sourced Fact-checking Websites however these rely on manpower which is costly and time-consuming. Since these are problems made by advancements in technology using technology to counter these problems is a good idea. Famous tech companies like Facebook and Google are trying to find a solution. There has been a little research on automating fake news detection some methods proposed are machine learning algorithms, deep learning techniques, Natural Language processing and finally there are a few fake news detection frameworks however, none of them have been very successful as fake news is still a very prominent problem. That is where this report comes into the picture.

This paper aims towards helping in eradicating fake new by proposing an automated, more effective, end-to-end fake news detection framework called 'HOAXIMETER' by evaluating the strengths and weaknesses of the preexisting fake news detection frameworks and incorporating deep learning and a knowledge based approach. The paper is arranged in 4 sections. Section 1 presents the introduction and the background which is followed by Sect. 2 having a detailed literature review. Section 3 presents the methodology and the novelty of work. Section 4 concludes the paper and throws light on potential future work.

2 Literature Review

In this section, a brief review of previously proposed fake new detection framework is provided. Murphy et al. [1] claims to be the very first to fill the void on the topic of fake news detection by suggesting a very simple framework. It doesn't rely on calculators or complex statistics rather it used three simple cues those are Lexical Structure, Simplicity and Emotion (LeSiE) as it states that these cues are the major difference between fake and real news. It used PolitiFact's dataset and uses turkeys test for evaluation.

Bedi et al. [2] categorizes news into two domains fake news on traditional media and fake news on social media. Bedi et al. [2] proposes a framework that connects authorized news websites to the social media feedbox of verifying whether a new is genuine or not and sends the link to the existing genuine news. Huxiao Liu et al. [3] segregates its features into two categories, 'author-based' and 'content-based' features. Author-based features are further categorized into certification status, average likes and number of follows and content-based features are further categorized into audience activity, shocked style and abstract statistical features. Huxiao Liu et al. [3] proposes the Fake News Detector based on Multisource Scoring (FNDMS) framework which goes as follows. First, the news event is taken and a keyword search is performed. Second Jaccard distance is measured. If $P < 0.5$ is removed (as unrelated) but if $P > 0.5$ (related) this is followed by Bert filtering of the related news which is followed by extracting features data leading to single-source scoring and DST which will determine whether the news is fake or real.

Singhal et al. [4] propose Spot fake a framework which also uses BERT to learn text features and VGG pre-trained on 'imagenet' for image features on a Twitter and Weibo dataset and compares its results to EANN and MVAE and come to the conclusion that their method works better than both of them. Ali Jarrahi et al. [5] propose FR-Detect: A Multi-Modal Framework for Early Fake News Detection on social media Using Publishers Features. It proposes two new and important features, 'Activity Credibility' and 'Influence' to investigate the publisher's role in spreading fake news. Parikh et al. [6] aims to detect fake (photo shopped and tampered) tweet images on social media by taking screen captures of tweets, extracts metadata from the tweet, populates and validates the metadata, queries twitter API, performs similarity to find a match, compares and validates using a timestamp, username, display name and content and computes a decision. Abdelminaam et al. [7] proposed CoAID deep, a framework to detect Covid-19 misinformation on Twitter using six machine learning algorithms and deep learning techniques on benchmark datasets. They preprocess the data through various methods like stemming, tokenization, stop word, lowercasing etc. After which they apply machine learning and deep learning approach to the data and get the result of accuracy, F measure, recall and precision.

Naeemul Hassan et al. [8] made 'Claim Busters', a framework that proposed NPL and supervised learning to spot claims and used and knowledge base approach to check them. Beer et al. [9] gives a literature review on approaches to identify fake news. It sheds light on different types of approaches that can be used to detect fake news. According to this report, there are five types of approaches for fake news detection i.e., language approach, machine learning approach, knowledge-based approach, hybrid approach and topic agnostic approach (from most to least used). Nicholas Diakopoulos's book 'Automating the News How Algorithms Are Rewriting the Media' [10] briefly talks about the topic of news verification in chapter two of the book. It also gives a table of journalistic uses of data mining with supporting capabilities and specific examples where it mentions the use of regression to predict the likelihood of a claim being true and the use of a knowledge base comparison.

From the literature review, it is quite evident that with every work proposed till date, there are significant limitations and tradeoffs, and the world can still benefit from a coherent, consistent, comprehensive, effective and efficient fake news detection framework. We have named it 'Hoaximeter' and have proposed it claiming that it will outperform the existing approaches ad frameworks.

3   Methodology

3.1   Analysis of Previously Proposed Frameworks

Language and Machine Learning approach is the most used approach [9] and this can be backed up by the literature review of this report as well as it can be seen that most of the frameworks proposed in the literature review propose. However, there are a few problems with this approach most important being the availability, reliability and accuracy of dataset. Data is the most primary source of information that the machine relies on for the machine learning process to make sense hence There are few standards a good dataset should maintain like good data quality and quantity (big data doesn't necessarily mean smart data however there should be enough data for the machine learning to be accurate) no missing data, no duplicates, no typos/errors etc. these ensure data sanity. Data sanity is of utmost importance for the accuracy of the machine learning approach however, finding data set with data sanity is not the easiest work due to many reasons like privacy, people unwilling to share, people/organizations putting out false data on purpose or simply unavailable data due to which data usually have to be altered and so that machine learning can take place. There are also problems with machine learning algorithms most common being understanding which process need automation a common example of what this means is machine learning might think that the music is dancing to people while common sense would tell us otherwise hence while machine learning is the future of technology there are a lot of problems that need to be dealt with for it to be perfect.

Bedi et al. [2] uses a knowledge based approach which is considered to be faster before there is a widespread of the false information and in that sense is better to the machine learning approach however, the knowledge bases should be accurate and reliable due to which this framework would not be considered very reliable with its approach since its definition of authorized news websites is not very accurate e.g., categorizing http://www.9gags.com in business and the cultural site would not be seen as very accurate. Huxiao Liu et al. [3] uses Bert filtering in the fake news detection framework for text summarization and is a good method however jacquard distance is not the best set similarity measure by there are many sets similarity measures Jaccard distance character and word based both seem to lack in accuracy [11]. Character-based Jaccard Index identifies every letter as a different element of the set hence it will identify 'abcdefghijklmnopqrstuvwxyz' and 'the quick brown fox jumps over the lazy dog' as the same thing. Word-based Jaccard Index gives a similar problem it can't identify paraphrasing or rephrasing for a thing that has the same meaning. However, there are many other alternative set similarity measures that can be used e.g., Tversky Index, smith waterman, Levenshtein, Bedfellows (which uses set of a six-association metrics) [12] etc. According to [13] cosine similarity gives a better and more accurate set similarity. Claim buster, a framework [8] which focuses more on claim spotting and then fact-checking. Even though this worked pretty well it didn't always reach an answer or even an approximate. This is because the knowledge base connected may not have an answer. Abdelminaam et al. [7] use a machine learning approach however it also uses a deep learning approach of LTSM and GRU which is a Lesser used and newer approach to tackle fake new this deep learning approach shows promising results. Singhal et al. [4] includes detection of not only fake text but images as well which is not seen much. However, it works only on Twitter and on 280-word limit of Twitter hence doesn't work well with bigger pieces of data. Parikh et al. [6] recognizes its limitations as only working on images of tweets and changes in the display name and deleted Twitter account will result in inaccurate results.

3.2 HOAXIMETER

3.2.1 System Overview

A new framework is essential to achieve the goal of eradicating fake news hence, this report proposes 'HOAXIMETER' new and improved framework for fake news detection. Hassan et al. [8] used a knowledge approach however while testing the framework it was noticed that there was no result at all on many queries and the framework couldn't answer all question or even give an approximation of a new being true or false in 'HOAXIMETER' framework and 'HOAXIMETER' mainly tackles that. 'HOAXIMETER' takes a hybrid approach of knowledge base and deep learning

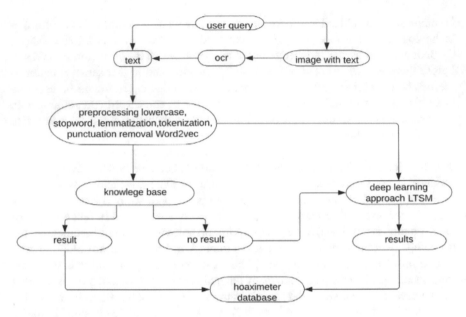

Fig. 1 The proposed Hoaximeter framework

to detect fake news and hence if the result of the Knowledge base is inconclusive the framework uses a Deep learning approach known as LSTM to give an approximation of the accuracy of the text (Fig. 1).

The functional framework can be described as below:

HOAXIMETER FRAMEWORK

User query of either text or image with text.

if Query = Image then

    USE Optical Character Recognition (OCR)
    EXTRACT Text from Image

else

    PRE-PROCESS Data
    APPLY
        Knowledge − based Approach
        DeepLearning − based Approach
    SAVE Results to database

3.2.2 The New and Improved Framework

The 'HOAXIMETER' framework first takes in user input of what a user may doubt is fake news. Most of the aforementioned frameworks work only on text or only on images but with the 'HOAXIMETER' framework, there is the option of inquiring about a text or textual data in an image. The user has the option of inquiring either about some text or an image with text of whose authenticity they doubt. If the user chooses to enquire about an image with text the text from the image is extracted through Optical Character Recognition (OCR) technology also known as text recognition technology. OCR technology is very effective and accurate in extracting text in images it does so by scanning through the image for dark and light patterns that make up letters and numbers. This technology was famously used to extract information in the Panama paper leaks. After the text is extracted/input it is preprocessed in numerous ways.

Text preprocessing is important as it makes the data cleaner and hence the approaches give better and faster results. Text preprocessing is done by Lowercasing which involves converting all the text to lowercase e.g., "FAKE" to "fake", punctuation removal involves removing all commas full stops exclamation marks question marks etc., stop word removal which involves removing of word that don't add much substance to sentence like articles, conjunctions, clothing's etc. example removing of words like "the" "a" "of" etc.

Lemmatization is considered an advanced approach to Stemming. Stemming is a faster method and involves narrowing down a word to its root/base. Example cares to care, cats to cat etc. However, Lemmatization actually simplifies word to a simpler meaning like saw to see, mouse to mice lemmatization is a slower process and harder to create for new languages but in this framework, we use it as it may increase the accuracy of the learning [14]. Tokenization involves breaking up text into words referred to as tokens.

'HOAXIMETER' uses Word2Vec instead of Bert filtering for Word embedding. BERT filtering has quite a few advantages over Word2Vec like context dependency that is it can understand where two same words are used in different context it supports out of vocabulary words that being said since Lemmatization is already used BERT filtering won't make much of a difference adding to which Word2Vec pre-trained word embedding's are available to use directly off-the-shelf [15].

After the data/text is preprocessed, the framework directly compares the query to the external knowledge base. Knowledge base is a set of facts. Knowledge base approach uses external sources for fact verification. Knowledge base fact checking can be manual and automated, 'HOAXIMETER' framework uses automated knowledge bases for fact checking the reason for this is manual fact checking wouldn't work very well with high volume of data it wouldn't be very fast either hence we use automated fact checking as it is faster and can deal with more data at a time hence is more efficient than the manual approach. Given a claim, it is compared to the knowledge base. The 'HOAXIMETER' use google API [16] and wolf gram API [17]. The query is compared to the knowledge bases to find evidence to support or debunk the query. Deep learning approach 'Long short-term memory' network also known as LSTM. They are in recent times used to solve various problems example speech recognition, handwriting recognition etc. They can be seen as an advanced RNN. These would be considered advanced as, unlike RNN, LSTM can learn long term patterns they are known for having good long-term memory. The process of LSTM has three gates and the process has a total of four steps as shown below [18, 19].

Step 1: Forget gate layer

$$ft = \sigma(Wf \cdot [ht - 1, xt] + bf)$$

In this step, the sigmoid decides what information to discard from the previous hidden state and current input. the sigmoid output is either 1 to keep the information or 0 to discard/forget information. Hence the name forget gate layer makes sense.

Step 2: Input gate layer

$$it = \sigma(Wi \cdot [ht - 1, xt] + bi)$$

In this step, the information of the previous hidden state and current input is passed through a sigmoid function known as input gate layer and a tanh function. The sigmoid function outputs a number between 0 and 1.

$$C \sim t = \tanh(WC \cdot [ht - 1, xt] + bC)$$

The tanh function produces an output between −1 and 1. the outputs of both of these functions are multiplied and the sigmoid then decides what information to keep.

Step 3: Update the old state

$$Ct = ft * Ct - 1 + it * C \sim t(5)$$

This step is just the implementation of calculation done in the previous steps. The information to be forgotten is forgotten and information from the second step is added. A new cell state is attained by is pointwise multiplied by the forget vector and pointwise adding the output of the input gate.

Step 4: Output the cell state

$$ot = \sigma(Wo[ht - 1, xt] + bo)$$

Finally in this step, the output gate decides the next hidden state this is important for further steps in the loop. A sigmoid and a tanh function is carried and the output of both is multiplied to decide the information to be carried in the next hidden state (Fig. 2).

The LSTM is trained on the dataset [20] and based on that makes estimates regarding the accuracy of the query. The framework applies both approaches to the query and presents the result to the user. It is likely the results come faster from the

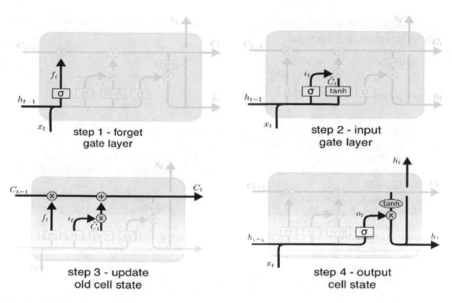

Fig. 2 LSTM cells and its operations [19]

knowledge base compared to the deep learning approach so even if the knowledge base has the result just in case there are no results the deep learning approach works on the back end. The results of both the approaches are saved in the 'HOAXIMETER' databases for future use and faster results of the same query.

4 Conclusion and Future Work

In conclusion, Hoximeter framework has quite a few advantages over the previously proposed frameworks like its deals with text and images with text, it doesn't only deal with one platform like Twitter and it always gives a result either from the knowledge base or an estimated result from the deep learning approach. However, that being said it has its flaws the deep learning approach gives an approximation if the query is true or false and may not always be accurate. The deep learning approach requires padding. Other than that, while the deep learning approach shows quite high accuracy it is suspected that its very specific to the particular dataset chosen [21] and thus the problem of not having proper data to train still exist. Example while LIAR dataset is a benchmark dataset and would have been good it's simply not big enough for proper training. There is still a dire need for a good dataset and the process is highly automated but still on the back end its manpower making these datasets and knowledge bases. Future work includes working on the flaws acknowledged as well trying to automate the fact checking process as much as possible. However right now this is still considered pretty difficult to achieve as the framework is limited by knowledge bases and datasets.

References

1. Choy M, Chong M (nd) Seeing through misinformation: a framework for identifying fake online news. https://www.arxiv.org/pdf/1804.03508.pdf
2. Bedi A, Pandey N, Khatri SK (2019) A framework to identify and secure the issues of fake news and rumours in social networking. IEEE Xplore. https://doi.org/10.1109/PEEIC47157.2019.8976800
3. Liu H, Wang L, Han X, Zhang W, He X (2020) Detecting fake news on social media: a multi-source scoring framework.. IEEE Xplore. https://doi.org/10.1109/ICCCBDA49378.2020.9095586. Accessed 28 Mar 2022
4. Singhal S, Shah RR, Chakraborty T, Kumaraguru P, Satoh S (2019) SpotFake: a multi-modal framework for fake news detection. IEEE Xplore. https://doi.org/10.1109/BigMM.2019.00-44. Accessed 28 Mar 2022
5. Jarrahi A, Safari L (nd) FR-detect: a multi-modal framework for early fake news detection on social media using publishers features. https://www.arxiv.org/pdf/2109.04835.pdf. Accessed 29 Mar 2022
6. Parikh SB, Khedia SR, Atrey PK (2019) A framework to detect fake tweet images on social media. In: 2019 IEEE fifth international conference on multimedia big data (BigMM). https://doi.org/10.1109/bigmm.2019.00-37. Accessed 29 Mar 2022
7. Abdelminaam DS, Ismail FH, Taha M, Taha A, Houssein EH, Nabil A (2021) CoAID-DEEP: an optimized intelligent framework for automated detecting COVID-19 misleading information on twitter. IEEE Access 9:27840–27867. https://doi.org/10.1109/ACCESS.2021.3058066. Accessed 29 Mar 2022
8. Hassan N, Arslan F, Li C, Tremayne M (2017) Toward automated fact-checking. In: Proceedings of the 23rd ACM SIGKDD international conference on knowledge discovery and data mining. https://doi.org/10.1145/3097983.3098131. Accessed 16 May 2022
9. de Beer D, Matthee M (2020) Approaches to identify fake news: a systematic literature review. Integr Sci Dig Age 136:13–22. https://doi.org/10.1007/978-3-030-49264-9_2. Accessed 16 May 2022
10. Diakopoulos N (2019) Automating the news: how algorithms are rewriting the media. Harvard University Press, Cambridge, Massachusetts
11. http://www.skillsire.com (nd) How do you perform unsupervised classification on fake news? https://www.skillsire.com/read-blog/347_how-do-you-perform-unsupervised-classification-on-fake-news.html. Accessed 16 May 2022
12. Iubel N (nd) Introducing bedfellows. https://www.source.opennews.org/articles/introducing-bedfellows/. Accessed 16 May 2022
13. Abhishek K, Sahoo S (2020) Role of text similarity in fake news classification regarding COVID-19. Research gate. https://www.researchgate.net/publication/348266889_Role_of_Text_Similarity_in_Fake_News_Classification_regarding_COVID-19. Accessed 16 May 2022
14. Beri A (2020) Stemming vs lemmatization. https://www.towardsdatascience.com/stemming-vs-lemmatization-2daddabcb221. Accessed 16 May 2022
15. Gupta L (2021) Differences between Word2Vec and BERT. The Startup. https://www.medium.com/swlh/differences-between-word2vec-and-bert-c08a3326b5d1#:~:text=Word2Vec%20will%20generate%20the%20same. Accessed 17 May 2022
16. toolbox.google.com (nd) Fact check tools. https://www.toolbox.google.com/factcheck/explorer. Accessed 17 May 2022
17. products.wolframalpha.com (nd) Wolfram|Alpha APIs: computational knowledge integration. https://www.products.wolframalpha.com/api. Accessed 17 Jun 2022

18. Olah C (2015) Understanding LSTM networks—colah's blog. http://colah.github.io/posts/2015-08-Understanding-LSTMs/. Accessed 17 May 2022
19. Phi M (2019) Illustrated guide to LSTM's and GRU's: a step by step explanation. https://www.towardsdatascience.com/illustrated-guide-to-lstms-and-gru-s-a-step-by-step-explanation-44e9eb85bf21. Accessed 17 May 2022
20. www.kaggle.com (nd) Fake and real news dataset. https://www.kaggle.com/datasets/clmentbisaillon/fake-and-real-news-dataset?select=True.csv. Accessed 17 Jun 2022
21. Abrahamson A (2020) Detecting fake news with deep learning. https://www.towardsdatascience.com/detecting-fake-news-with-deep-learning-7505874d6ac5. Accessed 16 Jun 2022

Cyber Resiliency in Electric Power Industry Based on the Maturity Model

Mohammad Ebrahimnezhad and Mehran Sepehri

Abstract Many risks always exist in the power industry, as the supply and demand are affected by many stochastic internal and external factors. The number and magnitude of technological, economic, social and environment changes in this industry are also accelerating. The power industry is faced with many uncertain dimensions, where other forms of energy come into play, such as oil and gas supplies. Cyber-attack is where the most drastic pressure is put on the industry. Local and remote attacks occur often, which require a comprehensive command system to deal with them. Cyber-attacks must be prevented and mitigated ahead of time by Resiliency strategies and be managed as they occur by crisis management procedures. Implications and damages are kept to a minimum for business continuity. In this paper, we review the background and existing literature of power industry to identify the main factors and relationships. We use Iran case (Ministry of energy) as a developing country with many sources of natural energies as well as an extensive network of electric power. A recent case of Staxnet is to illustrate the magnitude of issues and shortcomings in the current system of cyber-attack response. We use future study approach to consider several scenarios of plausible future for power industries. We utilize the standard C2M2-ES for systematic dealing with Cyber risks and uncertainties in electric power industries. Optimization and simulation models are then used to evaluate and compare various strategies to respond to various circumstances.

Keywords Cyber-attack · Resiliency · Power industries · Maturity model

M. Ebrahimnezhad
Ministry of Energy and Sharif University of Technology, 18365 Tehran, Iran

M. Sepehri (✉)
Northampton University, Northampton NN1 5PH, UK
e-mail: mehran.sepehri@northampton.ac.uk

© The Author(s), under exclusive license to Springer Nature Switzerland AG 2023
H. Jahankhani (ed.), *Cybersecurity in the Age of Smart Societies*,
Advanced Sciences and Technologies for Security Applications,
https://doi.org/10.1007/978-3-031-20160-8_22

1   Introduction

Today, due to the diversity of expectations from power distribution networks, there is a need to change the way they are managed, managed, and controlled. Intelligence helps to step in this direction in the right direction and by reducing risks as much as possible. The first step towards smartening is to develop a smart grid cyber security roadmap. The Smart Grid cyber security Maturity Model (SGMM), developed in collaboration with IBM, the APQC Institute and several companies providing electricity services from different countries, influences the path of roadmap development and the level of maturity of the organization in this direction with a comprehensive view of different areas [1].

In this regard, power distribution networks around the world are becoming an intelligent system. In the path of intelligence, many tools play a role as a driving force. At first glance, many see technology growth as the only means of moving toward intelligence, when it is not. Other tools that help distribute electricity networks are: Paying attention to strategic issues, management and laws, formulating business requirements to increase operational productivity, defining safety and environmental requirements, managing the organization's assets, monitoring and following changing behavioral patterns of energy consumption, creating and developing a competitive market and most importantly Cyber resilience is the main pillar [2].

Continuity and reliability of electricity supply in a country is vital to the operational continuity of the critical infrastructure and government. High levels of cyber security are required to achieve such key measures in the electricity industry. We propose solutions and models on how to improve the cyber security using the maturity model in order to achieve continuity of electricity supplies. Maturity levels start from the relative level (level 1) and continue to the adaptive level (level 4). Total of 10 domain are defined in the model. The practices of cyber security are clustered into domains. The model is used for assessment of the strength and weakness factors for the organizations from the model. The case to demonstrate the concepts is from Iran. To succeed in this path, which is inevitably considered a mandatory process, it is recommended to follow the cyber security roadmap of the smart grid [3].

2   Cyber Security Model of Smart Power Networks (SGMM)

There are various models for developing a roadmap for smart grids. One of the most authoritative and comprehensive of these models, which addresses all aspects of the issue from the perspective of the organization, subscribers and operating technologies, is the Smart Grid Maturity Model (SGMM). This model is a management tool to evaluate, develop and control the smart path of power generation, transmission and distribution industry, under the supervision of the Institute of Software Engineering (SEI) at Carnegie Mellon University. This model was developed by IBM

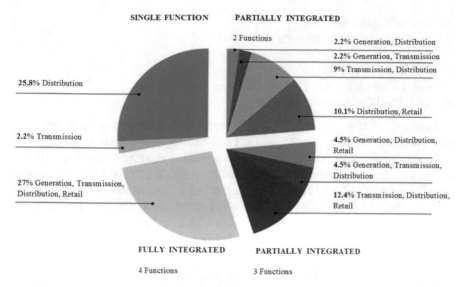

Fig. 1 Dispersion diagram of the use of SGMM model in the electricity industry in the world

Consulting in collaboration with APQC and an international group of power generation, transmission and distribution companies called GIUNC. Then its development and updating is entrusted to the Software Development Institute (SEI) [3].

In this model, special attention is paid to the requirements of cyber security intelligence at different levels of the electricity industry, especially the distribution and transmission sector, and a framework for understanding the current conditions of intelligent and smart grid development and capabilities needed to develop cyber security in the electricity industry is presented.

Figure 1 shows the distribution of the use of the SGMM model among different types of power companies [3].

3   Eight Aspects of Smart Grid Cyber Security Maturity Model

This model evaluates the current state of the network in terms of the degree of maturity of the organization Cyber Security in smartening the structure, processes and network using a questionnaire with a total of 175 questions in various aspects. The SGMM Model questions are addressed in the following eight aspects. These aspects, which cover all the main dimensions affected Cyber security by SmartGrid in an electrical energy service company, are:

1. Strategy, Management and Rules (SMR)
2. Organizational structure (OS)

3. Network operation (GO)
4. Work and Asset Management (WAM)
5. Subscribers (CUST)
6. Technology (TECH)
7. Value Chain Integrity (VCI)
8. Social and Environmental (SE)

Each aspect of this model pursues a specific goal in measuring the Cyber security maturity of the organization in the sustainability. The "strategy, management and rules" and "social and environmental" aspects of the goals address the organization's motivations. The aspects of "network operation", "labor and asset management", "technology" and "value chain integration" define the methods and programs of the organization in estimating goals. And determine the aspects of "organizational structure" and "subscribers" of stakeholders and the impact of cybersecurity implementation on it in the Smart Network Maturity (SGMM) model [3].

4 Levels of the SGMM Model

In this model, for each of these aspects, 5 levels are considered, which are:

1. **Introduction:** At this level it is important to have a vision of the smart grid and to understand what the interests of the organization and customers are. The organization takes the initial steps at this level.
2. **Empowerment:** At this level, the strategy for investing in the next steps is determined. Decisions are also made according to the leading path, at least at the operational level of cyber security.
3. **Integration:** At this level, the smart grid application begins to expand. Cyber security operational links are also established between two or more operational areas.
4. **Optimization:** The features and capabilities of the smart grid are widely used in the network, and signs of improvement from the smart grid have appeared. At this stage, cyber vulnerabilities are identified.
5. **Leadership:** The organization has entered new areas in its business and is among the top among similar partner organizations in the world. Also, the network is always ready to respond to cyber incidents, and has the ability to recover automatically.

5 The Process of Developing a Roadmap for Smart Cyber Security in Energy Distribution and Transmission Companies

To develop a cyber security roadmap for the smart power grid, as shown in the figure below, to start this process, the current situation is first assessed. In the continuation, meetings are held with managers and high-ranking individuals of the organization under the title of ideation meetings, and in these meetings, in each of the 8 aspects of motivations, the necessary measures to achieve these motivations and obstacles to achieve these motivations are discussed. Then, using the results of these meetings, the evaluation is performed, the business goals of the organization, information on the needs of stakeholders, legal requirements, vision, and mission are updated according to the objectives of cyber security. To achieve the vision and mission of cyber security, network intelligence, actions, and programs with the help of senior managers of each of the eight aspects are envisaged. Finally, after prioritizing these measures, a cyber security roadmap for the smart grid will be formed [4] (Fig. 2).

6 Assessment of the Current Situation

To assess the current situation of each organization, a questionnaire prepared and the SGMM model is used. This questionnaire consists of 175 questions, each of which has 8 levels in each, including different questions. A team of senior executives in all areas will answer multiple-choice questions. A simple or weighted average can be used to get the final score in each level, depending on the importance of the questions. According to the table below, if the average score is higher than 0.7, the current situation in the relevant aspect and level is consistent with the SGMM model. If this score is between 0.4 and 0.7, significant improvements have been made in the relevant aspect and level. If this score is lower than 0.4, it indicates that the initial steps have been taken, and if the score is 0 in the case, it means that no action has been taken in this aspect and at this level. In the SGMM model, typically, the implementation of study and pilot projects leads to the introduction of the basic level and empowerment in various aspects (Table 1).

6.1 Organizational Business Goals

The business goals of each organization are determined by its vision and mission. If you need more information and details, meetings can be held with senior managers of the organization in this regard.

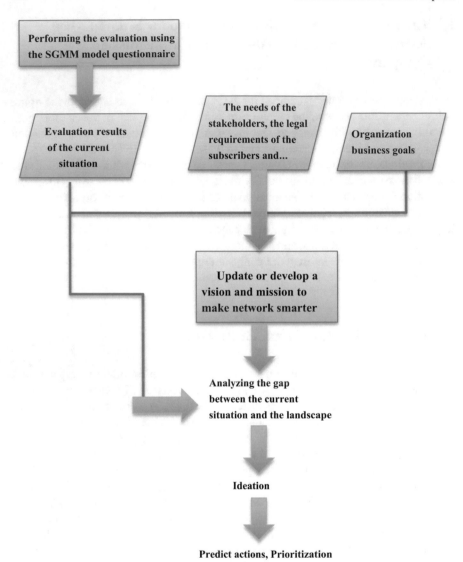

Fig. 2 Cyber security roadmap process for SmartGrid

Table 1 Table captions should be placed above the tables

Adaptation of the status quo to the model	≥0.7
Significant progress	0.7 < 0.4 ≥
Preliminary progress	0.4>
No action has been taken in this regard so far	0

6.2 Awareness of Stakeholder Needs and Regulatory Requirements

To know the needs of stakeholders, you can use various methods, including using polling tools in communication portals with employees and subscribers, holding brainstorming sessions, conducting face-to-face interview sessions, and so on.

6.3 Smart Cyber Security Outlook

The smart cybersecurity vision is developed using the three inputs mentioned above. At this stage, according to the opinion of senior managers of the organization, the optimal situation in each aspect of the SGMM model for the next 5 years is determined.

6.4 Gap Analysis

According to the developed perspective, it can be determined at what level the organization wants to be in each of the 8 aspects of the SGMM model. The results of the evaluation determine the current state of the organization and the outlook for what we expect from the future state of the organization. Therefore, by analyzing the gap between the current situation of the organization and the vision of Smart Cybersecurity, the needs of the organization can be identified. These needs are, in fact, the answers to the questions of the SGMM model that must be announced in to reach the desired level (Fig. 3).

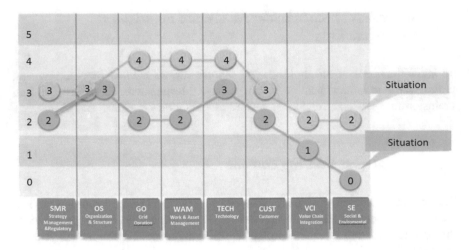

Fig. 3 Determining the optimal situation

6.5 Ideation

To achieve the above vision, meetings are held with experts and senior managers of the organization under the title of idea generation meetings. In these motivational sessions, ideas for achieving the motivations and the obstacles that exist in the way of achieving them are presented separately in each aspect.

7 Predicting and Prioritizing Actions

After the brainstorming sessions, separate meetings are held with the individuals and units in charge of each of the aspects in the SGMM model. In these sessions, the ideas that have already been put forward are discussed and re-examined, and from the heart of these ideas, steps emerge to reach the desired levels. The following list of measures is provided to senior managers of the organization to rate each of the measures in terms of feasibility and efficiency. In order of prioritization, using the diagram shown in the figure below, the actions are divided into 4 categories. These 4 categories are defined as follows according to the degree of feasibility and efficiency:

Area "A": The actions that are located in this area are well-suited to the cyber security of the company and are in good condition in terms of feasibility to be done. These actions should be followed up as soon as possible and action should be taken to implement them.

Area "B": The actions that take place in this area are the steps that are needed to become an intelligent and resilient cyber organization; however, it is less possible to

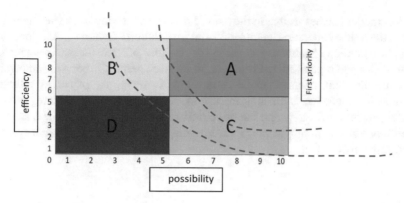

Fig. 4 Prioritization of actions

do them due to the potential of the organization. The organization must make every effort to enable these actions to take place as soon as possible.

Area "C": These measures are in good condition in terms of feasibility, but due to their lower efficiency, they should be of lower importance to the organization.

Area "D": The actions of this area, which have an almost low score in terms of feasibility and efficiency for the company's cyber security, should be considered in situations where the actions of other areas have been largely done. Some of these actions may be Remove in review.

In order to prioritize actions to be implemented, actions are divided into three categories of first priority, second priority and third priority using two lines. According to Pareto law, about 20% of actions should be in the first priority area.

Due to the dynamics and changes in technology, conditions and criteria of the country's electricity distribution industry, as well as changes in the level of capability of companies and the possibility of updating the SGMM model, their actions and priorities need to be reviewed after one year.

A review should be made during a reassessment of the organization's cyber maturity level. In the re-evaluation, the progress of the organization based on the proposed projects will be reviewed and if necessary, amendments will be made to the plans. The process is always repetitive (Fig. 4).

8 Conclusion

What can be summarized is to achieve an organization with a smart network with high cyber resilience, it is necessary to examine all the dimensions and aspects of the organization. The organization should know its needs and priorities in an integrated and comprehensive manner and strive to improve the situation in all aspects. One of the best tools to assess the current situation is to draw the desired conditions and develop a roadmap for the Cyber Security Maturity Model (SGMM).

This model teaches us that in the path of smart grid, trying to use new technologies in the field of operation, automation and installation of smart meters alone is not enough. Attention to other aspects of cyber security of network management such as strategies and organizational structure, work management processes and physical asset management, attention to customer empowerment, use of energy efficiency optimization methods, facilitating the use of new energy and prioritizing environmental protection Biology and the principles of sustainable development are other conditions for continuing to move towards the excellence of cyber security of the electricity network [5].

References

1. US Department of Energy and US Department of Homeland (2014) Electricity subsector cyber security capability maturity model, version 1.1
2. US Department of Energy and US Department of Homeland (2014) Cyber security maturity model version 1.1
3. IBM (2009) Smart grid maturity model: creating a clear path to the smart grid
4. AlShemeili A, Yeob Yeun C, Baek J (2016) PLC monitoring and protection for SCADA framework. In: Advanced multimedia and ubiquitous engineering
5. Xu X, Kang W, Fang Z, Sun B, Wang Y, Zhu T (2014) Global and initiative safety mechanism in industrial control system. Int J Comput Sci Eng 9(1/2):139–146
6. Anwar A, Mahmood AN (2014) Cyber security of smart grid infrastructure

Blockchain Technology in Cybersecurity Management

Mercy Ejura Dapel⬥, Mary Asante, Chijioke Dike Uba,
and Michael Opoku Agyeman⬥

Abstract Blockchain is a decentralised ledger used to secure digital currency, perform deals and transactions. A new transaction is validated when each member of network has access to the latest copy of the encrypted ledger. The features of blockchain are immutability, trackability, trustworthiness and decentralisation. This paper explores the concept, characteristics, and the need for blockchain in cybersecurity management. Blockchain is based on cryptography to ensure trust in transactions. Blockchain technologies made a remarkable contribution in Cybersecurity.

Keywords Blockchain · Cybersecurity · Cyberattacks · Cyberthreat

1 Introduction

Over the years the emergence of cyber threats and the increased frequency of cyber-attacks reinforced the need for cybersecurity initiatives. Solutions has been proposed to address cyber threats, one of such is blockchain technology. Blockchain is a technology that provides secure records validated by algorithms thereby encouraging trust on peer-to-peer networks [1]. Traditional security methods support preserving privacy and security however, their centralised nature with low computational capabilities makes them less efficient [2]. Blockchain technology has been adopted in many applications to strengthen security issues in a decentralised manner. It eliminates the need for password which is described as the weakest link in cybersecurity [3]. Blockchain is an emerging technology for cybersecurity, it has been the subject of many researches. Blockchain is a method for securing data through decentralized and peer to peer systems [4]. The use of blockchain has been applied to core business

M. E. Dapel · C. D. Uba · M. O. Agyeman (✉)
Centre for Advanced and Smart Systems (CAST), University of Northampton, Northampton, UK
e-mail: Michael.OpokuAgyeman@northampotn.ac.uk

M. Asante
University of Warwick, Coventry, UK

processes like banking [5], logistics [6], pharmaceutical industry [7], smart contracts [8] and most importantly cybersecurity.

Blockchain is an emerging technology for cybersecurity, it has been the subject of many researches in recent years. Blockchain is a distributed ledger that provides opportunity for data protection through decentralised identity. The ledger system makes information transparently available to members of the blockchain. It has gained traction in different application fields with focus on creating trust. Blockchain is used in cases where data requires trust without the need for third-party verification.

Blockchain use cases have emerged in areas such as healthcare. With patients data on blockchain, organisations can work together to improve care while patients privacy is protected [9]. In pharmaceutical industry, Blockchain can add traceability to drug supply thereby increasing the success rate of clinical trials. In a supply chain digitising paper based processes makes data shareable and trustworthy. It adds intelligence and automation to execute transactions [10]. In a loan process, consent for access to personal records is granted on the blockchain. Trust in the automated process for evaluating loan applications helps to drive closing faster and improve customer satisfaction [11]. Blockchain is a digital ledger that has the ability to be programmed to anything of value. The use of blockchain gained interest for its distinguished cybersecurity capabilities, resulting in many organisations becoming interested in using its security infrastructure to safeguard their information security systems [7]. With Blockchain, if an application needs to manage sensitive information it is solved by protecting a node, if the node is compromised, cybersecurity is threatened [12].

Blockchain technology has been applied in the use of Ethereum to advertise blacklisted internet protocols (IPs) that were suspected to be involved in Distributed Denial of service attacks (DDoS) [13]. Implementing security with blockchain is demonstrated in terms of confidentiality, integrity and availability. Blockchain technology offers a decentralized storage that can store data without the need of a single trusted party. Information is managed through a distributed ledger where nodes maintaining the ledger do not need to be mutually trusted, trust is distributed among all nodes. To add data to the ledger, a consensus is needed to be reached among all involved nodes. It possesses several features such as decentralization, immutability and validation [14].

Using blockchain technology for identity management links digital identity to a device IP address. Digital identities are secured using the principle of public and private key cryptography. A generated digital identity presents the verified identifier in the form of a QR-code or a digital certificate. Blockchain technology encourages secure communication thereby increasing trust when carrying out a transaction [15]. It utilizes peer-to-peer networks and distributed systems which include registers to store transactions. Blockchain security protects transactions against internal, malevolent and unintentional threats [16]. The core concept and unique property of Blockchain technology makes it attractive for business and cybersecurity. Protecting information through a decentralised technology tend to be more secure as it undergoes verification processes. When using blockchain the threshold of data veracity is higher [17]. Blockchain is an evolving technology that is finding traction in several areas such as banking [18], logistics [19], pharmaceutical industry [20], defence of IoT devices [21] and cybersecurity. It is a decentralized database that generates a digital log of trusted transactions that create reliability and reduce risk when a business is entered with an unfamiliar party. It can be shared across a public or private network that provides transparent and verifiable cybersecurity system. With blockchain, documentation of transactions can be verified by participating users in the Blockchain network [15]. Items can be tracked in a supply chain for traceability across companies using Blockchain. The operating process is similar to an IT process in a large organisation where compliance inspection and certificate issuing is performed. The supplier and receiver must trust the information along the supply route, this is an ideal Blockchain scenario. While using Blockchain and there is an attempt to incorporate unregistered software, it will be flagged [7]. The security qualities of blockchain are cryptography, software-mediated consensus, public and private keys, contracts and identity controls. This qualities offer integrity and data protection by verifying and authenticating transaction records and maintaining traceability and privacy [22]. The emerging trend of blockchain could enable decentralized applications without intermediaries, this could serve as a foundation for internet security.

2 Related Research on Blockchain in Cybersecurity Management

This study focused on existing approaches of Blockchain to provide cybersecurity. Blockchain was initially conceived as a financial protocol in the form of bitcoin. In view of its security capabilities researchers began to focus on blockchain to address privacy and security issues. Blockchain integrates several components such as a distributed data storage, consensus mechanisms and encryption algorithms. It separates data randomly and distributes them across an entire network of computers [23]. Features of blockchain technology include trustworthiness, trackability and immutability. According to Bansal et al.'s [24] survey on Blockchain for cybersecurity in the field of IoT, Blockchain can authenticate users by creating a decentralised system to provide interaction among users. By using blockchain technology security, reliability and transparency is provided to users through optimization and revolutionisation which is present in blockchain technology.

According to Vance and Vance [25] Blockchain solutions provide protection against persuasive phishing and social engineering using digital identities to enable safer IoT devices to prevent DDoS attacks. Alotaibi [26] explored cybersecurity relating to IoT and utilized end to end traceability, data privacy, anonymity, identity management, authentication, confidentiality, data integrity and availability. Initiatives like GUITAR and REMOWARE allow IoT devices to be updated in real-time. Blockchain supports a variety of functions, it can be installed in a smart home to enhance cybersecurity [26].

Alkadi et al. [21] discussed intrusion detection, Blockchain and data centralization in the cloud. Centralizing data in the cloud offers capabilities to protect consumer privacy. They proposed Blockchain implementation in information trust management in the cloud as the volume of information stored on the cloud is over-whelming, it cannot be processed by conventional methods alone. Lack of trust between IoT devices is solved by a decentralized Ethereum Blockchain that enables the survey of information from the Industrial Internet of Things to the cloud Fan et al. [27]. Wang et al. [9] reviewed a study on Blockchain for IoT and proposed a distributed and decentralized approach that promises IoT security.

Mittal et al. [18] explored cybersecurity enhancement through Blockchain training via a serious game approach and proposed an adaptive sandbox game which educates the players on the importance of Blockchain. Their approach provided skill advancement and a greater learning outcome which can be adopted by a large work force. Serrano [28] presented a blockchain random neural network for cybersecurity and proposed a method that enables a decentralised authentication method. Their validation result proved that adding blockchain random neural network provides user access control algorithm with increased cybersecurity resilience.

3   The Use of Blockchain to Improve Cybersecurity

Blockchain is a ledger technology with potential blockchain based characteristics such as decentralization and immutability that ensures authenticity, verifiability, reliability and integrity of data. When credibility of data can be ensured, trust worthy outcomes can be produced. This section gives a brief view on use cases of blockchain in data protection, privacy and security.

- Data Storage and sharing—Blockchain ensures that data stored in the cloud is resistant to unauthorized change. It utilizes public and private ledger to protect data from tampering, the hash list allows data to be securely stored. Blockchain ensures that exchanged data is verified as being the same from dispatch to receipt [29].
- Network security—Blockchain authenticates and store data in a decentralized and robust manner. Blockchain uses public and private architecture for point to point communication between nodes in the network to make blockchain appropriate to address network security issues [25].
- Navigation of the world wide web—Blockchain is used to improve the validity of the wireless internet access points thereby monitoring the access control on the local ledger. It ensures the validity of World Wide Web by navigating to correct web pages through accurate DNS records and communicate with others through secure encrypted methods [4].
- Intrusion detection—Blockchain can be used to detect malicious behaviour in a network environment [30].
- Securing transactions—Blockchain is used to secure electronic transactions particularly sensitive data. Blockchain utilises encryption and hashing to store immutable records [16].

4 Conclusion

In conclusion, Blockchain improves data storage by creating a decentralized network that uses client-side encryption such that data owners will have traceable control of their data. The innovative features of Blockchain make it ideal for today's cybersecurity needs, it can be used to prevent identity theft by verification through a decentralized identity system. Blockchain technology can be used to prevent data breaches, identity theft and cyberattacks by improving cyber defense through consensus mechanisms. Blockchain technology can be used in companies to authenticate users without password thereby eliminating human intervention and preventing potential stacks vectors. Blockchain is gaining traction in data assurance thereby implementing trustworthy and secure data infrastructures. Although blockchain provide advantages in cybersecurity, it also comes with disadvantages such high energy consumption required to keep a ledger and ensure transparency. The initial capital cost is high. Blockchain secures IoT through reliable authentication and data transfer. For a start, businesses could go for a private Blockchain that can serve as a platform for technology.

References

1. Carrara GR, Burle LM, Medeiros DS, de Albuquerque CVN, Mattos DM (2020) Consistency, availability, and partition tolerance in blockchain: a survey on the consensus mechanism over peer-to-peer networking. Ann Telecommun 75(3):163–174
2. Zhang R, Xue R, Liu L (2019) Security and privacy on blockchain. ACM Comput Surv (CSUR) 52(3):1–34
3. Kshetri N (2017) Blockchain's roles in strengthening cybersecurity and protecting privacy. Telecommun Policy 41(10):1027–1038
4. Taylor PJ, Dargahi T, Dehghantanha A, Parizi RM, Choo KKR (2020) A systematic literature review of blockchain cyber security. Digit Commun Netw 6(2):147–156
5. Hassani H, Huang X, Silva E (2018) Banking with blockchain-ed big data. J Manag Anal 5(4):256–275
6. Tijan E, Aksentijević S, Ivanić K, Jardas M (2019) Blockchain technology implementation in logistics. Sustainability 11(4):1185
7. Haq I, Esuka OM (2018) Blockchain technology in pharmaceutical industry to prevent counterfeit drugs. Int J Comput Appl 180(25):8–12
8. Cong LW, He Z (2019) Blockchain disruption and smart contracts. Rev Financ Stud 32(5):1754–1797
9. Wang Q, Zhu X, Ni Y, Gu L, Zhu H (2020) Blockchain for the IoT and industrial IoT: a review. Internet Things 10:100081
10. Agbo CC, Mahmoud QH, Eklund JM (2019) Blockchain technology in healthcare: a systematic review. Healthcare 7(2):56
11. Sylim P, Liu F, Marcelo A, Fontelo P et al (2018) Blockchain technology for detecting falsified and substandard drugs in distribution: pharmaceutical supply chain intervention. JMIR Res Protoc 7(9):e10163
12. Fraga-Lamas P, Fernández-Caramés TM (2019) A review on blockchain technologies for an advanced and cyber-resilient automotive industry. IEEE Access 7:17578–17598
13. Moniruzzaman M, Khezr S, Yassine A, Benlamri R (2020) Blockchain for smart homes: review of current trends and research challenges. Comput Electr Eng 83:106585

14. Vishwakarma R, Jain AK (2020) A survey of DDoS attacking techniques and defence mechanisms in the IoT network. Telecommun Syst 73(1):3–25
15. Viriyasitavat W, Hoonsopon D (2019) Blockchain characteristics and consensus in modern business processes. J Ind Inf Integr 13:32–39
16. Stephen R, Alex A (2018) A review on blockchain security. IOP Conf Ser: Mater Sci Eng 396(1):012030
17. Leng J, Zhou M, Zhao JL, Huang Y, Bian Y (2020) Blockchain security: a survey of techniques and research directions. IEEE Trans Serv Comput
18. Mittal A, Gupta M, Chaturvedi M, Chansarkar SR, Gupta S (2021) Cybersecurity enhancement through blockchain training (CEBT)—a serious game approach. Int J Inf Manag Data Insights 1(1):100001
19. Arjun R, Suprabha K (2020) Innovation and challenges of blockchain in banking: a scientometric view. Int J Interact Multimed Artif Intell 6(3)
20. Tijan E, Aksentijevic´ S, Ivanic´ K, Jardas M (2019) Blockchain technology implementation in logistics. Sustainability 11(4):1185
21. Alkadi O, Moustafa N, Turnbull B (2020) A review of intrusion detection and blockchain applications in the cloud: approaches, challenges and solutions. IEEE Access 8:104893–104917
22. Gimenez-Aguilar M, de Fuentes JM, Gonzalez-Manzano L, Arroyo D (2021) Achieving cybersecurity in blockchain-based systems: a survey. Futur Gener Comput Syst
23. Le DN, Khan AA (eds) (2022) Evolving software processes: trends and future directions. Wiley
24. Bansal P, Panchal R, Bassi S, Kumar A (2020) Blockchain for cybersecurity: a comprehensive survey. In: 2020 IEEE 9th international conference on communication systems and network technologies (CSNT). IEEE, pp 260–265
25. Vance TR, Vance A (2019) Cybersecurity in the blockchain era: a survey on examining critical infrastructure protection with blockchain-based technology. In: 2019 IEEE international scientific-practical conference problems of infocommunications, science and technology (PIC S&T). IEEE, pp 107–112
26. Alotaibi B (2019) Utilizing blockchain to overcome cyber security concerns in the internet of things: a review. IEEE Sens J 19(23):10953–10971
27. Fan K, Bao Z, Liu M, Vasilakos AV, Shi W (2020) Dredas: decentralized, reliable and efficient remote outsourced data auditing scheme with blockchain smart contract for industrial IoT. Futur Gener Comput Syst 110:665–674
28. Serrano W (2021) The blockchain random neural network for cybersecure IoT and 5G infrastructure in smart cities. J Netw Comput Appl 175:102909
29. Feng Q, He D, Zeadally S, Khan MK, Kumar N (2019) A survey on privacy protection in blockchain system. J Netw Comput Appl 126:45–58
30. Li W, Tug S, Meng W, Wang Y (2019) Designing collaborative blockchained signature-based intrusion detection in IoT environments. Futur Gener Comput Syst 96:481–489

The Impact of the Internet on Terrorism and Violent Extremism

Georgina Butler and Reza Montasari

Abstract The Internet has a significant impact on both the process and planning behind cyber and physical acts and attempts to mitigate these threats. This paper aims to highlight the ways in which the Internet is used in both of these processes by drawing upon the attractiveness of the Internet to terrorist organisations and how they, therefore, use it to their advantage. Instances where the Internet has been used for threats or acts of terrorism and extremism are elaborated with examples. The paper also aims to explore how the Internet and intelligence can be used in the process of countering these online threats. Artificial Intelligence (AI) is one of the most discussed phenomenon in relation to cyber security and countering cyberterrorism due to the increased technological advancements in this field. To this end, the paper briefly draws upon this use of AI in organisations such as the Global Terrorism Database (GTD) and the Institute of Strategic Dialogue (ISD) and outlines how this should be the focus of many governments in protecting their national security. The discussion of this paper revolves around the potential issues and impacts of the Internet in both the methods of conducting these attacks and in countering them.

Keywords National security · Artificial intelligence · Cyberterrorism · Cyber security · Violent extremism · Propaganda · Global immediacy · Radicalisation · Terrorism

1 Introduction

When considering terrorism, the Internet, and their combined effect, one thing becomes clear: both have catastrophic consequences from the local to the international and global levels [1]. The increasing interest as well as use of the Internet

G. Butler (✉) · R. Montasari
School of Social Sciences, Department of Criminology, Sociology and Social Policy, Swansea University, Richard Price Building, Singleton Park, Swansea SA2 8PP, UK
e-mail: 1908091@Swansea.ac.uk

R. Montasari
e-mail: Reza.Montasari@Swansea.ac.uk

© The Author(s), under exclusive license to Springer Nature Switzerland AG 2023   427
H. Jahankhani (ed.), *Cybersecurity in the Age of Smart Societies*,
Advanced Sciences and Technologies for Security Applications,
https://doi.org/10.1007/978-3-031-20160-8_24

has drawn the focus of multiple researchers and policymakers into the ways in which terrorists and violent extremists use it within the strategy of terrorist acts [2]. Although primarily used for commercial and academic purposes, the Internet is also a vital medium for personal communication and simple entertainment, and as technology has advanced, more people are online, resulting in more content [3]. This has bene-fited terrorist and extremist organisations by allowing them to become more active online without fear of being monitored. This is due to the mass of data as well as the increase in the amount of private and hidden areas of the Internet [4]. GCHQ refers to the Internet as an invisible, but a very much active, part of daily life within the UK, and this resulted in cyber threats being associated with the highest tier of threats alongside terrorism and war [5]. After reviewing the new Prevent strategy, the 2010 UK government abandoned the term "violent extremism" and stated that all forms of extremism are considered to be anything that draws vulnerable individuals towards terrorism [6].

The majority of studies in this field focus on the Internet's resources and capabili-ties in terms of creating a space for like-minded individuals to come together publicly and privately [4]. As a result, platforms like Twitter, Facebook, and YouTube have provided terrorist groups with a direct channel through which they can manipu-late and persuade vulnerable individuals by exploiting insecurities and self-identity issues [7]. This paper will introduce the concept of Internet connectivity to terrorist activities, how the Internet plays a direct or indirect role, and why the Internet is so appealing to terrorist organisations. Furthermore, once it is clear why these organi-sations choose this method, this paper will identify the various ways the Internet is used during the process of terrorist acts and how this may impact national security. Finally, because the Internet has had such an impact on terrorism and extremism, the latter section will examine how the Internet and technology can be used to combat this, and thus what the UK government is doing to combat cyberterrorism [8].

2   The Internet as an Attractive Option

The Internet in general has grown substantially over the past few decades and this has contributed to an increased involvement within terrorism among other threats to cyber security [1]. Along with this advancement comes multiple developments within the Internet and especially relating to the use of platforms such as Twitter, Facebook and YouTube [7]. Historically, information regarding terrorism and terrorist-related activities was rarely accessible and, therefore, was mainly shared through television or newspapers. Nowadays, the Internet has become a medium for terrorist organ-isations and those wishing to spread propaganda and communication has become significantly faster [2]. One of the main characteristics of terrorism and the main aim for the majority of terrorist groups is to encourage this sense of fear [9]. As a result, these groups target national security and governmental institutes as this is the fastest way to threaten a country and to instil this fear among society. Other signs of terrorism are the threat or use of violence and political motivation, as without these it cannot be considered a terrorist act, but rather just a common crime [7].

In terms of the link between terrorism and the cyberworld, cyberterrorism is defined by Pollitt as a "premeditated, politically motivated attack against information, computer systems, computer programs, and data which result in violence against non-combatant targets by sub-national groups or clandestine agents" [10, p. 8]. Aside from the fact that online communication is inexpensive and immediate, the usability and interactivity are also appealing options for many of these extremist groups [2]. The current cyberterrorism landscape places a high value on national security, which Grabosky defines as the "capacity to deter or to resist the invasion of one's territorial borders by foreign military, naval, or air forces". However, more recently, it involves threats to welfare, social cohesion and even public health [11, pp. 67–68]. Another factor to consider is that if a country is less technologically advanced, it will be more vulnerable to cyber-attacks because it lacks the resources and capabilities to defend against these advanced threats [12]. According to Weimann, modern terrorists value the Internet's anonymity and ability to inflict large amounts of damage quietly so that systems can be exploited [13].

While terrorist groups can be identified and their content removed online, governments and agencies must be aware that terrorists have learned and adapted to the secure nature of the Internet and can now avoid certain types of identification and removal [2]. This highlights the fact that as the Internet and technology advance, so will the knowledge and lengths to which these individuals will go to remain hidden. Social media platforms have provided a much more inexpensive channel for these groups to post and share information, allowing them to get their beliefs and messages across to anyone who wants to listen [14]. As a result, as more individuals post more personal and sensitive information on platforms such as Facebook and Twitter, believing they are sharing news and updating their family and friends, this information can be retrieved, manipulated, and used in criminal activity by terrorist groups [15]. Another reason why the Internet is appealing is that it can be accessed and used anywhere and has a sense of global immediacy [13].

The COVID-19 pandemic has had a significant impact on the global economy and national security, and terrorists have exploited this vulnerability even further [16]. According to INTERPOL, the global pandemic has resulted in a shift from individual attacks to threats against larger governments and critical infrastructure. Because of the pandemic, many corporations and businesses are working from home and rely on the Internet, but they are becoming lax in terms of security. Terrorist groups are developing more sophisticated methods, such as encouraging COVID-themed scams and campaigns, to increase their chances of success. For example, the majority of phishing and misinformation scams occurred when new COVID vaccinations became available, demonstrating that these terrorist organisations are taking advantage of their surroundings and continuing to find new ways to advance their strategies. Furthermore, another advantage of the Internet in these situations is that it is capable of breaking down multiple barriers erected by governments, and there is no need to be concerned about physical elements because they are not as advanced and intelligent as technology [14].

3   The Exploitation of the Internet in Cyberterrorism

It is clear why the Internet is appealing to these extremist groups; it is also useful to understand how they use it. The vast majority of research on this topic suggests that terrorists use the Internet for a variety of purposes, including recruitment, training, fund-raising, psychological warfare, and propaganda dissemination [2, 13]. According to the United Nations Office on Drugs and Crime, there are six major categories that identify how the Internet is used for terrorist gain: propaganda, finance, training, planning, execution, and the actual cyberattack [15]. Propaganda is commonly known as communication via the media that can provide "practical instruction, explanations, justifications and promotion of terrorist activities" [15, p. 3]. This can, therefore, come in the form of videos, photographs, voice memos, or any other form of media files that these groups use in order to help influence their victims [15]. Not only can these groups tailor the type of material they promote based on the target audience, but because of the breadth and depth of social media and the Internet, these groups can reach just about anyone, anywhere online [15]. As propaganda becomes more hidden and difficult to find, radicalisation on the Internet has become more private. This is especially challenging for security reasons as propaganda plays an important role in influencing the radicalisation process and is essential for the success of recruitment [17, p. 331].

Within propaganda, radicalisation is not the only major theme that is present, but both incitement and recruitment are also key elements of propaganda material [15]. Terrorists are known to use the Internet in order to find and groom individuals via chat rooms and social media platforms and to recruit them into their groups [13]. There are a number of different theories, such as the social learning theory, which explains how these groups promote violence and are able to do this [13]. Frieburger and Crane [18] imply that the social learning theory assumes individuals learn radical behaviour through other people which results in them developing similar, if not the exact same radical behaviour. So these organisations can use photographs or videos of fighting or extremist behaviour as a tool to facilitate the grooming process [7].

As a result, selecting a victim is not a random process for these organisations; terrorist groups will carefully select an individual based on their social media content and whether they exhibit vulnerable behaviours. Most of the time, it's because that person is struggling with their identity and does not mind if their anger is directed elsewhere [19]. Additionally, UNODC emphasise the importance of distinguishing between simple propaganda and that which is intended to incite specific acts of terrorism [15]. Under Article 19, paragraph 3 of the International Covenant on Civil and Political Rights, states, "preventing and deterring incitement to terrorism in the interest of protecting national security and public order are legitimate grounds for limiting freedom of expression" [15, p. 6]. Several groups in Syria and Iraq have turned to social media to incite and recruit individuals, but first they must educate them on their beliefs using propaganda and psychological manipulation. Once recruited, these groups must train these individuals online and help fund-raise to aid the attacks [2].

Terrorist organisations also use the Internet to fund-raise and finance their terrorist acts by exploiting certain infrastructures for financial gain [15]. They can target more financial infrastructures and companies to undermine national security, as well as a variety of government records, including medical records. Extremist groups frequently use the Internet for financial gain, for example, they easily exploit government-supported charities by diverting funds for their own use [15]. As a result, they may even impersonate charities in order to raise funds for the purchase of weapons or the training process [15]. Terrorist groups can also use social media chat rooms to solicit financial aid from their so-called supporters or victims, a practise known as direct solicitation. Intercepting online payments and utilising the tools to do so can be accomplished through various forms of fraud in which money is swiftly transferred through platforms such as PayPal [15].

Moving on from the financial aspect, terrorists can use the Internet to train individuals and in the planning process in a variety of ways. Social media has become more interactive as it fosters a sense of community by encouraging and enticing individuals to act on their beliefs, most of the time through violent or extremist acts [20]. Although instructional content will be available to anyone and will be able to reach any depth and breadth of the Internet, it will remain hidden from certain law enforcement agencies [15]. Not only are there detailed instructions in manuals and videos, but these Internet platforms can support multiple languages on a variety of topics ranging from how to plan these attacks to how to build and execute the creation of explosives or other weapons used [15]. As a result, researchers have "indicated that almost every case of terrorism prosecuted involved the use of Internet technology" with the commonality being the use of technology to plan the terrorist act [15, p. 8].

In terms of execution, the Internet can both assist the cyber-attack and contribute to the physical act itself [15]. Communication before and during the event is a critical component of successfully carrying out a plan. All of the previously introduced categories must have been successful to ensure the success of the execution, as the act will not be able to proceed smoothly without a fully trained and confident individual, as well as an effective plan with the necessary funding [15]. This is relevant to whether the attack is online or physical, as both require technological inputs and will leave an online trail [15]. However, the reason terrorist organisations use the Internet so confidently and frequently is not only because of its accessibility and reachability, but also because there is little chance of detection because these groups have improved their methods so that everyone involved remains anonymous [15].

4 Acts of Terrorism Using the Internet

There have been numerous instances where cyber-attacks have disrupted national security and had a significant impact on a local, national, and global scale. Although cyber and terrorist acts are not as well known or prevalent in the UK, they do occur and have an impact on national security and infrastructure. On March 24th, 2022, BBC News published an article about an English teenager accused of working for Lapsus$, arguably one of the most feared cybercrime gangs that targets major corporations and organisations [21]. The teenager in question was able to make a £10.6 million profit from hacking without his parents' knowledge, highlighting the hidden and secure nature of the Internet for these individuals or groups. As a result, anyone can go about their business undetected while causing severe damage to the security of companies and infrastructures for financial gain [21]. In fact, the article also states that the only reason the police were made aware of the issue is because a fellow online hacker had fallen out with the boy and released his personal details such as photograph, name, address and the fact he had committed these crimes [21]. It could be argued that if that person had not released this information, the teenager would have continued to exploit large corporations for money, potentially causing real harm.

Furthermore, on May 12, 2017, a ransomware attack globally impacted over 200,000 computer networks in approximately 100 countries [22]. This attack, known as 'WannaCry' was particularly successful with the NHS in the UK and remains one of the most popular stories about cyber-attacks in the country to date. Not only did the infection cause 80 of the 236 trusts to shut down, but it also resulted in approximately 20,000 cancellations in operations and general appointments due to the system' inactivity [22]. Due to the attack, approximately 5 accident and emergency departments across the country were unable to treat patients due to the blackout. As a result, patients were diverted to different hospitals, which in some cases worsened the patients' health and even worsened the situation [22]. The House of Commons Committee of Public Accounts stated that the NHS was not prepared for 'WannaCry' and that there was still a long way to go before advancements and precautions could be taken to improve the NHS' security [22]. The report also discovered that the UK government now knows more about the NHS' preparedness for a cyber-attack, highlighting that these trusts still would require a lot more support in terms of the necessary security standards [22].

Furthermore, the financial cost of this attack saw £21 million of funding to address the vulnerabilities within the NHS, and particularly in the Major Trauma Centres and Ambulance Trusts, whilst another £25 million was set to improve organisations response to attacks as well as supporting those who are most at risk [22]. Since then, there have been numerous attacks or threats against UK organisations and even members of the government, such as Downing Street. An article published on April 18, 2022 claimed that Foreign Office computer systems were suspected of being contaminated with spyware, specifically the Pegasus spyware [23]. Mobile phones inside Number 10, Downing Street, have also been reported to be infected with this spyware, which acts as a listening device and has previously been accused of targeting UK government officials [23]. This demonstrates the constant threat posed by cyber-attacks in the UK on a daily basis, and how the government is constantly under pressure to update and advance their counter-measures.

As a result of the impact seen here and in many other examples of cyber-attacks, intelligence and government agencies were required to improve traditional countermeasures and explore new methods and avenues in order to deter these cyber criminals [15]. According to multiple lines of research into the idea of policing and countering cyberterrorism, national governments are said to bear a significant amount of responsibility in these deterrent methods. To begin, one example of how the UK attempts to respond to these threats is through the assistance of organisations such as the GCHQ (the Government Communication Headquarters). The GCHQ aims to collaborate with academics and law enforcement to develop skills that can protect against cyber-attacks and identify online crime as quickly as possible [5]. The government can also assist in providing intelligence that can help prevent these threats from occurring, whether by educating individuals about Internet safety or by strengthening their national infrastructure and network security [5].

The UK accomplish this specifically by collaborating with the National Crime Agency (NCA) as well as numerous organisations and law enforcement agencies. In one case, the GCHQ, NCA, and FBI collaborated to dismantle a criminal conspiracy known as 'Gameover Zeus', which had exploited over 500,000 Internet users world-wide, including over 15,000 in the UK [5]. Within GCHQ, a division called the National Cyber Security Centre was established in 2017 with a strong emphasis on how to approach cyber threats, particularly educating and assisting businesses with cyber security issues [5]. Furthermore, the Ministry of Defence noted that the 2010 Strategic Defence and Security Review (SDSR) found that the United Kingdom's ability to defend against these cyber threats and attacks is critical to national security [24]. Not only that, but by developing these Defense Cyber Security Programs, the government will be able to educate organisations and communities on how to stay safe online and how to use such security systems [24].

5 The Use of the Internet to Counter Cyber Threats

One of the ways through which the government can strengthen their defence is through Artificial Intelligence (AI). Several lines of research indicate that Machine Learning (ML), a subset of AI, is one of the most recent and effective tools for combating cyberterrorism. Studies have found that manual detection of radical conversations and chat rooms on the dark web is nearly impossible, so automatic detection is becoming more appealing [25]. ML is defined as "the automated detection of patterns and recurrences in large data sets through various training techniques" [20, p. 2977]. ML algorithms are employed in order to take the mass amount of data online and classify radical or non-radical behaviour using classification techniques [25]. An example of classification techniques are Support Vector Machines and Decision Trees, both of which are used to classify the information into two groups and help determine what is radical and what is not [26].

There exist several different organisations and agencies which contribute to this, such as the Institute of Strategic Dialogue (ISD) and the Global Terrorism Database (GTD), specialising in countering online extremism and terrorist threats [27, 28]. Therefore, their methods take a more active approach as it looks into preventing the attack before it happens [28]. ISD have successfully removed a large number of Twitter accounts and Facebook pages that contained huge amounts of Islamist disinformation content [29]. The GTD collect and store data on all terrorist acts, both online and offline, and uses this information to monitor and, hopefully, predict future attacks [28]. Both of these organisations have successfully removed extremist content online and contributed to the countering of cyberterrorism. However, using AI has both advantages and disadvantages. For example, while it has the potential to be at the forefront of counterterrorism, issues concerning bias and a lack of privacy arise. This debate over privacy includes the intrusive qualities of AI and how using it to combat terrorism can actually exacerbate national security [30]. It also raises issues of accountability because it is difficult to determine who is to blame or who deserves credit for the outcomes produced by AI [31]. This not only highlights the need for additional research on the challenges and potential recommendations of AI but also implies that it has the potential to have an impact on the online counter-terrorism and extremist behaviour.

6 Discussion and Concluding Remarks

The Internet can play either an extremely active or a less interactive role in terrorism; in either case, it is still used in the majority of cases of violent extremism or terrorist acts. However, it should be noted that the Internet will always be used in the recruitment phase of terrorism, as it is unlikely that terrorist organisations will physically recruit individuals in the flesh nowadays. Since the Internet has become more widely used in the extremist process, propaganda, training, and planning have advanced, benefiting these groups significantly. This demonstrates how the Internet can either directly or indirectly contribute to extremism and terrorism. Because of its anonymity, accessibility, and global immediacy, the Internet is an extremely appealing option for many of these groups [13].

It has the ability to provide them with a platform to express their beliefs and goals, as well as a safe and private location where they can hide from the public and law enforcement. However, the Internet can also be used to combat online extremism through AI and ML. These algorithms are attempting to mitigate and prevent the spread of radical content online, thereby preventing terrorist groups from sharing their beliefs and committing these violent acts. Although the Internet has had a vital role in the development of terrorism as well as helping it to advance by exploiting national infrastructures, it can also have a significant impact on countering terrorism and extremism. If the government uses these methods effectively, they will be able to monitor and, hopefully, prevent this behaviour before it occurs. There are numerous challenges associated with this method of mitigation that governments should be aware of before proceeding, ranging from the possibility of bias to the issue of accountability. It can be argued that the Internet is growing fast, so are these terrorist organisations. Therefore, improvements and advancements are required to control the attacks and assist agencies in preventing them entirely.

References

1. LaFree G (2017) Terrorism and the Internet. Criminol Public Policy 16(1):93–98
2. Aly A, Macdonald S, Jarvis L, Chen T (2016) Violent extremism online: new perspectives on terrorism and the Internet. Routledge, Abingdon
3. Warf B, Grimes J (1997) Counterhegemonic discourses and the Internet. Geogr Rev 87(2):259–274
4. Gill P, Corner E, Conway M, Thornton A, Bloom M, Horgan J (2017) Terrorist use of the Internet by the numbers. Criminol Public Policy 16(1):99–117
5. GCHQ (2019) The cyber threat. GCHQ: Information. https://www.gchq.gov.uk/information/cyber-threat. Accessed 25 Apr 2022
6. Lowe D (2017) Prevent strategies: the problems associated in defining extremism: the case of the United Kingdom. Stud Confl Terror 40(11):917–933. https://doi.org/10.1080/1057610X.2016.1253941
7. Awan I (2017) Cyber-extremism: Isis and the power of social media. Society 54(2):138–149
8. Chen H (2007) Exploring extremism and terrorism on the web: the dark web project. In: Pacific Asia workshop on intelligence and security informatics. Springer, Berlin, pp 1–20
9. Marsili M (2019) The war on cyberterrorism. Democr Secur 15(2):172–199. https://doi.org/10.1080/17419166.2018.1496826
10. Pollitt M (1991) Cyberterrorism—fact or fancy? Comput Fraud Secur 2:8–10
11. Grabosky P (2015) Organized cybercrime and national security. In: Cybercrime risks and responses, pp 67–80
12. Lewis JA (2002) Assessing the risks of cyber terrorism, cyber war and other cyber threats. Center for Strategic & International Studies, Washington, DC, pp 1–12
13. Weimann G (2004) Cyberterrorism: how real is the threat? United States Institute of Peace, Special Report 119, vol 31. https://www.ethz.ch
14. Denning D (2001) Activism, hacktivism, and cyberterrorism: the Internet as a tool for influencing foreign policy. In: Networks and netwars: the future of terror, crime and militancy, pp 239–288
15. United Nations Office on Drugs and Crime (UNODC) (2012) The use of the Internet for terrorist purposes. United Nations, Vienna
16. Yu S, Carroll F (2021) Implication of AI in national security: understanding the security issues and ethical challenges. In: Montasari R, Jahankhani H (eds) Artificial intelligence in cyber security: impact and implications: security challenges, technical and ethical issues, forensic investigative challenges. Springer, Cham
17. Rowe M, Saif H (2016) Mining Pro-ISIS radicalisation signals from social media users. In: ICWSM. Association for the Advancement of Artificial Intelligence, pp 329–338
18. Freiburger T, Crane JS (2008) A systematic examination of terrorist use of the Internet. Int J Cyber Criminol 2(1)
19. Orsini A, Vecchioni M (2019) How does one become a terrorist and why: theories of radicalization. Department of Political Science. Luiss. https://tesi.luiss.it/id/eprint/24388
20. Verhelst HM, Stannat AW, Mecacci G (2020) Machine learning against terrorism: how big data collection and analysis influences the privacy-security dilemma. Sci Eng Ethics 26:2975–2984. https://doi.org/10.1007/s11948-020-00254-w
21. Tidy J (2022) Lapsus$: Oxford teen accused of being multi-millionaire cyber-criminal. BBC NEWS. https://www.bbc.co.uk/news/technology-60864283
22. House of Commons Committee of Public Accounts (HCCPA) (2018) Cyber-attack on the NHS: thirty-second report of session 2017–2019. House of Commons
23. Corera G (2022) No 10 network targeted with spyware, says group. BBC NEWS. https://www.bbc.co.uk/news/uk-61142687
24. Ministry of Defence (MOD) (2011) Combating cyber attacks. Defence and armed forces. GOV.UK. https://www.gov.uk/government/news/combating-cyber-attacks
25. Gupta P, Varshney P, Bhatia MPS (2017) Identifying radical social media posts using machine learning. Tech Rep. https://doi.org/10.13140/RG.2.2.15311.53926

26. Kitrum (2020) How to use machine learning in cybersecurity? Blog. https://www.kitrum.com
27. Jackson P (2017) If you want to understand anti-fascist movements, you need to know this history. HuffPost. https://www.huffpost.com/entry/anti-fascist-movements_b_599b11b 8e4b04c532f4348f4. Accessed 29 Mar 22
28. START (2022) Global terrorism database. National Consortium for the Study of Terrorism and Responses to Terrorism. https://www.start.umd.edu/gtd/about/. Accessed 12 Apr 2022
29. Institute for Strategic Dialogue (IDS) (2022) Institute for Strategic Dialogue. https://www.isd global.org/about/
30. Brey P (1998) Ethical aspects of information security and privacy. In: Petkovic M, Jonker W (eds) Security, privacy, and trust in modern data management. Springer, New York
31. Hussain Z (2017) The ABCs of machine learning: privacy and other legal concerns. Law Practices Today. https://www.lawpracticetoday.org/article/machine-learning-privacy-legal-concerns/#:~:text=%20The%20ABCs%20of%20Machine%20Learning%3A%20Privacy% 20and,officer%2C%20has%20said%20that%20%E2%80%9CAI%20will...%20More%20? msclkid=5604c6d8b8c811ec94d09eb8666f6e67

A Critical Review of Digital Twin Confidentiality in a Smart City

Alex Kismul, Haider Al-Khateeb, and Hamid Jahankhani

Abstract Digital twin technology is used to enable businesses to create efficiencies by modelling their physical counterparts. Use cases include modelling a physical device through its lifecycle to perform predictive maintenance, product training, future product development, product performance enhancement, or using the digital twin to control its physical counterpart to perform tasks on IoT or other connected devices. A digital twin leads to less downtime on a physical device as all the modelling or testing is conducted in a virtual environment meaning the physical device can continue to perform the tasks required of it. The digital twin and its physical counterpart are linked and synchronised through heterogeneous network connections. This poses a cyber security question of whether there is a risk of using a digital twin within a smart city. This paper aims to critically examine the confidentiality requirements for a digital twin in a smart city by performing a critical analysis of current literature.

Keywords Digital twin · Smart city · Cyber-physical systems · Confidentiality · Privacy · Data protection

1 Introduction

A digital twin (DT) can communicate bidirectionally with a physical entity (e.g., Internet of Things (IoT)) by use of network connectivity between the operational device and the DT [1–3]. This connectivity extends the data held within the physical entity to the DT and reciprocally the DT can send instructions to the physical entity. This brings a form of network convergence between Information Technology (IT) and Operational Technology (OT) systems.

A. Kismul · H. Jahankhani
Northumbria University London Campus, London, UK

H. Al-Khateeb (✉)
School of Engineering, Computing and Mathematical Sciences, University of Wolverhampton, Wolverhampton, UK
e-mail: H.Al-Khateeb@wlv.ac.uk

These factors may introduce contemporary and novel cyber security threats to the DT and its physical counterpart by connecting systems that historically are separate operational and information system networks. This research critically evaluates confidentiality threats to DTs in a smart city as they are likely to contain sensitive data that must be protected. The study will offer a critical review and assess the use of DT technology that processes information from a physical entity in a smart city and how it may pose a significant confidentiality and cybersecurity risk. A thematic narrative literature review is presented which identified numerous contemporary and novel threats by connecting a DT to a physical operational entity.

2   Background and Related Work

The literature review examines the evolution of DTs for contextual awareness. It also examines the uniqueness of smart cities in terms of searching for specific ethical requirements such as privacy, and citizen safety.

2.1   Digital Twin Computing Background

Cyber-Physical Systems (CPS) are created through digital transformation strategies which are predominantly associated with Industry 4.0 initiatives [4]. Connected physical assets such as the Internet of Things (IoT) devices, and big-data computing processing via heterogeneous network connectivity methods allow for the creation of CPS. CPS enables organisations to enhance business process efficiencies through the analysis of readings from the physical assets data outputs (e.g., sensor and actuator outputs) to greater understand asset performance and to make efficiencies through the asset's lifecycle.

A DT is a virtualised replica of a physical entity. This can be a digital replica of an IoT device, an engine, a human (e.g., transplanted organ), a city, or any physical thing that can be replicated in a digital form [4].

A DT can communicate in real time with the physical entity it replicates. Effectively, it can model it the physical counterpart by receiving and processing data from the physical entity and send communication signals such as instructions to the physical device. The state of the physical twin can be changed by altering the state of the DT and vice-versa [2, 3].

DTs can be created and used in sectors such as smart cities, healthcare, aviation, automotive, or any sector where there is a benefit in creating a digitised replica of a physical asset. Use cases include product design, device lifecycle planning, and optimisation of process controls. These use cases lead to improved business efficiencies, and cost reductions through strategies such as predictive maintenance [2, 3]. A DT is not a standalone entity in software. Often DTs will be used in conjunction with Artificial Intelligence (AI), Machine Learning (ML), and Augmented Reality

(AR). For example, the DT could pair a person's unique physical artefacts with digital models reflecting their status in real-time for analysis and representation within virtual and augmented reality systems [5, 6].

There are niche companies providing DT technology, either as a consumer service (SaaS, PaaS, etc.), or through providing bespoke services that are created based on the specific physical entity technology and its required use case. Often these services are provided on cloud-based locations away from the operational physical entity to enable cost savings through economies of scale. DTs can be deployed using open-source software or proprietary vendor provided solutions [7]. Along with other types of CPS, the concept of a DT introduces a form of convergence by connecting an operational physical asset to a digital replica over a shared network connection [8]. This literature review will concentrate on DTs within smart cities.

2.2 Smart City Background and Digital Twin Relevance

A connected place or smart city is defined by UK National Cyber Security Centre (NCSC) as "a community that integrates information and communication technologies and IoT devices to collect and analyse data to deliver new services to the built environment and enhance the quality of living for citizens" [9]. Smart city services include CCTV, traffic light management, waste management, street lighting management, transport services and other public services such as healthcare and emergency services [9].

Basic building blocks for a smart city are instruments (e.g., sensors, cameras, consumer devices, and other specific city digital devices), connectivity, and intelligent services used for analysis and decision making [10]. Smart city infrastructure is built by using technologies such as IoT, big data analytics computing, and network connectivity over existing (wired/wireless) and emerging deployment of network technologies (such as 5G/WIFI 6). IoT Sensors collect the data and stream it to computing data processors for a variety of applications such as Real-time reporting in Smart Homes [2, 3].

DTs can play a role in smart cities, for example, planning decisions within built environments can benefit from using near real-time 3D modelling. Collecting vast amounts of data can lead to the creation of modelling to enable smart city planners to adaptively change how the smart city operates [11].

2.3 Digital Twin Integration with Smart Cities

Deren et al. [11] assert there are two fundamental aspects to creating a DT city, a data foundation, and a technical foundation. The data foundation is created by the transmission of data from sensors, cameras, and other digital data collection devices. The technical foundation is the IoT, network transmission technologies (wired/wireless), cloud, fog services, big data processing, and integrated intelligence (e.g., AI) [11].

This raises two important considerations—the coexistence of multiple entities (service providers, city technology infrastructure, and consumer devices), and the convergence of data and information between operational networks and DTs likely located on separate networks (e.g., cloud infrastructure).

Elmaghraby and Losavio [12] previously investigated cyber security concerns with smart cities and concluded that the smart city building blocks need to be protected using the Confidentiality, Integrity, and Availability (CIA) triad (and authenticity), adherence to privacy laws, and consideration of the democratic social concept of the right to privacy [5, 6]. The social right to privacy is a complex subject and is taken to account within this document as a concept based on citizen trust and confidence, rather than examining various specific legal statutes [13].

Traditional industrial network hierarchical models such as IEC 62443 [14, 15] used in industrial zones is less relevant for connected places. The network is no longer hierarchical in the traditional sense of layers of rigid zones from the sensor to the application server. Emerging digital technologies such as multi-access edge computing (MEC), fog computing, and cloud computing will become the standardised technology stacks within smart cities. High speed, low latency connectivity will be critical for real-time representation as the technology use cases grow.

Cyber security considerations of the DT, physical twin, data flow, and the types of technology used need to be considered holistically when planning DT deployments. By processing potentially sensitive personal data, the risks and threats increase and could become uncontrolled. The risk of connecting DTs to their physical counterparts is if one of them is compromised by an attacker it could compromise the other.

A further complication occurs due to identifying the ownership of specific data and information. If a smart city is compared to a manufacturing business that uses DTs it can be assumed the intellectual property of the data belongs to the manufacturing business owner or stakeholders, in a smart city there are numerous groups of service providers, individual citizens, and civil departments. As smart cities grow, collaboration requirements will also grow and networks will continue to converge [7]. Data and information must be identified and categorised whether it is in the public domain, a closed domain, and is governed under a specific set of regulations (e.g., if defined as critical infrastructure or is an essential service). A further consideration is cloud servers storing and processing data (e.g., GDPR compliance). Cloud servers may not be in the jurisdictional areas required for compliance [16]. These considerations outline the importance of information governance for a smart city using CPS [5, 6].

Much work has been conducted to highlight a general requirement for cyber security and information security guidance or frameworks for using IoT and Smart Cities. Sookhak et al. [16] and Montasari et al. [17] discuss the security and privacy of smart city requirements based on IoT security and cloud security. Relevance is paid to security and privacy issues that should be considered when designing smart city frameworks.

Vitunskaite et al. [18] highlight the smart city threat landscape in their research. This is based on malicious activity and accidental activity. They discuss the risks to smart cities regarding third-, fourth- and fifth-party access. Of concern is connecting IoT devices which are not built with security in mind or by design. In addition, they discuss the lack of roles and responsibilities documented in standards concerning emerging smart city development and deployment of technology. Other key findings are requirements for maintaining the trustworthiness of data, malware mitigation (i.e., targeting IoT devices and connected application servers), and the impact of a Distributed Denial of Service (DDoS) attack. A key conclusion from their report is security by design is a major factor in ensuring secure and resilient smart cities. This is especially applicable when connecting DTs to smart city network infrastructure.

Vitunskaite et al. [18] report that threats to smart cities can be intentional or accidental. Of specific interest to the protection of confidentiality are eavesdropping and unauthorised access threats. However other threats can relate to the loss of confidentiality (e.g., malware resulting in data theft).

There is a common theme in literature for smart cities in terms of IoT protection, cloud computing protection, and secure network connectivity using encryption techniques. Limited information is currently available specifically aimed at DTs within smart cities. There are relevant similarities in terms of CPS and should be taken into consideration.

2.4 Relatable Standards and Guidance for Smart City Digital Twins

The European Union Agency for Cybersecurity (ENISA) provides guidance for "The Good Practices for Security of IoT" [19] which focuses on the secure development Lifecycle of IoT and "Baseline Security Recommendations for IoT in the context of Critical Information Infrastructures" [20] which focusses on IoT in mission-critical environments. This provides strong technical and procedural guidance to understand the IoT threat landscape, attack types, and mitigation techniques.

ENISA has produced a tool for smart environments (IoT, Smart Cities, Industry 4.0, Smart Cars, Smart Hospitals and more) that can be downloaded and used as guidance [21], this is broadly similar to control frameworks such ISO2700 series controls, NIST CSF, and other good practices.

The UK NCSC has put together a set of principles for connected places (smart cities), that is based upon three guiding principles:

- Understanding the connected place
- Designing the connected place
- Managing the connected place

These principles are broadly based on regulatory requirements such as the Network and Information Systems (NIS) and General Data Protection Regulation

(GDPR) in collaboration with the UK Centre for Protection of National Infrastructure (CPNI) (with relevance to critical infrastructures). CPNI has also provided guidance under a specification entitled PAS185:2017 for the development and operation of connected places [22–24].

The US Government has legislated "The Internet of Things Cybersecurity Improvement Act". This regulatory requirement overseen by the US NIST institute mandates US federal agencies are required to comply with the guidance published by NIST [25, 26]. This guidance is specifically intended to work with existing NIST frameworks (NIST 800-53—Security and Privacy Controls for Information Systems and Organizations) to tighten up cyber security for IoT. It also enables a baseline set of standards for manufacturers to comply with when developing and building IoT devices. It has some limitations as the regulatory framework only mandates US federal organisations to comply, others may do so on a voluntary basis. Additionally, its primary focus is securing IoT, rather than considering DT technology.

Applicable standards such as NIST 800-53 [27] and ISO 27001 [28] have relevance. Applicable sections of ISO 27001 include segregation of duties, access controls, user training and awareness, asset management, information classification, user access management (including privileged account management), user responsibilities, cryptographic controls, malware and vulnerability prevention, security in development, supplier relationships and legal and contractual requirements (including accessing and handling PII and IP). Whilst these controls are generic, they provide guidance to ensure access controls are robust and are used to protect business information (including PII). This requires an up-to-date inventory of assets and a solid understanding of the business information held within them.

2.5 Legal Requirements to Protect Private Data

Data protection is a regulatory requirement in most modern economies. The data protection act (DPA) in the European Economic Area (EEA) is GDPR (the UK has also adopted this DPA after leaving the bloc). This law applies anywhere to anyone who targets or collects personal data of a citizen in the European Union (EU) [29].

GDPR Article 25 states data protection must be by design and by default [29]. Data protection by design is a requirement for organisations to design privacy handling requirements into new systems that process personal data. This includes technical and organisational requirements. This principle is designed to reduce privacy risks and promote trust.

Data protection by default mandates an organisation to protect privacy technically by choosing the most protective settings [29]. Furthermore, only personal data that achieves the specific business purpose should be processed [30].

3 Organisations Concerned with the Development of Smart Cities

The G20 of developed nations Smart Cities Alliance was formed in 2019 [31] to bring together industrialised countries to work in a public–private sector partnership to enhance smart city standards through collaboration. This partnership is currently developing its standards. Currently released is a policy for cyber resilience based on the NIST Cybersecurity Framework (CSF) [32], the NCSC Cyber Assessment Framework (CAF) [33] aimed at critical infrastructures, or requirements for UK/EU NIS compliance, or organisations associated with public safety. Other currently released policies include a Privacy Impact Assessment (PIA) which takes into account legal and ethical considerations for data and information privacy and a Cyber Accountability Model.

At the time of writing this document, numerous policies on the roadmap are not yet released. It is still currently incumbent on the technology owners and local policymakers to ensure specific technical controls comply with the standards and frameworks within their region.

Whilst all these policies align with security good practice guidelines and standards, they are generic and high-level. From the examination of the standards and frameworks from the UK, EU, US and G20, there is a lack of maturity concerning specific technical cyber security controls that are relevant to DTs that connect and control their physical counterpart. The policies released are generally based on contemporary controls and do not take account of the novel threats and risks posed by introducing DTs into smart cities. This is where the current gap in knowledge exists.

4 Current Considerations for Digital Twin Cyber Security

A key cyber security consideration for a DT is the synchronous bi-directional communication with its physical counterpart [8]. The protection of information held within a DT, and resultant data flow between the CPS systems is paramount. Maliciously attacking a DT could result in loss of information such as intellectual property, privacy loss such as Personally Identifiable Information (PII), degrading the quality of service, or in extreme cases the safety of employees or citizens. Holmes et al. [8] advocate the CIA triad as a method to secure DTs.

The Industrial Internet Consortium (IIC) outlined the requirement to protect the DT itself, (i.e., protect the intellectual property within), and to protect the physical asset(s) it connects to. When considering security controls, it must be done holistically and with consideration of the strength of the security of the IoT device, the network and any other connected device or system. Consideration needs to account that a compromise of the physical asset could lead to compromise of the DT, and conversely, compromise of the DT could lead to compromise of the physical asset. The article states there must be a security culture within an organisation that is led from the top. A

Table 1 Digital twin risks

Risk	Summary	Aspect of CIA triad
System access	The unauthorised access or unauthorised elevation of privileges could lead to IP theft, non-compliance, and loss of information integrity	Confidentiality/integrity
IP theft	Theft of IP (i.e., theft of business information, PII) enabling an attacker to sell on IP, or use to reproduce the DT for further malicious activity	Confidentiality
Non-compliance	Compromise of the DT could lead to loss of privacy compliance requirements such as DPA's (e.g., GDPR). Resulting in fines or reputational loss	Confidentiality/integrity
Information integrity	If information integrity is compromised, the DT could compromise the physical counterpart (e.g., degrade service, produce unexpected results, endanger safety, compromise privacy)	Integrity

secure by-design methodology is needed that is incorporated into a Software-Defined Lifecycle (SDLC). Software hardening should occur, and encryption to protect data at rest and in transit [34].

A publication by the consulting firm Royal HaskoningDHV [35] highlights four contemporary cyber security risks to DTs that directly impact the confidentiality and integrity of a DT. This is summarised in Table 1.

Lomax Thorpe [35] asserts the risks covered in Table 1 should be understood by the business and governed by an enforceable policy.

Nonetheless, the US federal standards organisation NIST recently produced a draft guidance of cyber security considerations concerning DTs [1]. This literature states there is a requirement to address novel threat considerations that DTs introduce when connecting CPS. Confidence in DTs is of paramount importance, especially to ensure information and data privacy is maintained. Fundamentally, connecting DTs introduces novel threats which may not be fully addressed by existing contemporary controls. Table 2 illustrates the threat considerations posed in the NIST document.

The document produced by Voas et al. [1] refers to numerous NIST standards that exist as potential frameworks that can be used. These suggestions follow IoT-specific controls, standard risk controls, privacy considerations, the NIST CSF, encryption at rest and in transit, strong authentication, physical security controls, and fault tolerance. A Zero Trust approach is also suggested [1].

Table 2 Novel cybersecurity challenges

Statement	Summary consideration from document
"Massive instrumentation of objects (usually IoT technology)"	Vulnerabilities in IoT devices Low-level computing capability of IoT technology combined with limited upgradeability or patching of vulnerabilities Trusted sourcing and deployment of IoT devices Remote control of IoT devices at a mass scale and within different environments (e.g., botnet attack) IoT device secure communication (e.g., PKI, symmetric/asymmetric encryption of link)
"Centralization of object measurements"	Containment of potentially sensitive information within a central location Unauthorised system access compromise could lead to loss of IP from all data collated from sensors sent to the DT (e.g., information theft, ransomware) A large volume of technology (IoT and DTs to protect) where a vulnerability in one could lead to compromise of the other Taking control of DT leads to control of physical devices Interconnected system vulnerabilities
"Visualization/representation of object operation"	Manipulation of how information is presented through malicious access to the DT Loss of data integrity leading to system integrity loss (including connected systems)
"Remote control of objects"	Compromise of the physical objects a DT remotely controls Presentation of false readings Safety of sensor, human safety Loss of IP/data theft
"Standards for digital twin definitions"	Removes the proprietary aspects of the technology Enables attackers to build their own DT definitions (also consider phishing techniques) Could become part of an attacker toolkit

Source [1]. Novel threat—"NIST Draft NISTIR 8356 Considerations for digital twin technology and emerging standards"

5 Risks and Architectural Requirements

Overall, the question around the protection of the DT in a smart city requires further consideration. For example, the inclusion of techniques and controls such as privacy by design, privacy enhancing technologies (PETs) and other techniques discussed earlier above [1, 8, 35]. As a smart city grows so will the convergence and coexistence of different technology and data. Fundamentally the specific types of technology, usage of the technology and the sensitivity of the information should drive the specific controls (i.e., through a risk-based methodology). Continuous development of controls to counter contemporary and novel threats will enable a future framework of controls that will be state-of-the-art.

The protection of data and information must be included when designing and deploying DTs to ensure it is not subject to theft, accidental loss, or misuse. This is especially important when privacy is a requirement. Technology, processes and authorised people need to be identified to ensure correct cyber security governance is in place. This is an especially difficult undertaking. For example, a smart CCTV system will capture multiple thousands of people in a city per day. One of the most important undertakings is ethically protecting the right to privacy, whilst still fulfilling the purpose of the system.

Examining the confidentiality perspective (of the CIA triad) and making consideration for the privacy requirement of citizens' data is closely linked to data protection regulations [13]. Information anonymisation or encryption may be a regulatory requirement to protect privacy where a sensor has retrieved Personally Identifiable Information (PII) which is processed in a DT or connected system. Business intellectual property information is also highly likely to be contained within a DT and its physical counterpart and requires protection. To promote confidence and trust in the use of DT technology, data and information must be used ethically. The benefits of using DTs should be clearly defined to ensure it is not perceived as an information gathering service of citizen data (e.g., for unethical surveillance purposes [36].

Some work has commenced on the creation of a framework required to protect DTs and their physical counterparts in the industry. Gehrmann and Gunnarsson [37] has produced a set of requirements (Table 3) that address aspects of system security, performance, and accuracy of the data exchange between DTs and dependant devices for industrial automation and control system security. The table below summarises the findings.

From a smart city perspective, these requirements in the table above are valid and comparable. A smart city has a strong mandate to protect citizen safety and keep mission-critical services running for many people. Numerous high-profile attacks have recently occurred against city services (and utility providers providing critical services). Furthermore, the protection from supply chain attacks must also be mitigated to ensure the propagation of malware [38] is limited throughout the smart city digital infrastructure. A recent global supply chain attack in 2020 illustrated the threat by using techniques including infiltration, lateral movement across systems, and implemented malware to eventually steal information [39].

Table 3 Architectural requirements for industrial digital twins

Requirement	Summary consideration
"Synchronization security"	The starting state and idle state between the DT and its counterpart are always in alignment The DT provides confidentiality protection The DT provides synchronisation protection
"Synchronization latency"	Message exchanges between the DT and physical twin must not affect time-critical control functions
"Digital twin external connections protection"	Authentication of connections between the DT and any external entities must occur to protect confidentiality and integrity
"Access control"	Access controls are applied to the DT and any entity that requests access to the DT. Includes third parties and information exchanged with other DTs
"Software security"	The physical twin software should be trustworthy and must be resilient to zero-day attacks (e.g., a backup available)
"Local factory network isolation"	Network controls are required to be in place to ensure only valid synchronisation traffic is permitted and can mitigate against DoS attacks on the physical twin
"Digital twin Denial-of-Service (DoS) resilience"	The DT should be protected from DoS attacks, whilst ensuring this doesn't restrict synchronisation between the DT and the physical counterpart (see "synchronization latency")

Source [37]

6 Conclusion

Any digital replica of a physical entity that can communicate and control its physical counterpart could be prone to the contemporary and novel threats discussed earlier, either accidentally or maliciously. This could result in defects, IP theft, or even threaten human life. Therefore, the protection of the DT is at least of equal importance to protecting the connected operational physical assets.

Key considerations for DTs within smart cities are citizen safety and privacy are of paramount importance, and the protection of intellectual property is also of key importance. What is required is a specific framework to protect the DT that takes account of novel threats and that understands the risk landscape for a DT within a smart city (and other related sectors). Consideration should be made to the protection of data and information. Where feasible technical controls should be state-of-the-art to build on existing contemporary controls.

All the considerations here are valid and critical; however, they are generally based on generic contemporary controls, rather than specifically targeting the protection of a DT with the novel risks and threats it may encounter (including the environment it is used in). The considerations above are predominantly related to IoT and CPS. This highlights there is a gap in the literature concerning the protection of confidentiality of data and information for a DT in a smart city.

Conflicts of Interest The authors declare no conflict of interest.

References

1. Voas J, Mell P, Piroumian V (2021) (Draft) Considerations for digital twins standards. NIST Database (Draft). https://nvlpubs.nist.gov/nistpubs/ir/2021/NIST.IR.8356-draft.pdf
2. Singh M, Fuenmayor E, Hinchy E, Qiao Y, Murray N, Devine D (2021) Digital twin: origin to future. Appl Syst Innov 4(2):36. https://doi.org/10.3390/asi4020036
3. Singh R, Al-Khateeb HM, Ahmadi-Assalemi G, Epiphaniou G (2021) Towards an IoT community-cluster model for burglar intrusion detection and real-time reporting in smart homes. In: Montasari R et al (ed) Challenges in the IoT and smart environments, a practitioners' guide to security. Advanced sciences and technologies for security applications. Springer International Publishing, Cham, pp 53–73. Print ISBN 978-3-030-87165-9. Electronic ISBN 978-3-030-87166-6. https://doi.org/10.1007/978-3-030-87166-6_3
4. Fuller A, Fan Z, Day C, Barlow C (2020) Digital twin: enabling technologies, challenges and open research. IEEE Access 8:108952–108971. https://doi.org/10.1109/ACCESS.2020.2998358
5. Ahmadi-Assalemi G, Al-Khateeb HM, Maple C, Epiphaniou G, Alhaboby ZA, Alkaabi S, Alhaboby D (2020) Digital twins for precision healthcare. In: Jahankhani H et al (ed) Cyber defence in the age of AI, smart societies and augmented humanity. Advanced sciences and technologies for security applications. Springer International Publishing, Cham, pp 133–158. ISBN 978-3-030-35746-7. https://doi.org/10.1007/978-3-030-35746-7_8
6. Ahmadi-Assalemi G, Al-Khateeb HM, Epiphaniou G, Maple C (2020) Cyber resilience and incident response in smart cities: a systematic literature review. Smart Cities 3:894–927. https://doi.org/10.3390/smartcities3030046
7. Mylonas G, Kalogeras A, Kalogeras G, Anagnostopoulos C, Alexakos C, Muñoz L (2021) Digital twins from smart manufacturing to smart cities: a survey. IEEE Access 9:143222–143249. https://doi.org/10.1109/ACCESS.2021.3120843
8. Holmes D, Papathanasaki M, Maglaras L, Ferrag MA, Nepal S, Janicke H (2021) Digital twins and cyber security—solution or challenge? In: 2021 6th South-East Europe design automation, computer engineering, computer networks and social media conference (SEEDA-CECNSM), pp 1–8. https://doi.org/10.1109/SEEDA-CECNSM53056.2021.9566277
9. NCSC (2021) Connected places cyber security principles. NCSC. https://www.ncsc.gov.uk/collection/connected-places-security-principles
10. Ibm.com (2009) A vision of smarter cities. https://www.ibm.com/downloads/cas/2JYLM4ZA
11. Deren L, Wenbo Y, Zhenfeng S (2021) Smart city based on digital twins. Comput Urban Sci 1:4. https://doi.org/10.1007/s43762-021-00005-y
12. Elmaghraby A, Losavio M (2014) Cyber security challenges in smart cities: safety, security and privacy. J Adv Res 5(4):491–497. https://doi.org/10.1016/j.jare.2014.02.006
13. Benedik R, Al-Khateeb HM (2021) Digital citizens in a smart city: the impact and security challenges of IoT on citizen's data privacy. In: Montasari R et al (ed) Challenges in the IoT and smart environments, a practitioners' guide to security. Advanced sciences and technologies for security applications. Springer International Publishing, Cham, pp 93–122. Print ISBN 978-3-030-87165-9. Electronic ISBN 978-3-030-87166-6. https://doi.org/10.1007/978-3-030-87166-6_5
14. Iec.ch (2021) Understanding IEC 62443. https://www.iec.ch/blog/understanding-iec-62443
15. Sans.org (n.d.) Introduction to ICS security part 2. SANS Institute. https://www.sans.org/blog/introduction-to-ics-security-part-2/

16. Sookhak M, Tang H, He Y, Yu FR (2019) Security and privacy of smart cities: a survey, research issues and challenges. IEEE Commun Surv Tutor 21(2):1718–1743. https://doi.org/10.1109/COMST.2018.2867288

17. Montasari R, Jahankhani H, Al-Khateeb HM (2021) Challenges in the IoT and smart environments—a practitioners' guide to security, ethics and criminal threats. Advanced sciences and technologies for security applications. Springer International Publishing. Print ISBN 978-3-030-87165-9. Electronic ISBN 978-3-030-87166-6. https://doi.org/10.1007/978-3-030-871 66-6

18. Vitunskaite M, He Y, Brandstetter T, Janicke H (2019) Smart cities and cyber security: are we there yet? A comparative study on the role of standards, third party risk management and security ownership. Comput Secur 83:313–331. https://doi.org/10.1016/J.COSE.2019.02.009

19. European Union Agency for Cybersecurity (2019) Good practices for security of IoT: secure software development lifecycle. European Network and Information Security Agency. ISBN 978-92-9204-316-2. https://doi.org/10.2824/742784

20. European Union Agency for Cybersecurity (2017) Baseline security recommendations for IoT in the context of critical information infrastructures. European Network and Information Security Agency. ISBN 978-92-9204-236-3. https://doi.org/10.2824/03228

21. Enisa.europa.eu (n.d.) ENISA good practices for IoT and smart infrastructures tool. https://www.enisa.europa.eu/topics/iot-and-smart-infrastructures/iot/good-practices-for-iot-and-smart-infrastructures-tool/results#Smart%20Cities

22. NCSC (2021) Connected places cyber security principles. https://www.ncsc.gov.uk/files/NCSC-Connected-Places-security-principles-May-2021.pdf

23. Cpni.gov.uk (2021) Security-minded approach to open and shared data. https://www.cpni.gov.uk/security-minded-approach-open-and-shared-data

24. Cpni.gov.uk (2022) Security-minded approach to developing smart cities. https://www.cpni.gov.uk/security-minded-approach-developing-smart-cities

25. Congress.Gov (2020) H.R.1668—IoT cybersecurity improvement act of 2020. https://www.congress.gov/bill/116th-congress/house-bill/1668

26. NIST (2020) NIST releases draft guidance on Internet of Things device cybersecurity. https://www.nist.gov/news-events/news/2020/12/nist-releases-draft-guidance-internet-things-device-cybersecurity

27. NIST (2020) Security and privacy controls for information systems and organizations. https://nvlpubs.nist.gov/nistpubs/SpecialPublications/NIST.SP.800-53r5.pdf

28. Bsigroup.com (n.d.) ISO 27001—information security management (ISMS). https://www.bsigroup.com/en-GB/iso-27001-information-security/

29. European Commission, Directorate-General for Justice and Consumers (2018) The GDPR: new opportunities, new obligations: what every business needs to know about the EU's General Data Protection Regulation. Publications Office. Print ISBN 978-92-79-79453-7. https://doi.org/10.2838/6725. PDF ISBN 978-92-79-79430-8. https://doi.org/10.2838/97649

30. Ico.org.uk (n.d.) Data protection by design and default. https://ico.org.uk/for-organisations/guide-to-data-protection/guide-to-the-general-data-protection-regulation-gdpr/accountability-and-governance/data-protection-by-design-and-default/

31. Globalsmartcitiesalliance.org (2020) About the alliance—GSCA v2. https://globalsmartcitiesalliance.org/?page_id=107

32. Barrett M (2018) Framework for improving critical infrastructure cybersecurity version 1.1, NIST cybersecurity framework. https://nvlpubs.nist.gov/nistpubs/CSWP/NIST.CSWP.04162018.pdf

33. NCSC (2019) NCSC CAF guidance. NCSC. https://www.ncsc.gov.uk/collection/caf/caf-principles-and-guidance

34. Hearn M, Rix S (2019) Cybersecurity considerations for digital twin implementations. https://www.iiconsortium.org/news/joi-articles/2019-November-JoI-Cybersecurity-Considerations-for-Digital-Twin-Implementations.pdf

35. Lomax Thorpe B (n.d.) Risk mitigation in digital twins. https://global.royalhaskoningdhv.com/digital/resources/blogs/risk-mitigation-in-digital-twins

36. Mehta A (2022) Facial recognition technology 'will turn our streets into police line-ups', campaigners say. Sky News. https://news.sky.com/story/facial-recognition-technology-will-turn-our-streets-into-police-line-ups-campaigners-say-12572433
37. Gehrmann C, Gunnarsson M (2020) A digital twin based industrial automation and control system security architecture. IEEE Trans Ind Inform 16(1):669–680. https://doi.org/10.1109/TII.2019.2938885
38. Irshad M, Al-Khateeb HM, Mansour A, Ashawa A, Hamisu M (2018) Effective methods to detect metamorphic malware: a systematic review. Int J Electron Secur Digit Forensics 10(2):138–154. ISSN 1751-9128. https://doi.org/10.1504/IJESDF.2018.090948
39. Mandiant.com (2020) Highly evasive attacker leverages SolarWinds supply chain to compromise multiple global victims with SUNBURST backdoor. Mandiant. https://www.mandiant.com/resources/evasive-attacker-leverages-solarwinds-supply-chain-compromises-with-sunburst-backdoor

IoE Security Risk Analysis in a Modern Hospital Ecosystem

Sadiat Jimo, Tariq Abdullah, and Arshad Jamal

Abstract Internet of Everything (IoE) and Internet of Things (IoT) paradigms emerged in the recent years as key elements of the infrastructures in business, industry and everyday life. This has, created new challenges, including those related to privacy and security, in the pervasive computing area. The Internet of Things is recognized for allowing the connection of virtual and physical worlds by giving processing power to "things". The Internet of Everything goes beyond that by connecting people, data, and processes to the Internet of Things, thereby making a connected world. For any technology to be successful and achieve widespread use, it needs to gain the trust of users by providing adequate privacy and security assurance. Despite the growing interest of the research community in IoT and IoE, and the emergence of vibrant literature addressing its architecture and its elements, the security and privacy of these systems and the consequential ways in which the varying capabilities of constituent devices might impact it, are still not fully understood. In this paper, a modern hospital ecosystem is used as a case model for the IoE security risk analysis. This model is used for understanding the nature of cyber-attacks against the healthcare industry with a focus of first identifying the threat actors that attack the health industry, why they do so, and how they do so. To answer these questions, an analysis was carried out on medical-related systems and devices used in the healthcare industry using Shodan IoT search engine. A DREAD threat model exercise is then used to carry out a qualitative risk analysis on healthcare networks to understand where, among various threats, the greatest risk lies. This analysis also included a focus on supply-chain attacks and the way this translates to the healthcare network. Finally, results from the DREAD threat model are used to recommend technical and non-technical

S. Jimo · T. Abdullah (✉)
Department of Computing and Mathematics, University of Derby, Derby, UK
e-mail: t.abdullah@derby.ac.uk

S. Jimo
e-mail: s.jimo1@unimail.derby.ac.uk

A. Jamal
Northumbria University London Campus, London, UK
e-mail: arshad.jamal@northumbria.ac.uk

© The Author(s), under exclusive license to Springer Nature Switzerland AG 2023
H. Jahankhani (ed.), *Cybersecurity in the Age of Smart Societies*,
Advanced Sciences and Technologies for Security Applications,
https://doi.org/10.1007/978-3-031-20160-8_26

451

measures that would help in providing security and assuring privacy within healthcare industry utilizing IoE technology.

Keywords Internet of Everything · Security · Attacks · Threats · Privacy · Preventive measures

1 Introduction

The Internet of Everything (IoE) is a concept that intelligently connects people, data, processes and things (IoT) through the Internet [1]. IoE is designed to improve how industries, cities and people live today. Everyday objects like smart home appliances, wearable devices and so on are assigned digital identifiers which gives users new way to interact, share information and connect [2]. The IoE provides solutions for a wide range of applications such as: health care, retail, emergency service, intelligent cities, waste management, traffic congestion, safety, logistics, and industrial control. According to Cisco, IoE is defined as [1]: *"Internet-of-Everything (IoE) is a valuable and relevant networked connection that allows the interaction and integration of people, data, process and things (IoT) over the internet for rich experience, unprecedented economic opportunities and turning information into acts that seats new capabilities for individuals' businesses and countries"*.

1.1 Overview of Cyber-Attacks in the Healthcare Industry

The global life expectancy has steadily increased due to the improvement in medicine and health technologies. The heart of any modern hospital is the technology that helps with the transfer and sharing of patient data to enable quick, effective and cooperative patient care by the different disciplines in the medical field. Hospital's operations such as medical (admission and discharge, diagnostic, treatment, and life support, etc.), administrative, financial, legal, records and all other information processing aspects are usually managed by a Hospital Information System. This means that every device and application running on a modern hospital network represents a possible entry point for a cyber-attack against a technology-driven modern hospital. The critical nature of hospitals puts them in a position whereby cybercriminals have realized that there is a high probability of a pay-out if the IT environment gets compromised using ransomware. Apart from this, valuable information such as personal identifiable information (PII) and financial data can be monetized in the deep web market or used for identity theft. Healthcare breaches can specifically be serious because personal data can, in some cases, mean the difference between life and death. For example, medication could get mixed up, or people might fail to get treatment for critical conditions such as diabetes. At the same time, the increasing number of IoT devices being used in healthcare to power everything, from pacemakers to wearable location trackers for elderly, are adding to the risk.

The hospital environment has various pathways for several threat actors and many vulnerable areas. Although ransomware has been in the spotlight for public attention and media coverage, it is not the only threat. Based on our research into cyber threats against hospitals, three broader high-risk target areas by cybercriminals are hospital operations, data privacy and patient health. These high-risk areas are also the focus of this paper. In the remainder of this section, we explore who is attacking healthcare systems, attack motivations, attack methods and the supply chain attacks for a better understanding of the rest of the paper.

(1) **Who is attacking the healthcare industry**: There are different categories of attackers that might want to attack, steal and abuse the healthcare system for a variety of reasons. They range from highly skilled criminal gangs that are controlled and funded or hackers using different methods such as ransomware and phishing to generate illegal revenue for their gang or even politically driven malicious actions.

Software espionage and customized malware tools have been used to steal intellectual properties, gather intelligence, cause social engineering attacks or gain competitive advantage. For instance, in 2014, a foreign government sponsored an attack on the second largest healthcare insurance provider in the United States [3]. Another type of attacker are internet activists known as Hacktivists who tend to attack highly profiled and visible targets or cyber assets to draw attention to their political causes. Lastly, script kiddies are attackers with low-level programming skills that use automated tools for hacking and are mostly motivated by attention on social media sites from peers [4].

Inside threat is another possible category of attackers. These attackers are usually motivated by money, revenge, politics, ego, ideology, and coercion, or a disgruntled employee who steals equipment or data or keeps old employee and admin accounts active for snooping purposes. Also, insider threats can be out of negligence, like opening a phishing email by mistake [5].

(2) **Motivations for Attacking the healthcare industry**: While majority of cyber-attacks nowadays are motivated by money, that's not always the case for attacks on healthcare industry. Healthcare providers such as hospitals are highly visible targets and an attack against them will be a high impact, which is a key motivator for many of these perpetrators. Vulnerable patient's health and safety will suffer whenever there is disruptive attack that can sabotage, disable or knock offline critical systems inside a hospital.

Stolen data from Intellectual properties such as PII, drug trial data, financial/insurance data, medical records, research data and the likes, can be used for financial fraud, identity theft, privacy violation, blackmail, industrial espionage, and for sale in the dark web. Insider attacks are mostly an act of revenge because perpetrators have physical access to the system or expert knowledge of their use. Meanwhile, Hacktivists, draw attention to their social and/or political industry causes by defacing high-profile targets.

(3) **Attack Methods on the healthcare industry**: The healthcare industry is a complex and massive ecosystem with thousands of endpoints, users and systems. Unpredictable attacks often surface in a hospital ecosystem due to the function, complexity and size. The following attack vectors can be used by threat actors to sabotage or infiltrate a system.

- *Distributed denial-of-service (DDoS) attacks*: Normal traffic of a targeted service or network is overwhelmed with a flood of internet traffic, thereby brining the service or network down. Multiple compromised systems are used to launch this type of attack.
- *Malware*: Malicious software code specifically designed to disable, compromise, disrupt, damage or steal data from computers. There are numerous examples where malicious code such as ransomware, Trojan, keyloggers, worms, and others have been used to attack healthcare networks.
- *Spear phishing*: Deceptive emails are sent to specific individual or organization usually from a known source or trusted sender seeking unauthorized access to confidential information.
- *Exploitation of software vulnerabilities*: Weaknesses in software are used for malicious purposes. The U.S. Food and Drug Administration (FDA) in August 2017, recalled over half a million pacemakers due to the vulnerabilities found in the firmware that could allow hackers access to the device and allow them to manipulate battery and pacing strength [6].
- *Privilege Misuse*: A situation where an unauthorized person or an insider uses legitimate permissions for malicious activities. In one example, a hacker accessed a healthcare supplier's network by installing a 3rd party software that had weak passwords and allowed administrative access.
- *Data manipulation*: involves altering data or the digital images. In 2015, FDA warned that certain infusion systems contained a vulnerability that could allow a hacker to manipulate the data in infusion pumps used for dosage calculations, thus putting patients' lives at risk [7].

(4) **Healthcare Supply Chain Attacks**: The supply chain is often one of the overlooked aspects of hospital operation. There are a set of people who can attack a hospital through the supply chain that delivers technology to the hospital. These attackers are commonly worried about and are hard to detect and protect against. Through the supply chain, perpetrators can introduce an unwanted design or function, manipulate data, extract sensitive and confidential information, introduce counterfeit devices, disrupt daily operations, install malicious software, and affect business continuity. The risk in a hospital supply chain can be from numerous potential entry points such as medicine and medical product suppliers. Threat actors might have tampered with the manufacturing process of a product been delivered to the clinic/hospital. Other entries are through vendors, medical equipment contractors, HVAC, telephony, ISP or even hospital staff.

Several types of attacks can be launched through a hospital supply chain, the most common is through 3rd party vendors. These vendors have access to sensitive credentials like logins, passwords, and badge access, all of which can be compromised. In 2019, a night shift security guard, with a work access badge to a building, installed a botnet in the hospital's HVAC controllers [8]. Also, a hospital in Illinois stored paper copies of their patient records in the office after transitioning to EHR, these paper copies were sold without their knowledge [9]. These breach examples show that clinics/hospitals can be at risk through supply chain attacks. A risk-based vendor management program under a comprehensive enterprise risk management/governance framework will assist in minimizing supply chain threats.

Any of the above-mentioned attacks can be used to launch a major cyber-attack against hospitals. According to the statistics and data provided by Privacy Rights Clearinghouse (PRC), a non-profit California based corporation, the snapshot below shows the common attack types of the cyberthreat landscape of the health industry. It also shows that the number of reported breach incidents in hospitals resulting from malware attacks and hacking are increasing.

Ransomware has been affecting the whole cyberthreat landscape for a long time. However, WannaCry incident to date was the highest profile case among other healthcare-related ransomware attacks.

In most case, ransomware uses email phishing as the main infection vector, but the WannaCry ransomware have a built-in worm-like function. Ransomware encrypts databases, folders, and documents on the victim's computer, making them inaccessible, and demands a ransom payment in the form of digital currency like Bitcoin to decrypt and give back control of the data.

2 Related Work

The Internet of Everything is faced with privacy and security issues [10]. Due to this, there has been a significant amount of research to address these challenges associated with devices. There has been a major focus on the Internet of Things which is the foundation the Internet of Everything is built on. Various researchers have tried to address the security challenges in IoT as an isolated subject without our attempt to bring it into the context of a more comprehensive system including IoE entities such as people, data, and processes. The thought that these harmless internet-connected devices (smartphones, wearables devices, tablets, routers, switches, surveillance camera, healthcare smart devices etc.) can be accessed by an unauthorized person to conduct various cyber-criminal activities like ransomware, video recording, deactivating or manipulating device capabilities for black mail purposes, steal useful information, create a havoc and disrupt network or business continuity is quite worrying and scary. The emergence of the IoT and IoE has increased security importance for device makers, software vendors and organization that depends on interconnected network of smart devices [11] to support operations and serve customers. Several cybersecurity incidents related to IoT have been reported in various industries and sectors (steel mills, energy grids and water supplies). The amount of damage is expected to increase by 32% by 2021 [12].

A large retailer in 2013 suffered a major privacy breach when hackers gained access to their network through their heating, ventilation and air-conditioning system. Another German iron plant suffered a fire damage when their control system was breached by hackers causing a furnace to shut down which led to a fire outbreak. A joint research team from the University of California, San Diego and University of Washington have recently shown that hackers can gain remote access to a vehicle's critical systems by using connected applications that enable roadside assistance. Through the music system's CD drive, hackers were able to take control of the car, this highlights the potential risks in the development processes and supply chain for companies manufacturing the cars; application creators and wireless technology; and for the automotive industry.

An IoT threat-based security analysis for smart healthcare system [13], power management and smart cars was conducted by developing a threat model consisting of sources of threats, classes of attack vectors and attack impacts to determine where efforts should be invested to secure IoT network. Other researchers [14, 15] also address similar security issues facing IoT implementation.

Researchers at a French technology institute, Eurecom, conducted an analysis of approximately 32,000 firmware images from potential IoT devices manufactures and discovered 38 vulnerabilities across 123 products. About 140,000 internet facing devices were confirmed to have some of these vulnerabilities during the research. The vulnerabilities discovered includes poor encryption and backdoors that can be exploited by an unauthorized person, opening hundreds of thousands of devices on a network with potentially serious consequences.

Many wireless IP cameras are susceptible to cyberattacks. This study also showed how data from millions of Internet-connected devices are available and readily accessible on the search engine called Shodan. Their research compliments this research as we will see in subsequent chapters how internet-connected medical-related systems and devices are exposed to the internet. There was an incident in early 2012, whereby hackers compromised the software running on SecurView IP cameras. With search engine like Shodan that provides insight into data, IoE security needs to be heightened to ensure data privacy and mitigate virtual and physical risk to IoE devices.

Fell and Barlow [9] discussed that addressing the connectivity, security, mobility and reliability of IoT challenges through IPv6 can be used to prepare IoE for the future. IPv6 adds a layer of security (authentication and encryption) by design when implemented which is an advantage over IPv4, but this not necessarily prepare IoE for the future. The Internet of Everything security requires a layered approach where each stakeholder, vendor, user and even mobile and internet providers contribute to security and data safety, which is what this research focuses on [16].

A privacy enhancement protocol over Bluetooth Low Energy (BLE) advertising channels which are based on a three-way Handshake protocol between the peripheral and the gateway for nonce R's deployment [17]. This enhancement is however not practical as it requires changing both peripheral and protocol.

An IoT driven system was proposed by Banga [18], the security system utilizes a web camera to detect the motion of an intruder in a camera range when the owner is not at home. Whenever an intrusion is detected, a security alert is sent to the owner. This security system relies on SMS using GSM technology.

In a recent study, it was lent that more than 35% businesses using IoT technology do not change the default password of their devices and 54% of these businesses do not use 3rd party security tools to protect their network and devices from both inside and outside security threats which is both alarming and shocking. The year 2013 Target attack was as a result of the vulnerability found in the company's HVAC system through which hackers gained unauthorized access to, and cause havoc on the company's network [19, 20]. In 2017, a wide scale Distributed Denial-of-Service (DDoS) attack was mounted through IoT devices which affected a large percentage of TalkTalk routers in the United Kingdom and Deutsche Telekom in Germany.

Wearable devices, smart utilities, Industrial IoT, and smart factory initiatives are already being used in many organizations. But securing devices, data and networks, tackling security complexity and complying with regulations are major challenges. The risks are no longer theoretical: On average, some responding organizations claimed they have suffered at least three attacks on IoT devices over 12 months, while just a quarter (27%) have not experienced any [14]. Yet cybersecurity remains a crucial challenge and hindrance to progress. If IoT endpoints are left unsecured, they could be hijacked to gain access to corporate networks, conscripted into bonnets or sabotaged to disrupt key processes which makes security a big challenge for IoE. More connection means more data vulnerability, there are those who want to take advantage of the vulnerability. The critical success factor for businesses is data protection—we must secure our data.

3 Methodology

For this research, the IoT device search engine—Shodan was extensively used to collect raw data that helped to identify unpatched vulnerabilities in the exposed cyber assets. Shodan helps to conduct an Open-Source INTelligence (OSINT) gathering for different organizations, geographic locations, services, devices etc. Information from Shodan can also be used to perform detailed surveillance and gather intelligence about an attack by an adversary [21].

The basic unit of data that Shodan gathers is called the banner, textual information that describes a service on a device. The content of the banner varies depending on the type of service. In addition to the banner, Shodan also grabs metadata about the device, such as its geographic location, hostname, operating system, and more [21].

On Shodan, because searching for exposed devices can't always be tracked back to the organization to which they belong, devices are oftentimes registered to the Internet Service Provider (ISP). For this research, it is more desirable to have high-confidence verifiable results vs. larger data sets that may contain false positives. To avoid false positive results, the Shodan searches were narrowed down using the following search filters: "org:hospital", "isp:hospital", "org:clinic", and "isp:clinic". Using these search filters, the result shows many hospitals and clinics around the world who have registered their devices or systems using their own organization names or marking themselves as the ISP. Shodan also has built-in modules (_shodan.module) that can identify protocols. The _shodan.module classification was extensively used in my analysis.

Exposed healthcare system software screenshots in this paper were collected from Shodan image database (https://images.shodan.io/). All included screenshots were found in Shodan using search term 'rfb authentication disabled', i.e., they came from VNC servers that have authentication disabled, and thus, are open to all to access.

After identifying the exposed medical-related systems and devices over the internet, they were then classified into three critical broad area of interests—information systems, medical devices and hospital operations. The DREAD threat model is used to perform qualitative risk assessment and analysis [22]. Qualitative risk analysis is opinion based [23, 24]; it uses rating values to evaluate the risk level. For this research, data were collected from the Health Information Trust Alliance (HITRUST) threat catalogue to determine the rating of these threats. The risk rating of each threat is arrived at by asking the following questions.

- Damage potential: rates the extent of damage that occurs if a vulnerability is exploited.
- Reproducibility: rates how easy it is to reproduce the attack.
- Exploitability: rates how easy it is to launch an attack. This also considers the preconditions such as whether the user must be authenticated.
- Affected users: rates a rough estimate of number of users affected if an exploit became widely available.
- Discoverability: rates how easy is it to find an exploitable weakness if found by hackers.

A rating scale of 1–3 is used to rate each category, with 3 being highest probability of the occurrence happening with a serious damage penitential and 1 being the exact opposite. Each risk is calculated using the DREAD algorithm shown below.

$$\text{Risk\_DREAD} = \frac{(D + R + E + A + D)}{5}$$

Based on the results gotten from the threat model risk analysis [25, 22], some observations were drawn to understand where, among various threats, the greatest security and data privacy risk lies. Recommendations on how to prevent or mitigate these risks were given.

A. Critical Systems Inside Hospitals

For this paper, three broad categories of systems and devices that are used in the hospital are used for this threat modelling exercise. This is by no means a comprehensive list of all the systems and devices present inside a hospital, but rather a broad cross-section of them. Threat assessment for these three "broad" categories of devices and systems is expected to provide a good understanding of the everyday cyber-threats hospitals face and help the IT staff and senior management prioritize and develop cyber defense strategies.

In this paper, data collected from the Health Information Trust Alliance (HITRUST) threat catalogue to determine the rating of these threats. A rating scale of 1–3 is used to rate each category, with 3 being the highest probability of the occurrence happening with a serious damage potential and 1 being the exact opposite. Each risk is calculated using the Risk_DREAD equation.

B. Cyberattacks Risk Measurement Against Hospitals and Clinics

Six cyber-attack vectors are most likely to be used by cybercriminals against critical hospital systems, based on the analysis of the past cyber-attacks against hospitals. The threat model exercise will help to identify critical areas that need improved defenses and active monitoring. While specific cases may have different ratings, this exercise will help us understand where the real areas of weakness are and therefore where focus should be on 'Medical Devices', 'Information Systems' and 'Hospital Operations'. The following observations are drawn based on the categories of healthcare systems and devices (Table 1):

- Most hospital network are mostly set up as a flat network with no additional layer of security such as DMZ. This means that if systems and devices that are directly not use in patient care are compromised, then this could serve as a point of entry in disrupting the hospital network.
- DDoS attacks are serious threats that are very easy to execute through compromised devices and systems. Specialized knowledge about the systems and devices are not required to launch a DDoS attack. Information about exposed systems and devices are available online using IoT search engines like Shodan, perpetrators can DDoS these exposed devices and knock them offline.
- Medical devices on their own have specialized software, which are in most cases not connected directly to the hospital network, therefore the probability of them getting compromised is very low. Except the threat actor spends time to investigate and launch a targeted attack, medical devices are not likely to be the avenue

Table 1 Categories of healthcare systems and devices

Category	Systems and devices
Information systems	Picture archiving and communication systems (PACs), mobile health applications, EMR/HER systems, laboratory information systems, radiology information systems
Medical devices	Respiratory ventilators, infusion pumps, imaging, e.g., X-ray, ultrasound, MRI, CT heart–lung machines, anesthesia machine, robotic surgical tools, dialysis machine, active and passive monitoring systems, radiotherapy systems
Hospital operations	Office applications (databases, fileservers, email, payroll), work order and staff scheduling systems, hospital paging systems, drug and equipment inventory systems, building control systems. Automated drug dispenser, barcode scanners and printers, pneumatic tube transport system

of choice for cybercriminals. Having said that, the controllers of these medical devices are connected to the hospital network and hence open to cyber-attacks. As an example, if the MRI scanner controller is compromised with ransomware, all MRI activities will stop.

- The greatest threat on medical devices as per the above assessment is DDoS attacks. The other three attacks apart from data manipulation are rated medium risk because the probability that the medical devices will be connected to the hospital network is very slim. It very difficult to manipulate data on these devices, threat actors would target the device controllers instead of the devices themselves to modify data (Table 2).
- Medical devices and information systems are not designed to send or receive emails therefore spear phishing is not directly possible. On the other hand, if a POE on the hospital network is successfully compromised, this can cause serious damage inside the hospital ecosystem.
- Hospital Operations and Information Systems threats are mostly categorized as high risk because the damage potential if these devices/systems are compromised

Table 2 Threat ratings table for hospital risk analysis

	Rating	High (10)	Medium (5)	Low (0)
D	Damage potential	The attacker disrupts the system which can cause serious damage to hospital operations thereby putting patients' lives at risk	Attacker disrupts the system and can cause moderate damage to hospital operations (lasting a short period of time and only few departments are affected)	Attackers disrupt the system and can cause minor damage on the hospital operations that does not affect patients' health directly
R	Reproducibility	Attack is easy to reproduce at anytime	Attack is easy to reproduce (only within set limitations)	Attack not easy to reproduce (even with security loopholes knowledge)
E	Exploitability	Little or no knowledge of the system is required for exploitation	A fundamental knowledge of the system is required by a skilled operator to exploit it	Attack requires a very skilled operator with in-depth knowledge of the system for exploitation to occur
A	Affected user	Many users will be affected by the attack	A good number of users will be affected by the attack	A small number of users will be affected by the attack
D	Discoverability	Attack info published and readily available. Vulnerabilities found in most used apps and systems	Applications and systems vulnerabilities are less common. It requires skills to discover exploitable weak	Unlikely to discover vulnerabilities, hence attack to applications and systems is difficult

is very high. Almost everyone inside the hospital including nurses, doctors, technicians, and patients will be directly impacted. These systems mostly use off-the-shelf software like MYSQL, Windows, etc., the vulnerabilities of these platforms are well-known and as seen in the case of WannaCry, we saw that the systems are not regularly patched.

The examined the moving parts in the healthcare ecosystem in the previous and current sections, this will give hospital IT teams a broader perspective of the existing weaknesses in healthcare networks. In the next section an IT defense strategy is provided including technical and non-technical recommendations to address the issues raised here.

4 IT Defense Recommendations for Hospitals

Prevention strategies for data breach and cyber-attacks should be part of the daily business operations at the hospitals. Although data breaches and cyber-attacks are inevitable but having mitigation processes [26], effective alert and containment are important. The key principle to defense is to assume the systems and devices are compromised and take countermeasures:

- Prevent attack preemptively by securing all avenues that can be exploited.
- Ongoing security breaches should be quickly responded to and contained to stop loss of sensitive data.
- Lessons should be learned from past incidents in order to strengthen defenses.

Relevant standards and regulations pertaining to the healthcare industry such as ISO (International Organization for Standardization), NIST (National Institute of Standards and Technology), PCI-DSS (Payment Card Industry Data Security Standard), HIPAA (Health Insurance Portability and Accountability Act), and even GDPR (General Data Protection Regulation) are put into consideration while making the recommendations to prevent and detect the kinds of attacks that can hit healthcare organizations in the next section (Table 3).

Table 3 DREAD results for medical devices

Attack vectors	D	R	E	A	D	Rating
Spear phishing	–	–	–	–	–	Not applicable
DDOS	3	3	3	2	3	High
Vulnerability exploitation	3	1	1	2	1	Medium
Malware infection	3	2	2	2	2	Medium
Privilege escalation and misuse	3	1	2	2	2	Medium
Data manipulation	3	1	1	1	1	Low

A. Technical Recommendations for Hospitals

Each organization, even within the healthcare industry have a different set of challenges when it comes to preventing and mitigating cyber-attacks but based on the cyber threats faced by a hospital from my research findings, the recommendations below are defensive strategies that are considered mandatory minimum for the hospital IT ecosystem.

(1) *Network segmentation*: Splitting the network into multiple sub-networks called VLANs (Virtual Local Area Network) to improve network performance and security. Medical devices can be assigned to a different network separate from the corporate network, by doing this, overall security is improved, and risk reduced.

(2) *Firewalls*: These are network security systems that control ingress and egress traffic based on an applied rule set. It serves as a barrier by preventing incoming traffic to a trusted network from unknown and bad domains. It can either be host-based or a network firewall.

(3) *Antimalware and Antiphishing solutions*: Antimalware products can detect, block and remove malicious software such as worms, viruses, Trojans, worms, keyloggers and so on, from files in a computer system. Antiphishing products can detect and block phishing and spam emails.

(4) *Breach Detection Systems (BDS)*: Security solutions focused on detecting intrusions caused by targeted attacks and other sophisticated threats designed to harvest information from the compromised systems. BDS analyses complex attacks out-of-band, detecting, rather than preventing, network breaches. BDS can analyse network traffic patterns across multiple protocols, identify malicious domains, and uses emulation-sandboxing to model the behavior and impact of malicious files that are being dropped or downloaded.

(5) *Intrusion Prevention System/Intrusion Detection Systems*: Are network security systems that examine and monitors the entire network traffic flow to detect and prevent suspicious traffic by doing deep packet inspection. While an IDS only detects bad traffic, IPS blocks the bad traffic when it is identified.

(6) *Patch management (virtual or physical)*: This software keeps servers, endpoints, and remote computers updated by applying the latest security patches and software updates. Virtual patch management uses a security enforcement layer to prevent malicious traffic from reaching vulnerable systems. In a large environment where patches need to be thoroughly tested before applying, virtual patching provides the stopgap measure of filtering out malicious traffic and attempting to exploit known vulnerabilities (Table 4).

(7) *Vulnerability scanner*: Automated tools that scan endpoints, servers, networks, and applications for security vulnerabilities that an attacker can exploit. One of the tried-and-tested ways malware does lateral movement is by exploiting vulnerabilities on the target machine it wants to infect. A vulnerability scanner scans and identifies unpatched vulnerable endpoints, servers, and applications, which the IT administrator can then patch.

Table 4 DREAD results for information devices

Attack vectors	D	R	E	A	D	Rating
Spear phishing	–	–	–	–	–	Not applicable
DDOS	3	3	3	3	3	High
Vulnerability exploitation	3	2	2	3	2	High
Malware infection	3	2	2	2	2	High
Privilege escalation and misuse	1	1	1	3	2	Low
Data manipulation	2	1	2	3	1	Medium

(8) *Shodan scanning*: Shodan is a search engine for internet-connected devices. Shodan provides an easy one-stop solution to conduct Open-Source INTelligence (OSINT) gathering for different geographic locations, organizations, devices, services, etc. Software and firmware information collected by Shodan can potentially help identify unpatched vulnerabilities in the exposed cyber assets. Hospitals should monitor their IP ranges in Shodan to ensure their managed devices and systems are not exposed to the internet.

B. Non-Technical Recommendations for Hospitals

An important step in ensuring that any organization can formulate a good defense strategy for their network is by training their IT staff and making sure they have access to adequate security resources. Since new threats are always emerging, it is crucial that they have the means to experiment and learn from mistakes so the organization can be saved from the next big cyber-attack or data breach. Also, an incident response process that prescribes the behavior desired from any employee that either discovers the security breach or receives the report should be in place. An incident response team that consists of members from various departments such as technical, human resources, threat intelligence, public relations, executive management and legal should be formed and identified for quick mobilization. To complement the technical recommendations already given, regular social engineering training for all staffs and third-party partners should be enforced. These training will keep their insights learned.

C. Managing Supply Chain Threats

As supply chain attacks continue to grow, hospitals are recommended to develop and improve their risk management programs for their third party and vendors. The following activities are recommended:

- Vulnerability assessment should be carried out on all new medical devices before connecting them to the hospital network. This assessment is essential as it helps determine if they pose any cyber risks and if the device has not been compromised on the manufacturer's end [27]. Also, vulnerability and security testing of 3rd party software should be performed to ensure they are free from hackers (Table 5).

Table 5 DREAD results for hospital operations

Attack vectors	D	R	E	A	D	Rating
Spear phishing	3	2	3	2	2	High
DDOS	3	3	3	3	3	High
Vulnerability exploitation	3	2	2	3	2	High
Malware infection	3	2	2	3	2	High
Privilege escalation and misuse	1	1	1	3	1	Low
Data manipulation	2	1	2	3	1	Medium

- Medical devices should be purchased from manufacturers who go through rigorous security assessment of the products during design and manufacture. Such device will pose a low risk when connected to the hospital network.
- A plan for updating and patching firmware/code for hospital medical equipment and devices implanted in patients should be developed.
- Risk assessment should be carried out on all vendors and suppliers in the supply chain. Thorough background check should be carried out on all employees who have both physical and remote access to medical devices and computers.
- Risk assessment should be carried out on all vendors and suppliers in the supply chain. Thorough background check should be carried out on all employees who have both physical and remote access to medical devices and computers.
- Authentication should be required before any devices can access the hospital network [28].

5 Conclusions

IoE networks security is about implementing an end-to-end solution and not just about securing individual devices. IoE environment security requires a layered approach where each stakeholder, vendors, users and even mobile and internet providers contribute to security and data safety.

Striking a good balance between securing the hospital network and ensuring efficient hospital operation through modern technology is a struggle that the healthcare IT teams constantly face. Regulations set in General Data Protection Regulation (GDPR) amongst others have to some extent helped to ensure and emphasized the importance of keeping healthcare systems and data secure by enforcing fines and penalties for non-compliance. It is crucial that the healthcare industry is more vigilant in ensuring that the systems and devices on their network are not searchable on the public internet. Surprisingly, a high number of exposed medical systems amongst others were found while searching on Shodan. While exposed cyber assets do not necessarily mean they are compromised, they do point cybercriminals in a specific direction if they want to find weaknesses in a target institution.

One component in the hospital operation that can also lead to security compromise but is often overlooked is the supply-chain. The many moving parts that ensure that hospitals can deliver life-preserving services can also endanger those very services such as non-core services which have physical access to the healthcare consoles, vendors who are allowed to access the internal networks and also software developer for devices that are eventually shipped to and used in diagnosing patients or even raw material suppliers. To cover the weakness in supply-chain, healthcare IT team must develop and regularly review a strategy that identifies all vendors and third parties that the clinic or hospital directly interacts with. When an IoT device is manufactured it should be issued a CA certificate. Certificates are for data encryption and communication between device and vendor system. CA certificate must be installed on the device itself so that the device is able to determine if a software update that it received is good to install through code signing certificates. One aspect where vendors still struggle with is securing user access to the device they have manufactured. Most devices are shipped with a default password and changing the default password to completely up to the user. There is nothing on the device preventing users from using poor choice of passwords like 'password1'. The more the IoE technology grows, the more the number of connected devices which then means that we must create and maintain a bunch of passwords for each of our connected devices too which can be frustrating.

The threat modeling proposed for this research is an essential foundation for defining security requirements of computer systems. Without identifying threats, it is impossible to provide assurance for the system and justify security measures taken. The DREAD threat modeling done on the healthcare industry network in this research show that DDoS attack is the greatest threat for medical devices and can be easily executed. While on the hospital information system, threats like DDoS, vulnerability exploitation, and malware infection directly impacts all hospital users and are easy to implement given the systems are typically off-the-shelf platforms hence why they are categorized as high risk.

In modern hospitals the daily operations are supported by the exchanged data among a wide array of interconnected systems, devices, and application. Technology plays a critical role in patient care, the primary objective of hospitals, and thus, it strongly recommended that the prevention strategy of data breach and cyber-attack become an integral part of daily hospital operations. Adequate priority should be given to cybersecurity because it is important that patients' health is not jeopardized by the actions of malicious and/or profiteering hackers.

A blend of security technology and threat response protocol, including employee/partner awareness and education is strongly recommended. Creating, enforcing, and frequently reviewing a risk management system and governance framework that are related to resources transfer from and to any entity outside a trusted network circle to minimize supply chain attack risk is strongly recommended. The smooth operation of daily hospital services makes a life-or-death difference for patients—IoE security should be an enabler, never an obstacle, to delivering these life-preserving services.

References

1. Capgemini (2016) https://www.uk.capgemini.com/resources/securing-the-internet-of-things-opportunity-putting-cybersecurity-at-the-heart-of-the-IoE
2. Li S, Xu LD (2017) Securing the Internet of Things. Safari Books Online. Syngress. https://my.safaribooksonline.com/book/networking/security/9780128045053. Accessed 23 Apr 2020
3. Rosner G (2016) Privacy and the Internet of Things. Safari Books Online. O'Reilly Media, Inc. https://my.safaribooksonline.com/book/hardware/9781492042822
4. Basu SS, Tripathy S, Chowdhury AR (2015) Design challenges and security issues in the Internet of Things. In: IEEE region 10 symposium, pp 90–93
5. Morrow S (2016) Insider threats at hospitals. Infosec Resources. http://resources.infosecinstitute.com/insider-threats-at-hospitals
6. Russell B, Duren DV (2016) Practical Internet of Things security. https://my.safaribooksonline.com/book/networking/security/9781785889639
7. Garun N (2017) Almost half a million pacemakers need a firmware update to avoid getting hacked. https://www.theverge.com/2017/8/30/16230048/fda-abbott-pacemakers-firmwareupdate-cybersecurity-hack
8. Dissent (2017) Smart physical therapy hacked by TheDarkOverlord. https://www.databreaches.net/ma-smart-physical-therapy-hacked-by-thedarkoverlord/
9. Fell G, Barlow M (2016) Who are the bad guys and what do they want? Safari Books Online. O'Reilly Media. https://my.safaribooksonline.com/book/networking/security/9781492042464
10. Gilchrist A (2017) IoT security issues. Safari Books Online. De Gruyter. https://my.safaribooksonline.com/book/networking/security/9781501505621. Accessed 23 Apr 2020
11. Batalla J, Mastorakis G, Mavromoustakis C, Pallis E (2017) Beyond the Internet of Things: everything interconnected. Springer
12. Khakurel J, Melkas H, Porras J (2018) Tapping into the wearable device revolution in the work environment: a systematic review. Inf Technol People 31(3):791–818
13. Atamli A, Martin A (2014) Threat-based security analysis for the Internet of Things. In: Proceedings of the international workshop on secure Internet of Things. IEEE, pp 35–43
14. Kumar SA, Vealey T, Srivastava H (2016) Security in Internet of Things: challenges, solutions and future directions. In: 49th HICSS, pp 5772–5781
15. Andrea I, Chrysostomou C, Hadjichristofi G (2016) Internet of Things: security vulnerabilities and challenges. In: IEEE symposium on computers and communications
16. Jara J, Ladid L, Gomez-Skarmeta A (2013) The Internet of Everything through IPv6: an analysis of challenges, solutions and opportunities. 4(3):97–118
17. Rerup N, Aslaner M (2018) Hands-on cybersecurity for architects. Safari Books Online. Packt Publishing. https://my.safaribooksonline.com/book/networking/security/9781788830263
18. Ullah K, Shah MA, Zhang S (2016) Effective ways to use Internet of Things in the field of medical and smart health care. In: International conference on intelligent systems engineering (ICISE)
19. Jara J, Zamora MA, Skarmeta AF (2010) An initial approach to support mobility in hospital wireless sensor networks based on 6lowpan (hwsn6). J Wirel Mob Netw, Ubiquitous Comput, Dependable Appl 1:107–122
20. Krebs B (2014) Target hackers broke in via HVAC company. https://krebsonsecurity.com/2014/02/target-hackers-broke-in-via-hvac-company/
21. Yang N, Zhao X, Zhang H (2012) A noncontact health monitoring model based on the Internet of Things. In: Proceedings of the 8th international conference on natural computation, pp 506–510
22. Czagan (2014) Qualitative risk analysis with the DREAD model. http://resources.infosecinstitute.com/qualitative-risk-analysis-dread-model/
23. Wu YL, Yin G, Li L, Zhao H (2017) A survey on security and privacy issues in Internet-of-Things. IEEE Internet Things J 99:1–1

24. Abomhara M, Koien GM (2014) Security and privacy in the Internet of Things: current status and open issues. In: International conference on privacy and security in mobile systems (PRISMS), pp 1–8
25. Omotosho A, Haruna BA, Olaniyi O (2019) Threat modeling of Internet of Things health devices. J Appl Secur Res 14(1):106–121
26. Ashraf QM, Habaebi MH (2015) Autonomic schemes for threat mitigation in Internet of Things. J Netw Comput Appl 49:112–127
27. Groden (2015) Hackers could go after medical devices next. http://fortune.com/2015/08/04/hackers-medical-devices/
28. Marin-Lopez R, Pereniguez-Garcia F, Gomez AF, Ohba Y (2012) Network access security for the internet: protocol for carrying authentication for network access. IEEE Commun Mag 50(3):84–92
29. Kouicem E, Bouabdallah A, Lakhlef H (2018) Internet of Things security: a top-down survey. Comput Netw

The Future Era of Quantum Computing

Galathara Kahanda, Vraj Patel, Mihir Parikh, Michael Ippolito,
Maansi Solanki, and Sakib Ahmed

Abstract This paper will cover the impact of quantum computing and how it will
affect the entire landscape, from security to ecommerce and even technologies such
as the blockchain. First, it introduces the basics of encryption and compares it to
classical computing and how it will differ from the quantum era of computing. The
older style of encryption, when used in classical computing, will be defenseless in
the age of quantum computing. We go over different types of protections and other
theories to protect such attacks if and when they occur from a quantum computer. Our
research paper also covers decryption comparing how classical computing decryption
can be done manually or even automatically. Still, with the introduction of quantum
computing, it would have to utilize quantum mechanical principles to perform cryptography. In addition this paper will also go over how big companies such as IBM and
their technology strive to be ahead of their competitors in regards to advancement
in the quantum space. One of the biggest companies, such as IBM, is investing in
the quantum computing field with a power of 64 Qubits, and it is still in progress of
advancement in Qubits. Quantum computing will also bring many benefits, such as
fixing old security protocols and encryption issues. It will also solve problems such as
key encryption or even make older encryptions stronger when running on a quantum
system. In addition, cyber security with quantum computing will come with some
negatives which will weaken all classical types of security, such as RSA, DSA, and
ECDSA. It can keep trying to compute until it cracks through these classical security
protocols. Ecommerce in the future will be handled through quantum computing and
more advanced security protocols. Blockchain technology emerged with Bitcoin and
is used for secure ways to do transactions and communication between two users. The
technology relies more on classical cryptography protocols, and emerging quantum
computing will create security issues including areas of blockchain data transparency.
As the conclusion on quantum computing which requires future work to be done for
implementation of effective strategies to push the classical computing technologies
to the quantum era.

G. Kahanda · V. Patel (✉) · M. Parikh · M. Ippolito · M. Solanki · S. Ahmed
Department of Computer Science and Mathematics, Rutgers University, Newark, NJ 07102, USA
e-mail: vratel7@gmail.com

Keywords Cyber security · Encryption · Quantum computing · Classical computing · Cryptography · Decryption · Qubits

1 Introduction

Cryptography protects data from individuals who can bring the Internet and privacy on the digital platform to its knees. It is the most crucial step in the modern era of new technologies where most of our data is stored in cloud services and requires more secure protection. With newer technologies developing at such a fast rate, quantum computing is the next giant leap into future technology. Quantum computing harnesses the collective properties of quantum states, such as superposition, interference, and entanglement, to perform calculations. Currently, classical computing uses 0's and 1's known as bits, while compared to Quantum computers that use qubits which are 0's, 1's, and a combination of both 0 and 1 at the same time. Quantum computing is a mixture of quantum physics and computer science. It can process more data at faster rates than modern computers. Quantum computing will reach a point where anyone can access it, not just giant tech companies with a big budget in research and development. The problem lies in the fact that even though it has great potential, it may be used in a malicious way against modern technologies.

Our personal information and security may be at risk in the quantum age of computing. With the fast-changing world of computer science and hardware development, it can be challenging for security protocols to keep up. Security and the ways of encryption and decryption will be affected by this change. Quantum computing has both positive and negative effects. The positive aspect is that it will have better forms of security protocols, and the negative aspect is the sense that quantum computing can exploit and abuse classical computing security protocols such as brute force breach. When using quantum computing, a significant amount of processing power is needed to accomplish multiple tasks simultaneously in a fraction of the time. Shift to quantum computing will also be difficult for most companies to catch up and update their security protocols that minimize cyber-attacks.

2 Encryption and Decryption

First, we need to understand encryptions in security protocol and their differences in classical and quantum computing. As the world becomes increasingly dependent on technology every day, security becomes more critical to protect users' privacy. Technology has become an integral part of our lives, however, there are risks that come with the devices and technologies, especially when our personal data is involved. That is where cyber security and the aspect of encryption come into the picture. Security through encryption is fundamental as we must protect our personal data and information where hackers dominate the current arena. Encryption in classical computing, specifically, standard technologies primarily consist of the aspects of classical computing. Encryption is when a plaintext is turned into ciphertext (encoded version) so it is hard for humans to read or understand. However, decryption is converting ciphertext into plaintext (decoded version) for humans to read or understand. The point of encryption in classical computing explicitly protects users' data with specific security measures. Different encryption techniques take place according to those preferred security circumstances and schedules. There is high scalability of sensitive information which is stored in cloud systems, and encryption plays a crucial role in protecting that information from cyber-attacks. Encryption takes its course with quantum computing when there is a computed solution to the quantum computing language and analog that wires the set of quantum computations to be outputted on the encrypted quantum data. When it comes to the quantum gates of the encryption processes, photons, and linear optics are used to encrypt the encryption analogs and schemes on the set of the quantum gates sufficient for arbitrary quantum computations [1, p. 1].

Classical cryptography relies on the high complexity of the mathematical problems allowed for a large number of operations provided. Quantum cryptography relies on quantum mechanics and uses qubits compared to the standard classical cryptography that relies on ones and zeros. Hence, the level of encryption is faster to process and generate. For instance, classical cryptography would use a substitution cipher, whereas a quantum would use a homomorphic cipher because a quantum computer is much quicker. Cryptography, meaning scrambling words, is a practice to enable secure communication.

In earlier versions of cryptography, the decryption key was the same as the encryption key, and keys were exchanged between trusted and recognized parties. Current encryption methods no longer share the private key and are much stronger because, without the private key, the use of extensive computing requirements is necessary to decrypt. There are many methods of classical/conventional cryptography; one of the most important and popular methods is the Hill cipher encryption and decryption, which generates a random matrix. An inverse of the matrix in the Hill cipher is required to decrypt, which is challenging because the inverse of the matrix does not always exist. If the matrix is not invertible, the encrypted content cannot be decrypted. The most significant advantage of conventional cryptography over classical computing is that decryption is not computationally easy, and the introduction of quantum computing nullifies this advantage [2].

In Quantum computing, cryptography methods use quantum mechanical principles to perform cryptographic tasks. This approach applies in several areas requiring accurate quantum computers, which is not yet a reality. Hence, the applications are currently theoretical. The theoretical advantage over classical computers is that quantum computers do not follow the same logic. The main effects of quantum mechanics leveraged by quantum computers include superposition, interference, and entanglement. Superposition allows quantum bits (qubits) to exist in both an on and an off state simultaneously. This functionality differs from classical computers, which are bound to using bits of binary data that are either on or off. Superposition allows for testing multiple solutions simultaneously, which is why cryptography is broken so easily. With each qubit added to the system, more operators become available to solve a problem. The increase of this processing power is known as quantum parallelism. Shor's algorithm takes advantage of quantum parallelism. To factor a 1,000-digit number, it would take a classical computer over 10 million billion years [3, p. 21093]. Quantum computers could perform math much faster.

Concerns about Quantum computing's ability to break current cryptography algorithms have increased as innovations have been made. Currently, IBM has a working and fully programmable quantum computer with 64-qubits. IBM's progress is impressive, but it is essential to realize that classical computers are able to emulate quantum computers. Today's practical full-state simulation limit is 48 qubits because the number of quantum state amplitudes required for the full simulation increases exponentially with the number of qubits, making physical memory the limiting factor. Regarding security encryption, it is estimated that about 1500 qubits are required to hack Bitcoin private keys. However, today's quantum computers have impractically high error rates and can operate only in lab conditions at temperatures near absolute zero. It isn't known whether it works practically since these larger-scale quantum chips can't be simulated. In addition, Quantum computers are not necessarily going to follow Moore's law, making it challenging to forecast breakthroughs. Also, it is a complex engineering problem to keep scaling up, which is why, so far, none of the chips have been practically useful or stable. In late 2021, IBM released a 127-qubit processor. Although this chip isn't practically useful for anything, it is still an impressive milestone towards a commercial quantum computer.

As we move towards the era of quantum computing, it becomes more important than ever to understand the impact quantum computers will have on the security of current classical protocols. Quantum computers rely heavily on principles of quantum mechanics and use superposition and entanglement to store and process information. This, in turn, allows quantum computers to solve problems outside the range of classical computers.

Codebreaking and Beyond, discuss that one problem that can be solved on a quantum computer is integer factoring. The article explains that Shor's algorithm, "breakthrough insight was that a quantum computer can sample from the Fourier spectrum of functions such as modular exponentiation in a way that is exponentially more efficient than the best-known classical algorithms" [4, p. 24]. As a result, Shor's algorithm can effectively break the RSA cryptosystem. However, since we "rely on RSA's security against digital, classical attacks—that is, both the promise and potential threat of quantum computing to classical cryptography are immediate: on the one hand, a promise to solve some of the world's hardest problems quickly, on the other hand, a threat to dismantle our security methods and reveal secrets" [4, p. 22]. So, while quantum computers can provide solutions outside the range of classical computers, they also pose a potential threat to the security protocols we use today like RSA. Moreover, quantum computers also provide a solution to the key distribution problem in cryptography as well. The problem with using symmetric and asymmetric cryptography is that they provide a weak key distribution. In other words, "the most important problems in cryptography are concerned with the security and authenticity of exchanged messages. Assume that two parties Alice and Bob wishing to communicate over the insecure (public) channel want to share a secret key. It is very important to Alice and Bob to make sure that any potential intruders did not successfully achieve the information of the key. This is where the key distribution step is used for Alice and Bob to establish a secret key prior to exchanging any message within the public channel" [5, p. 19]. However, Quantum Key Distribution (QKD) uses the principles of quantum physics to generate secure keys. This is beneficial because the security in QKD is achieved by providing a totally secure key distribution technique. For instance, if Jerry and Sue wish to exchange secret information, Jerry must initially create a secret key KS using random numbers as the seed value. KS can be of any desired bit length, where the length of the key is directly proportional to the strength of the encryption. Once created, KS is converted into qubits, which are then sent across the quantum channel. The latter can be formed with an atomic particle that follows the laws of quantum physics, such as the currently implementable photons [6, p. 58]. Therefore, QKD provides a means of secure communication and can be applied to many of today's technologies like blockchains.

Although quantum computing does provide several benefits such as solutions to problems beyond the scope of classical computers and the ability to perform calculations faster than their classical computers, there are drawbacks as well. The biggest drawback is quantum computing's ability to break many of today's security protocols, like RSA and DSA.

In "*Cyber Security in the Quantum Era*" they discuss the significant potential threats of quantum computing in cyber security. They assert that it is very important to look at this system now and to create the necessary defenses against cyber-attacks using quantum computing. A con for quantum computing can be that quantum computing is very consistent. In the Wallden and Kashefi article, it states, "The development of large quantum computers, along with the extra computational power it will bring, could have dire consequences for cyber security. For example, it is known that important problems such as factoring and the discrete log, problems whose presumed hardness ensures the security of many widely used protocols (for example, RSA, DSA, ECDSA), can be solved efficiently (and the cryptosystems broken), if a quantum computer that is sufficiently large, "fault-tolerant" and universal, is developed" [7, p. 121]. RSA which stands for Rivest-Shamir-Adleman is a type of security that makes it very difficult to factor in large numbers that are the product of two large prime numbers and in order to find out the original prime numbers is where it is very difficult. DSA which stands for a digital signature algorithm is an algorithm that encrypts data to RSA. EDSA stands for Elliptic Curve Digital Signature which is used to ensure transactions are being spent by their original owners. All these different classical security protocols will not be able to withstand quantum fault-tolerant behavior. Fault-tolerant means after failure it continues to solve the problem, it is consistent. It will continue to try until it can crack the encryptions. Another disadvantage is that quantum computing can have the ability to crack today's classical security as well. Wallden and Kashefi's article states, "A quantum attacker can use Quantumness in various ways, not only in order to solve some classical problems quicker" [7, p. 123]. Since classical securities are less complex it provides quantum computing an easy feat because quantum computing can process enormous data and can easily crack classical cryptosystems or securities that we use today.

Quantum can have an impact on encryption schemes. In, *"The Impact of Quantum Computing on Present Cryptography"* by Vasileios Mavroeidis, Kamer Vishi, Mateusz D. Zych, and Audun Josang a chart was displayed on page 4 of their research report. The cryptographic algorithms they were comparing are AES-256, SHA-256 and SHA-3, RSA, ECDSA and ECDH, and lastly DSA. They compare the types, utilities, and most importantly how they withstand quantum attacks. According to their articles chart, AES-256 is a symmetric key that is used for encryption. It uses a hidden key to cipher and deciphers the information. It is secure against quantum impacts for now. SHA-256 and SHA-3 has no type but it is a hash function that compresses numerical inputs to values. But for SHA-256 its values are 256 bits. SHA-3 is similar but not as bits as SHA-256. Followed by RSA, ECDSA, and DSA which are public keys and are used for signatures and key exchange except for RSA which is for key establishment. But they all are not secure against quantum attacks.

In Mavroeidis, Vishi, Zych, and Josang's article, it states, "In addition, they emphasized that the relatively small keyspace of ECC compared to RSA makes it easier to be broken by quantum computers. Furthermore, Proos and Zalka explained that 160-bit elliptic curves could be broken by a 1000-qubit quantum computer while factoring 1024-bit RSA would require a 2000-qubit quantum computer. The number of qubits needed to break a cryptosystem is relative to the algorithm proposed" [8, p. 4]. The more bits that are used to compute the more likely they are able to break standard classical cryptographic security protocols. Since they are probabilistic and fault-tolerant they are able to use qubits that are not defined and use superposition which can find a value that's close to the actual value and it will keep trying to use qubits to get to closer values, eventually reaching the exact value needed to crack security system protocols. Another downside of classical securities is that they cannot solve problems compared to quantum computers. In the article, *"Cybersecurity Challenges Associated with the Internet of Things in a Post-Quantum World"* it explains: "In a parallel development, quantum computers are emerging which are known to be able to solve problems classical computers cannot. Because of the tremendous combinatorial speed of the quantum computers, which act in a superposition state where the state can be zero and one simultaneously (quantum bit), combinatorial problems are solved much quicker" [9, p. 157356]. Quantum computing is so quick compared to classical securities that even classical securities cannot solve problems. Security breaches are inevitable if quantum technologies are in the wrong hands.

3   Quantum Computing Implementation

With any outstanding achievement, there comes a significant complication. Quantum computing has been in discussion for the past decades, while only limited companies have built quantum computing in its early stages, no existing operating systems can run quantum computing. The computer cannot be placed in an office, like classical computers, as it requires a cooling system to operate and generates loud noise and heat. Within *Quantum computing: The Risk to Existing Encryption Methods*, Kirsch discusses the Environmental risk factors and other risks such as accuracy and Phase error. The research paper states, "Higher accuracy can be done with numerous trials of the same problem, but this diminishes the speed advantage of quantum computing" [10, p. 8]. This type of speed disadvantage will slow down the time it takes for the quantum computer to break todays' encryption. Even if the accuracy issue has been addressed, there is a significant risk to existing encryption, classical computers do not hold the capability of fighting an attack from quantum computers.

Another major complication that arises with implementing quantum computers is the classical encryption algorithms. For a classical computer to crack specific encryption, it would take minutes, days, or even years, whereas a quantum computer can do the same job faster than classical computers. "However, quantum computers can approach the same problem and attempt multiple combinations at once, significantly reducing the amount of time to find the correct combination" [11, p. 9]. Since quantum computers can try multiple combinations at once, it would take quantum less time to crack the password. Before we implement quantum computers, mathematicians and computer scientists need to develop more robust algorithms that can take a quantum computer at the same time as it would take a classical computer to decipher the text.

A classical computer uses bits to process information, whereas a quantum uses qubits to process its information. *Quantum Computing and the Threat to Classical Encryption Methods* cites another author, "Classical computers process information in bits, which have two states, on (1) or off (0); the state of the bit cannot be both on and off at the same time" [12]. However, quantum computers can be on and off at the same time. "Quantum computers use quantum bits, also known as qubits, to process data. These qubits use a method known as superposition. With superposition, the qubits can be in multiple states (on and off) at the same time. Superposition increases the problem-solving abilities of quantum computers" [13]. The Quantum bits makes it possible for the computer to process information faster than a classical computer, as it holds the capability of doing multiple things at once. This means that when quantum computing becomes publicly available in the near future, it will have the ability to attack systems at a faster rate.

The amount of heat and noise quantum is producing is not only creating environmental and health issues, in fact, it will also create a major problem for the quantum computer itself. In the article, *The Impact of Quantum Computing on Present Cryptography* it states, "Qubits are susceptible to errors. They can be affected by heat, noise in the environment, as well as stray electromagnetic couplings" [8, p. 2]. In quantum computing qubits hold most of the information, so when factors are affecting the qubits its resulting in missing information.

A core issue that rises, on quantum capability, is the reduction of its speed. Due to power quantum has the ability of computing, multiple possibilities of answers. As described in the article, *The Impact of Quantum Computing on present cryptography*, "Quantum algorithms are mainly probabilistic. This means that in one operation a quantum computer returns many solutions where only one is the correct" [8, p. 2].

When quantum is used as an additional security layer for E-payment, it will create a possibility of quantum computers to run slower than average, this means that during a cyber-attack, a quantum computer will act like a normal computer, rather than a supercomputer.

4   E-Commerce and E-Payments

We currently live in a time where most transactions take place online. With Amazon leading the way for people to purchase things with ease virtually and have the item at their doorstep before the end of the day in some cases. E-payments and transactions also skyrocketed due to quarantine implements caused by the global pandemic that hit during 2020. Now with all of these transactions taking place online and payment information being stored to a user's account there would have to be security built in to protect these purchases and transactions. There would also have to be secure connections from the buyer and seller on both ends. Now people with bad intentions could sometimes break these connections creating a fork in the road that will receive a payment from the user over the actual seller or even get the buyer's card information or account and transfer money that way. But in the modern day of classical computing there are ways to protect these transactions. Now with quantum computing on the horizon this can lead to new exploits or even more secure options for both buyer and seller transactions.

E-payments and third-party payment systems like PayPal, Apple Pay, and Alipay are all used in everyday life when it comes to making transactions online or in person. With this being said there are security protocols in play that protect the buyers and sellers' information. "However, the security of E-payment has not reached a perfect level" [14, p. 2]. With our old classical computers it does the job to provide a level of security but for something such as money being exchanged one would think that this security would be air tight with the least number of exploits. So here comes the question of how we can make tighter security to protect the user and their money and accounts. Quantum computing will be the key to this protection.

The current issue with security is that it is expensive due to having more complex algorithms for it to run. "However, in recent years, security is no longer the only requirement for E-payment, efficiency and convenience have become increasingly important" [14, p. 2]. So, with the power of quantum computing the plan would be to bring a cheaper and more efficient way to secure transitions and E-payments. The plan would be to optimize the schemes that we have and reduce their costs at the same time. Quantum mechanics with communication is called quantum entanglement. This is where two systems are in a state of entanglement, regardless of the distance between each system, also known as quantum mechanical nonlocality. These will be the steps used to create a more secure connection and create anonymity between both seller and buyer during the process. There are various quantum cryptography protocols that can be utilized in this approach. Quantum bit commitment and quantum key distribution or QBC and QKD respectively. QBC is based on BB84 and "declared that the scheme is unconditionally secure" [14, p. 2]. Now, these two protocols can work together because the QKD will add to the QBC and make this protocol stronger. "These quantum Cryptography protocols guarantee that Quantum communication can resist the attack of QTM (Quantum Turing Machine) and a PTM (Probabilistic Turing Machine) which makes it a more secure communication mode than the traditional" [14, p. 2]. This style of encryption using quantum cryptography will maintain the four following things: First would be fairness, whereas the buyer and the seller cannot see the security score of each other. This is put into play to protect each party in case there is foul play. The second is binding. Although neither party is made aware of each other's security score, this score will remain on each person's account and will give the score numbers to the attached user so it can be used to identify who is who. The third is untraced ability and with this, the payment processor will obtain both security scores that were bonded to the users and then get the sum to evaluate the transaction as it is taking place. This is to create a level of security for when users hand over the data to the platform and not worry about a data leak. The final part is resistance. This is if there is an eavesdropper trying to steal the data this process will be made aware that the attempt was made. But the information provided will never make it to the actual data that they are looking for in it.

The age of quantum computing will be approaching faster than we think. A lot of major tech companies are investing in this technology such as IBM and Google. Now, these types of technologies tend to grow at such rapid rates. Within a short time, this quantum technology will be fully operational and can lead to security issues with all the classical systems we are currently using. "The security of classical E-payment schemes depends on the difficulty of solving hard mathematics problems, which will be threatened by quantum computation" [7, p. 2]. These current e-payment processors use mathematical formulas to secure and encrypt the process of the transactions but with quantum computing figuring out these formulas and secret keys, it is a much easier task due to the processing power of quantum computing.

To combat the issue with quantum computing breaking down the classical security protocols, some researchers are trying to come up with different protocols that may be able to protect the information during the purchasing process. The first technique is a multi-proxy blind signature. "Proxy signature allows a designated person, called proxy signer, to sign on behalf of an original signer" [7, p. 2]. Now with this type of security, it can add a level of protection that can sign off that the purchase was authenticated. The next is controlled Quantum teleportation. "Our quantum multi-proxy blind signature is based on controlled teleportation, which takes the entangled four-qubit Cluster state as it's quantum channel" [7, p. 2]. This technique has three parts to it: a sender, a controller, and a receiver. For example, if someone was to make an online purchase, they would be considered the sender sending money to the seller, who would be considered the receiver. So in this instance there are 4 particles where the sender has the 1st and 3rd particle and the controller and sender has the 2nd and 4th particles. Then the sender would perform a Von Neumann measurement on the Nth particle as well as 1–4 particles as long as it can satisfy a quantum state. Once this operation is performed, the outcome from the sender will send the results to the receiver and controller. At that point, the Controller will send particle 4 through a

Hadamard gate where the 4th particle can change states. Now if the controller passes the sender and receiver, without incident, this will be communicated and will open up a line of communication between the sender and receiver, allowing the transaction to proceed.

5   Blockchain

In the world of computer science and quantum computing, hackers and cyber-attacks are very common and are understandably feared as many people would not want personal info to be revealed or stolen in the intentions of hacking systems to do so. One of the most vital forms of implementing a system to avoid any type of information leaks or hacks consist of the block chain. Blockchain is a system of recording information that makes the chances of hacking quite a bit more difficult compared to no implemented system. This is used in hopes to lead to the prevention of cyber-attacks, changing the information that is outlined in the sensitive case records or prevention of hackers being able to cheat the system in some way. Although these block chains are used to prevent any type of unwanted hacks and attacks, there are some factors that make this system vulnerable itself while trying to avoid the recordings of information or cheating the system.

The idea of quantum computing does pose a good number of threats in regards to efficiency and speed. Blockchain technology is vulnerable using quantum computing since they rely on classical cryptographic protocols. This is the exploit that will be used when a quantum attack occurs. With the advances of quantum computing and the level of efficiency that is present, quantum computers are able to predict any security measures and protocols that are used to protect the vital information of the users. Confidential and sensitive information can be leaked or hacked as this information is being stored in these blockchains. This raises a question to those who intend on using quantum computing for security purposes within the blockchains because it is never guaranteed that the security protocols are enough to keep information from getting leaked or hacked.

Throughout the last couple of years blockchains and distributed ledger technologies or DLTs have evolved considerably, and their implementation has been said to be helpful for various applications. This is because blockchains and DLTs have the ability to provide transparency, redundancy, and accountability. Distributed ledgers are a type of database that can be shared or replicated with members in a decentralized network. Primarily used for recording transactions, "the data is replicated and synchronized so that a consensus is achieved between the participating nodes" [15, p. 112].

Blockchains are a technology that emerged with Bitcoin and are used for secure communications, data privacy, resilience, and transparency [3, p. 21091]. They are a type of distributed ledger that links units of data using hash functions. Additionally, many of the security advantages that come with blockchain technology relies heavily on the use of classical cryptography and security protocols. This includes leveraging public-key/asymmetric cryptography, which is essential for authenticating transactions and hash functions which "allow for generating digital signatures and for linking the blocks of a blockchain" [3, p. 21091].

However, the progress towards quantum computing grants potential attacks derived from Grover's and Shor's algorithms on pre-quantum blockchains. The algorithms threaten the integrity of classical security protocols and force the redesign of present blockchain to withstand quantum attacks. In blockchains, hash functions such as SHA-256 or Scrypt are used frequently since they are easy to implement and hard to forge. These are essential for the generation of digital signatures which blockchains use to authenticate transactions. So a problem arises with quantum attacks making use of Grover's algorithm to accelerate brute force attacks by a quadratic factor and can be used to attack blockchains in two ways.

The first is by searching for hash collisions and replacing the entire blocks in the blockchain. This can be achieved by the use of "Grover's algorithm to find collisions in hash functions, which concludes that a hash function would have to output $3*n$ bits to provide a n-bit security level" [3, p. 21094]. Arriving at a conclusion like this means that our current hash functions are no longer valid in the post-quantum era and algorithms like SHA-2 and SHA-3 would have to increase their output size. The second is by using Grover's algorithm to speed the generation of nonces. This would mean that Grover's algorithm "makes it viable to insert a modified block into the chain without compromising the sequential consistency of the blocks…Thus the faster miners can take control of the content of the blockchain…it is feasible for a party with a quantum computer to rapidly outstrip competitors, who have only classical computing capacity, in generating additional blocks on the chain" [16, p. 5823]. While this would make recreating a blockchain faster, it would also undermine its integrity.

Moreover, blockchains use public-key cryptography for securing information exchanges through digital signatures. "Digital signature schemes are critically used to sign and validate transactions" [15, p. 112]. Bitcoin for example uses an elliptic digital signature algorithm that relies on a private key to sign a message and a public key to check the signature. However, a problem arises using Shor's algorithm on quantum computers since the algorithm has the potential to crack public-key cryptography and break digital signatures. Furthermore, "it is estimated that in the next 20 years such a kind of computers will be functional enough to be able to break easily current strong public-key cryptosystems. In fact, organizations like the NSA have already warned on the impact of quantum computing on IT products and recommended increasing the ECC (EllipticCurve Cryptography) security level of certain cryptographic suites" [3, p. 21093]. Therefore, with public-key cryptography being threatened by quantum computers there is a growing need to make blockchain technologies adaptable for the post-quantum era.

6 IBM

IBM which stands for International Business Machines is a company that provides hardware, software, cloud-based services, and cognitive computing throughout the entire world. IBM is currently working on improving quantum computing. The question is when will it come to the hands of the average user? When can we expect to see what quantum computing can do in our environment? It may come sooner than you think.

Quantum computers are being in use today; however, it is not in the things you use today. That is because there are still things that need to be improved with quantum computing. According to IBM, they state that the year 2023 will be where quantum computing takes its first steps into high-performance computing. Quantum systems are super computers that use quantum physics. IBM has a development roadmap in which it states in the year 2022 the number of qubits in the quantum system are 433 qubits and the quantum system or supercomputer is called "Osprey". IBM states they would need to increase the number of qubits to have an impact in the high-performance computing we have today. As we go further in the road map, it shows that by the end of 2023 the number of qubits will reach 1121. That would be a substantial increase from the current year of qubits.

By 2023 we will be seeing improvements and implementations of quantum computers to even replace some supercomputers. But with all these improvements, what were the challenges? What were some hurdles that needed to be jumped over? For quantum computing what is the one thing we expect from it? In the short term, they are results. The performance of the computational speeds and how well it outputs data with qubits is the challenge. According to IBM, they divide the performance into 3 metrics. They are scale, quality, and speed.

Scale is measured by the number of qubits that indicate the amount of information or data IBM can encode into the quantum systems. Scale is all about high coherence, high reliability, and lower costs for resources. It is important for quantum computers to scale up because more data is being processed and more bits are being generated. To handle all these things quantum computing must go up in processing more qubits.

Quality is measured by Quantum Volume which shows the quality of the circuits and how those circuits are implemented in the hardware that is being used for quantum systems. Quality is about having fewer operation errors. The fewer errors, the more you can improve on the accuracy of these quantum computers which will have better results.

Speed is measured by CLOPS which stands for Circuit Layer Operations Per Second. This shows how many circuits can run on a piece of hardware in a given amount of time. Speed is about the synchronization of quantum and classical circuits to increase the execution rate.

Another problem with these quantum computers is the amount of heat the computer will generate undergoing so much processing power. The heat itself can destroy many circuits and ruin other significant hardware components. According to Jerry Chow who is the Director of Quantum Hardware System Development, his team has developed what is called the "Super Fridge" which will help regulate temperatures by 15 millikelvins and help his team reach over 1000 qubits in future quantum systems. As more qubits are scaled the size of the Super Fridge needs to increase as well as it will ensure that extremely hot temperatures will not destroy the quantum computers.

Further developments in quantum physics theory and hardware are required to meet the ambitious quantum technology implementation goals companies/organizations have set for themselves. Theoretical physicists advance research by exploring the advantages of solving a particular quantum physics problem. These theorists are essentially creating the quantum computing rulebook and the ecosystem for the technology. In addition, theorists also develop strategies to mitigate errors and improve device performance in the short term while also figuring out how to correct errors in the long term, realizing fault-tolerant quantum computers.

Currently, quantum processors are disrupted by noise and errors because quantum circuits and their qubits are connected and manipulated by sequences of microwave pulses. As a result, during operation, qubits quickly forget their quantum states due to external disruptions. A big challenge is figuring out how to control large systems of qubits for long enough, and with limited errors, to run the complex quantum circuits. Fault-tolerant quantum hardware or resilient quantum computers immune to disruptions are required to implement future quantum applications.

Due to hardware and theoretical constraints achieving fault-tolerance and resiliency is currently impossible. Therefore, researchers are working to develop quantum error mitigation techniques to mitigate the effect of noise in quantum circuits without requiring additional qubits or hardware resources. These methods can be implemented today but don't scale well and only apply to a restricted class of quantum algorithms. Without a long-term solution, error mitigation serves to bridge the gap before we can implement full-scale quantum error correction. In their research paper, Piveteau et al. attempt to split the difference by applying error correction only to the low-overhead Clifford gates while applying error mitigation techniques to noisy T-gates [17]. The implication is that this allows us to produce circuits beyond the reach of classical computers without paying the full price of realizing a fault-tolerant universal gate set.

To justify further developments in quantum computing, researchers need to identify the applications where quantum is superior to classical. This leads to the question, what benefits are provided by quantum technology in a real-world implementation. A subfield of theoretical computer science, computational complexity attempts to classify and compare the practical difficulty of solving problems and the resources required, focusing on time and memory utilization. Allowing researchers to compare and study the ability and efficiency of classical or quantum computers in solving and/or checking various algorithms. In their research paper, Bravyi et al. explore the computational complexity of simulating quantum systems in thermal equilibrium and the problem of calculating statistical properties of this system, such as the average energy and entropy. This problem is complex for both classical and quantum computers, and so far, nobody has been able to quantify the difficulties in terms of the known complexity classes. Bravyi et al. [18] take the first step by showing that simulating a system of qubits with two-qubit interactions in the thermal equilibrium is as complex as the "Quantum Approximate Counting" problem.

Even as theorists continue to develop quantum theory and hardware, they continue to deepen their understanding of how qubits work together to handle errors via fault tolerance. Theory and robust quantum hardware continue to be the blockage towards quantum implementation. Despite the difficulties, it's helpful to remember that it took classical computing many decades to go from individually programmed logic gates to the sophisticated cloud-based services of today.

All the topics this paper has gone over are just the tip of the quantum iceberg that is to come. For society to switch over to quantum computing this will be another great leap and society reacting to it in a skeptical way will come along with it. Humans are naturally skeptical beings and are usually afraid of change and especially changes that most people wouldn't understand from the beginning. Once quantum computing goes from research and development labs to actual users and can be brought down and compact enough for regular people to purchase, this will surely send a shock to society. It will be in comparison to how classical computers used to take up an entire lab just to run small processes and now we have the power of a computer in the palm of our hands due to current efforts in microchip technology. Now moving that over with quantum computing we can determine that a lot of what has happened in the past may play out again on how society will react. It would seem like some technological advancements move too fast for people to keep up with and on the other side of the coin it might strike fear into people like how the year 2000 also known as "Y2K" scared people due to society running more on technology then it used to be. Therefore, the impact on society will be quick and it will also shock companies that will need to change with the new technology as well. At the speed of how fast technology grows, as a society humans will just have to embrace the changes brought upon us and utilize it to the best of our abilities to further the human race as a whole.

We are in a current time period that uses technology on a daily basis and coexists with it as a tool to help with completing tasks at such extreme rates. With this in mind, our security is and will continue to be under attack with data miners and hackers just trying to get some personal data information. With quantum computing on the horizon, these security protocols will be outdated and ready to be attacked if a user with quantum computing chooses to do so. Quantum computing will help the technology landscape in many ways such as making transactions a safer and easier process and offering much more protection to users. Everything great will come with pros and cons but with most new technology it will be a while before we see public use but it is good to prepare for all the cons that can come with quantum computing to defend against cyber-attacks. It will be very interesting to see quantum computing more normalized and see how nations will react and renovate this new technology.

7 Conclusion

As described in previous sections, quantum computing changes current technology drastically, however there are certain issues that need to be addressed. It is important to find alternatives for classical encryption methods that will fail with quantum computing. Major corporations will have to implement strategies for protecting their data and possibly implementing a quantum computing infrastructure. The crypto space and blockchain technology will be affected making it easier to make transactions on the block chain as well as speed up crypto mining. Ecommerce will change as well, creating a safer space for online transactions to take place. This can change the entire technology space and all the old systems will be exposed to exploits and need to keep up with this new tech to protect themselves from attacks.

8 Future Work

Future work is necessary to create a quantum computing environment with moderate temperatures. Extremely low temperatures that are required for quantum computing right now, discourage more organizations. Everything great will come with pros and cons but with the new technology it will take a while for the public to accept, but it is good to prepare for all issues that can come with quantum computing. It will be very interesting to see quantum computing more normalized and see how nations will react and renovate this new technology.

References

1. Fisher KAG, Broadbent A, Shalm LK, Yan Z, Lavoie J, Prevedel R, Jennewein T, Resch KJ (2014) Quantum computing on encrypted data. Nat Commun 5(1)
2. Rahman MNA, Abidin AFA, Yusof MK, Usop NSM (2013) Cryptography: a new approach of classical Hill cipher. Int J Secur Appl 7(2):179–190
3. Fernandez-Carames TM, Fraga-Lamas P (2019) Towards post-quantum blockchain: a review on blockchain cryptography resistant to quantum computing attacks. IEEE Access 8:21091–21116
4. Roetteler M, Svore KM (2018) Quantum computing: codebreaking and beyond. IEEE Secur Priv 16(5):22–36
5. Kartheek DN, Amarnath G, Reddy PV (2013) Security in quantum computing using quantum key distribution protocols. In: 2013 international multi-conference on automation, computing, communication, control and compressed sensing (iMac4s)
6. Nanda A, Puthal D, Mohanty SP, Choppali U (2018) A computing perspective of quantum cryptography [energy and security]. IEEE Consum Electron Mag 7(6):57–59
7. Wallden P, Kashefi E (2019) Cyber security in the quantum era. Commun ACM 62(4):120. Zhang J-L, Hu M-S, Jia Z-J, Bei-Gong, Wang L-P (2019) A novel E-payment protocol implemented by blockchain and quantum signature. Int J Theor Phys 58(4):1315–1325
8. Mavroeidis V, Vishi K, Zych DM, Jøsang A (2018) The impact of quantum computing on present cryptography. Int J Adv Comput Sci Appl 9(3)

9. Althobaiti OS, Dohler M (2020) Cybersecurity challenges associated with the Internet of Things in a post-quantum world. IEEE Access 8:157356–157381
10. Kirsch Mentor Z, Chow M (2015) Quantum computing: the risk to existing encryption methods
11. Rosales M (2019) Quantum computing and the threat to classical encryption methods. Dissertation/Thesis, p 50
12. Bone S, Castro M (1997) A brief history of quantum computing. Surv Present Inf Syst Eng (SURPRISE) 4(3):20–45
13. Buchanan W, Woodward A (2017) Will quantum computers be the end of public key encryption? J Cyber Secur Technol 1(1):1–22
14. Cai D-Q, Chen X, Han Y-H, Yi X, Jia J-P, Cao C, Fan L (2020) Implementation of an E-payment security evaluation system based on quantum blind computing. Int J Theor Phys 59(9):2757–2772
15. Khalifa AM, Bahaa-Eldin AM, Sobh MA (2019) Quantum attacks and defenses for proof-of-stake. In: 2019 14th international conference on computer engineering and systems (ICCES)
16. Cui W, Dou T, Yan S (2020) Threats and opportunities: blockchain meets quantum computation. In: 2020 39th Chinese control conference (CCC)
17. Piveteau C, Sutter D, Bravyi S, Gambetta JM, Temme K (2021) Error mitigation for universal gates on encoded qubits. Phys Rev Lett 127(20)
18. Bravyi S, Chowdhury A, Gosset D, Wocjan P (2021) On the complexity of quantum partition functions. arXiv e-prints arXiv:2110
19. Lipman P (2021) Council post: how quantum computing will transform cybersecurity. Forbes
20. Niu X-F, Zhang J-Z, Xie S-C, Chen B-Q (2018) A third-party E-payment protocol based on quantum multi-proxy blind signature. Int J Theor Phys 57(8):2563–2573
21. Sumathi M, Sangeetha S (2018) Enhanced elliptic curve cryptographic technique for protecting sensitive attributes in cloud storage. In: 2018 IEEE international conference on computational intelligence and computing research (ICCIC)

Printed in the United States
by Baker & Taylor Publisher Services